Power Systems

H.M. Hashemian
Maintenance of Process Instrumentation in Nuclear Power Plants

H. M. Hashemian

Maintenance of Process Instrumentation in Nuclear Power Plants

With 131 Figures

Springer

H.M. Hashemian
Analysis and Measurement
Services Corporation, AMS
Cross Park Drive 9111
37923 Knoxville, TN
USA
hash@ams-corp.com

ISBN 978-3-642-07027-3 e-ISBN 978-3-540-33704-1

Springer is a part of Springer Science+Business Media
springer.com

Softcover reprint of the hardcover 1st edition 2006

Final processing: PTP-Berlin Protago-TEX-Production GmbH, Berlin (www.ptp-berlin.com)
Cover-Design: deblik, Berlin
Printed on acid-free paper 62/3141/Yu – 5 4 3 2 1 0

This book is dedicated to my wonderful daughter, **Nikki Hashemian**.

H.M. Hashemian

Knoxville, Tennessee

USA

Preface

This book is written for the instrumentation and control engineers, technicians, and managers in nuclear power plants. It focuses on process temperature and pressure sensors and the verification of these sensors' calibration and response time. It also provides examples of typical problems and solutions with temperature and pressure measurements in nuclear power plants.

Contents

List of Figures

List of Tables

1 Introduction

Signals from sensors in nuclear power plants can be monitored while the plant is operating to verify the performance of the sensors and associated instrumentation and to diagnose process anomalies. This introduction provides some examples of this procedure. It also presents a review of computer-aided maintenance technologies as well as active methods for employing test signals to measure sensor performance and to identify problems in their cables and connectors. The remainder of the book will focus on nuclear plant temperature and pressure sensor operation and maintenance as well as active and passive techniques for remotely testing these sensors' performance after they are installed in an operating plant.

1.1 Reference Plant

Fig. 1.1 illustrates a loop of a pressurized water reactor (PWR), which will be used as the reference plant throughout this book. The figure shows the reactor vessel, a primary coolant loop, a steam generator, a pressurizer, and the secondary loop. Typically, a PWR plant consists of two to four of these loops, with the exception of some Russian PWR models, which have six loops. The sensors typically found in a PWR plant are indicated in Fig. 1.1 by small circles. More specifically, the figure shows neutron flux detectors on the outside of the reactor vessel, core-exit thermocouples on the top of the core inside the reactor vessel, narrow-range and wide-range resistance temperature detectors (RTDs) in the hot-leg and cold-leg pipes, and pressure, level, and flow transmitters in the primary and secondary loops.

A PWR plant was selected as the reference plant for this book because most of the nearly 500 nuclear power plants in the world today are PWRs. In addition to PWRs, however, most of the material in this book also applies to other conventional and advanced nuclear power plants such as boiling water reactors (BWRs); heavy water plants like Canadian deuterium (CANDU) reactors; Russian PWRs, which are referred to as *VVERs*; liquid metal fast breeder reactors (LMFBRs); and high-temperature gas-cooled reactors (HTGRs).

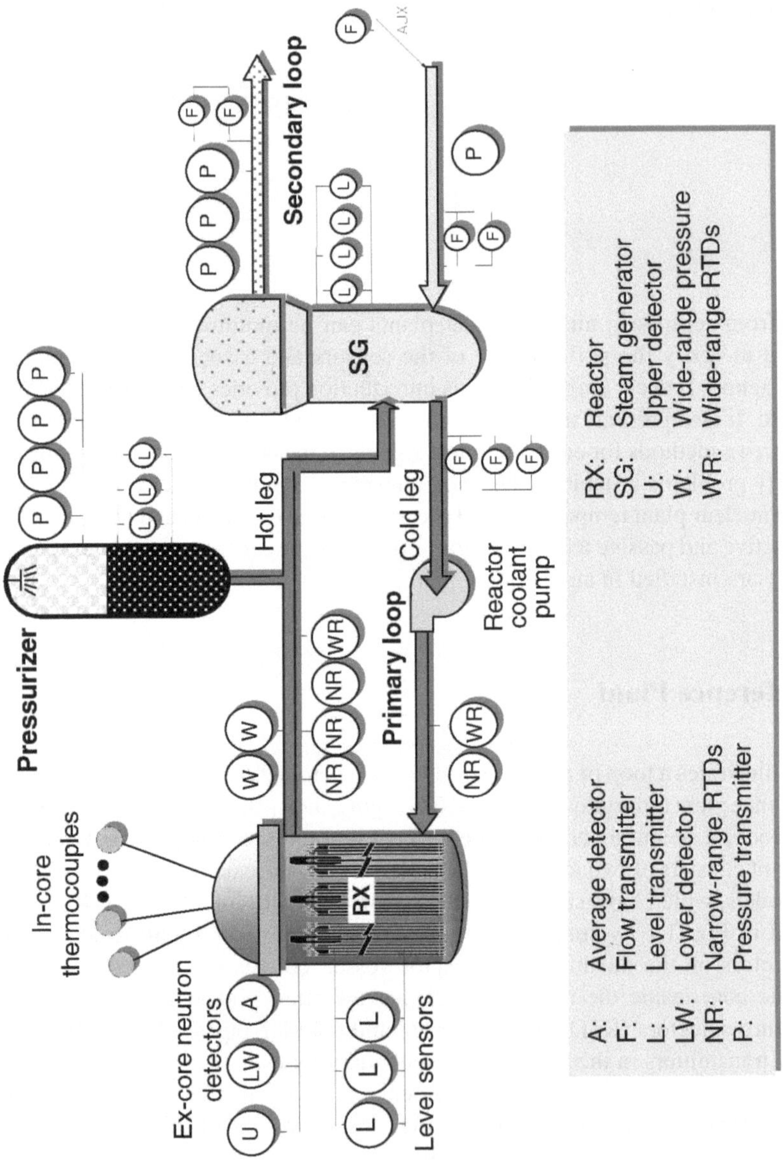

Fig. 1.1. A loop of a PWR plant and its typical sensors

1.2 On-Line Monitoring of Process Instruments Calibration

Fig. 1.1 shows that two to four sensors are typically used to measure each process parameter in a nuclear power plant. This redundancy improves the plant's availability and protects it from the operational or safety problems caused by the failure of single sensors. Although instrument redundancy is built into nuclear power plants mainly to enhance plant safety and availability, the nuclear industry has in recent years exploited this redundancy for other purposes, such as for verifying the calibration of process instruments. For example, a test called *cross-calibration* is performed on the primary coolant RTDs in PWRs in order to verify that these sensors remain accurate as they age in the plant.

The primary coolant system of a PWR plant typically has about 16 to 32 RTD elements. At isothermal conditions, these RTDs are exposed to essentially the same temperature. Therefore, the reading of the RTDs under isothermal plant conditions is recorded at several temperatures during plant startup or shutdown, and these temperatures are then compared to identify the outliers. Subsequently, cross-calibration data points from three or more widely spaced temperatures are used to generate a new calibration table for any outlier that is found.

For pressure transmitters that are not as redundant as RTDs, on-line monitoring—in which transmitter output signals are averaged or modeled—is used to identify calibration drift. Fig. 1.2 shows on-line monitoring data from four steam generator level transmitters in a PWR plant. Each graph represents each transmitter's deviation from the average of the four transmitters plotted over time. The data encompasses two years, which corresponds to a full operating cycle. It is apparent from this data that these transmitters did not drift over this operating cycle and do not therefore need to be calibrated. This example illustrates the principle of on-line calibration monitoring for process instruments in nuclear power plants.

The data in Fig. 1.2 corresponds to a one-point calibration check of the four transmitters. To cover a transmitter calibration over a wide range, on-line monitoring data are sampled not only during process operation but also during plant startup and shutdown periods. Fig. 1.3 shows the results in a nuclear power plant of on-line calibration monitoring for a nuclear plant pressure transmitter as a function of the transmitter's operating range. This indicates that the drift of the transmitter is contained within 0.5 percent of its span over the approximate range of 7.5 to 75 percent of its span.

1.3 Dynamic Testing of Pressure Transmitters and Sensing Lines

For dynamic testing of sensors and transmitters, on-line monitoring requires rapid data acquisition. The upper half of Fig. 1.4 illustrates the installation of a level transmitter at the end of a sensing line in a nuclear power plant. In this particular plant, on-line measurements are made once every fuel cycle to determine each transmitter's response time and to identify any significant blockages in the pressure sensing lines. For this example, data was sampled from the output of the transmitter once every millisecond and analyzed to examine the transmitter's dynamic characteristics.

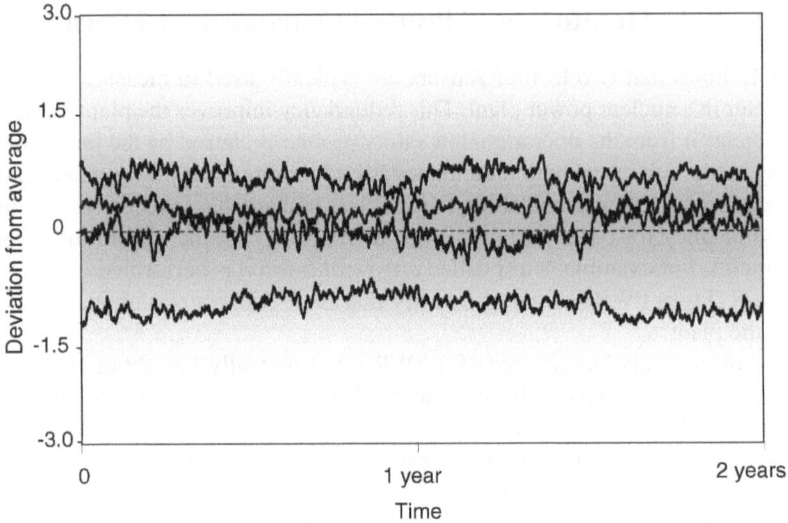

Fig. 1.2. On-line monitoring data from four redundant transmitters

The analysis entailed performing a fast Fourier transform (FFT) of the data in order to obtain its power spectral density (PSD), which is then used to determine the transmitter's response time. At first, the transmitter was found to be slower than expected, and its PSD did not compare well with previous baseline PSD. The plant

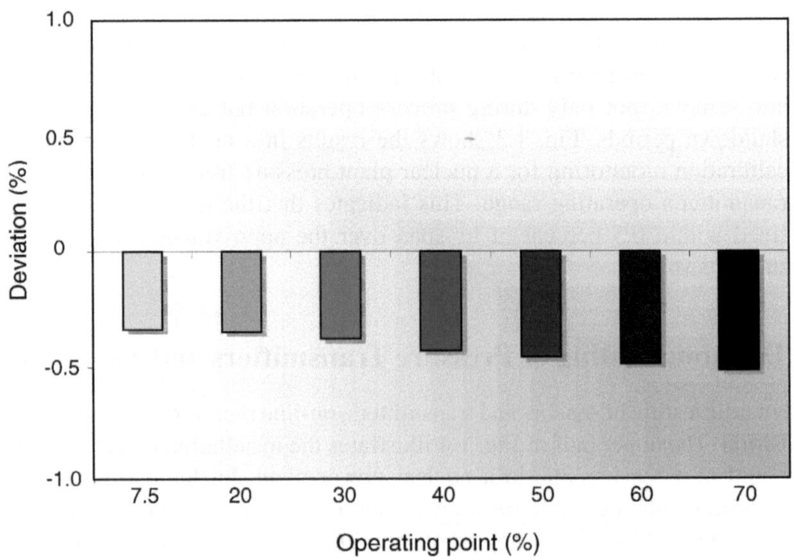

Fig. 1.3. Results of transmitter calibration verification over a wide range

was therefore notified that either the transmitter was sluggish or its sensing lines were partially blocked, or both. As a result, the plant maintenance crew examined the transmitter and its sensing lines during the plant outage and determined that crud from the reactor coolant water was obstructing one of the sensing lines. They therefore purged the sensing line. Subsequently, the dynamic tests were repeated to verify that the transmitter performance was restored. The lower half of Fig. 1.4 shows

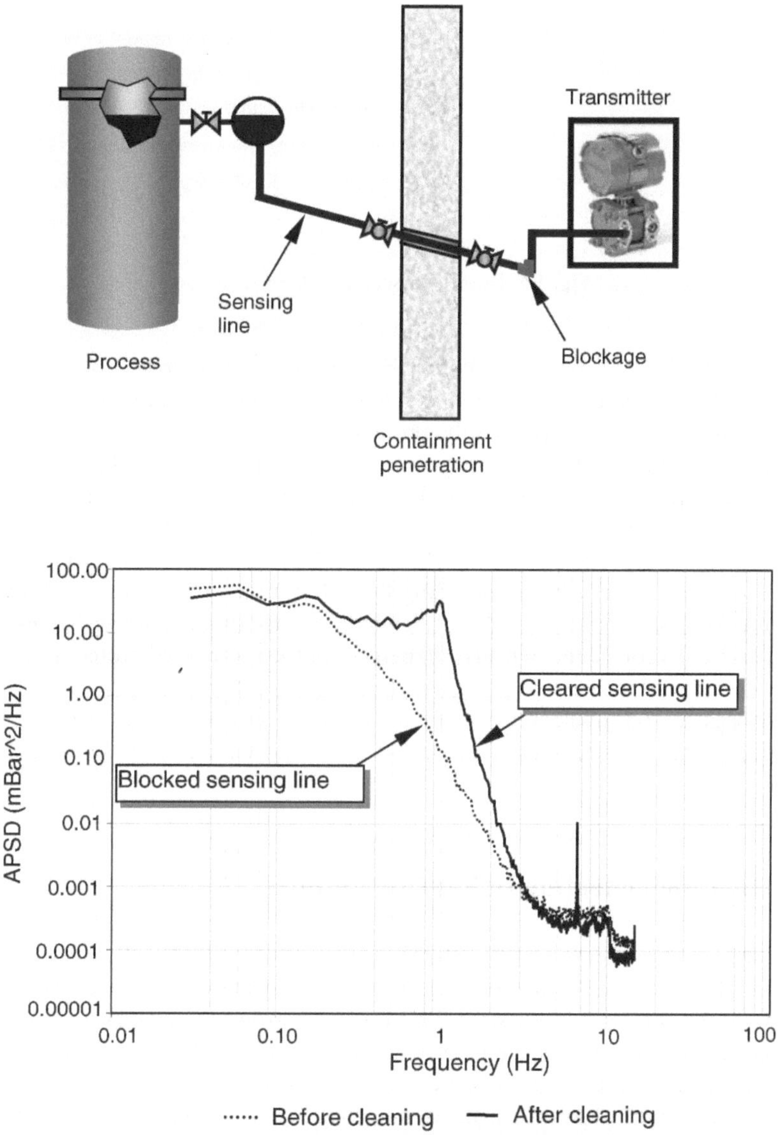

Fig. 1.4. On-line detection of sensing-line blockages

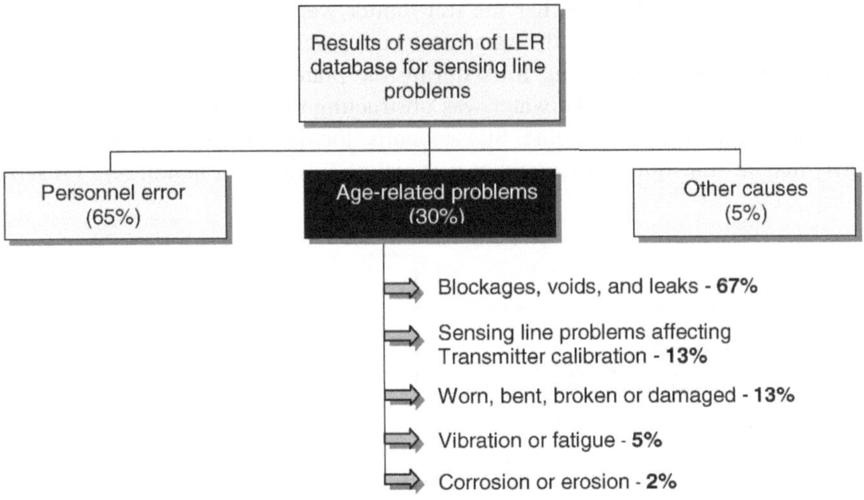

Fig. 1.5. Results of search of LER database

the transmitter's PSDs before and after the blockage was removed from the sensing line. It is clear that the blockage reduced the transmitter's dynamic performance and that purging the system corrected the problem.

Nuclear power plants have encountered many events involving blockages, voids, and leaks in pressure sensing lines. Fig. 1.5 shows the results of a search of the Licensee Event Report (LER) database. This database is maintained by the U.S. Nuclear Regulatory Commission (NRC) to track the failure of important equipment in U.S. nuclear power plants, including the safety-related pressure, level, and flow transmitters. The information in Fig. 1.5, which covers 10 years, shows that blockages, voids, and leaks contribute to nearly 70 percent of the age-related problems in sensing lines.

For this reason, nuclear power plants perform on-line testing of the dynamics of pressure transmitters, including sensing lines, to ensure safety and operational efficiency.

1.4 On-Line Detection of Venturi Fouling

In the secondary system of PWRs, the feedwater flow is traditionally measured using a venturi flow sensor. An inherent problem in venturi flow sensors, however, is the fouling of the venturi flow element. This fouling narrows the diameter of the sensing section of the venturi flow element and causes erroneously high indication of the feedwater flow. Through calorimetrics, the higher-than-actual flow that is measured because of venturi fouling translates into higher-than-actual indication of reactor power. In this case, the plant loses the ability to generate as much power as it is allowed. Experience has shown that flow uncertainties due to venturi fouling can

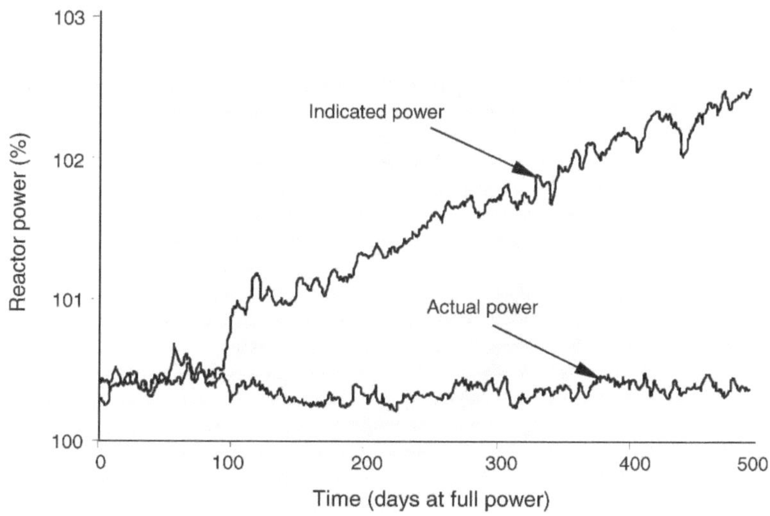

Fig. 1.6. Example of on-line monitoring results for detecting venturi fouling

cost a plant nearly 3 percent of power output. Because of this problem, many plants have installed ultrasonic flow sensors, which do not suffer from the fouling problem. Ultrasonic flow sensors are also more accurate than venturi flow sensors in most cases and have been approved by the NRC as a way to uprate plant power by up to 3 percent.

For this 3 percent gain in plant power output, plants must pay about $2 million (in 2006) to implement an ultrasonic flow sensor. This investment is obviously justified, and many plants have exploited ultrasonic flow sensors to reduce the uncertainty of their feedwater flow measurements and thereby increase the amount of power they are allowed to generate. On the other hand, using ultrasonic flow sensors, some plants have learned that their venturi flow elements have been reading lower than the actual flow. These plants have had to reduce power after installing ultrasonic flow sensors. Overall, the number of plants that have increased power production by using ultrasonic flow sensors has been much more than those who have had to decrease power.

The venturi fouling problem can be monitored on-line by using existing plant signals from upstream and downstream of the venturi flow sensor and from elsewhere in the plant. Fig. 1.6 shows an example of on-line monitoring results to examine the extent of venturi fouling and its effect on reactor power. The data covers 500 days, which corresponds to a complete operating cycle in the plant from which this data was retrieved. Fig. 1.6 shows two graphs: (1) the reactor power as calculated from analytical modeling using on-line monitoring data; and (2) the reactor power as indicated by the plant's instrumentation. It is apparent that the indicated power and the calculated (actual) power begin to diverge at about 100 days into the plant's operating cycle. More specifically, the indicated power climbs to about 2.5 percent above the actual power in 500 days. As a nuclear power plant is not normally allowed to operate beyond 100 percent power, this 2.5 percent error in reactor power indication is normally taken from the allowable power output of the plant.

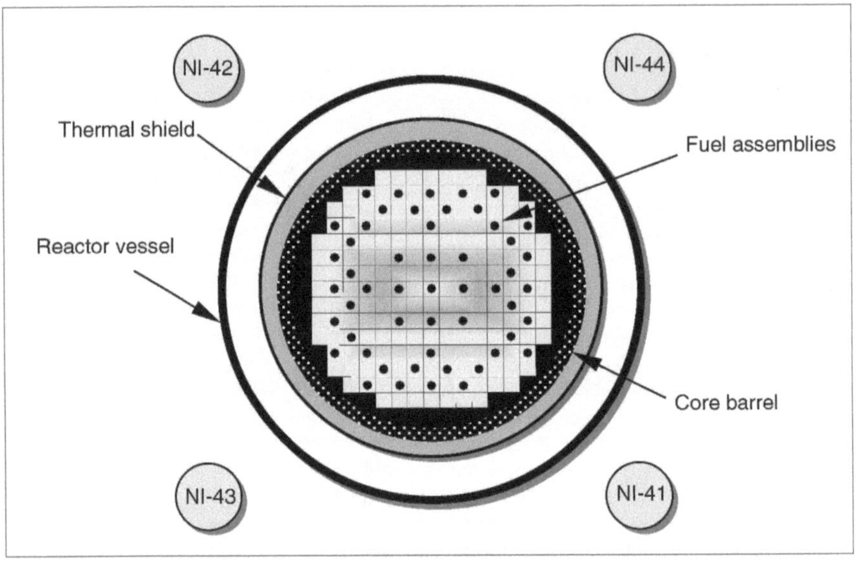

Fig. 1.7. Cross-sectional view of a PWR plant

1.5 Measuring the Vibration of Reactor Internals

Fig. 1.7 shows a simplified cross-sectional view of a PWR plant including the reactor vessel, core barrel, fuel assemblies, and thermal shield. Outside the reactor vessel, four neutron detectors, labeled NI-41, NI-42, NI-43, and NI-44, are shown. These detectors are referred to as *ex-core neutron detectors*, *neutron instrumentation (NI) sensors*, or *power range neutron flux monitors*. Their main purpose is to measure neutron flux as a way of monitoring reactor power. In addition, these detectors can serve to measure the vibrational characteristics of the reactor vessel and its internal components.

Typically, vibration sensors (e.g., accelerometers) are located on the top and bottom of the reactor vessel to sound an alarm in case the main components of the reactor system vibrate excessively. However, neutron detectors have proved to be more sensitive in measuring the vibration of the reactor vessel and its internals than accelerometers. This is because the frequency of vibration of reactor internals is normally below 30 Hz, which is easier to resolve using neutron detectors than accelerometers. Accelerometers are more suited for monitoring higher-frequency vibrations.

Fig. 1.8 shows the PSD of the neutron signal from an NI detector in a PWR plant. This PSD contains the vibrational signatures (i.e., amplitude and frequency) of the reactor components, including the reactor vessel, core barrel, fuel assemblies, thermal shield, and so on. It even contains, at 25 Hz, the signature of the reactor coolant pump rotating at 1,500 revolutions per minute, which corresponds to 25 Hz. Clearly, neutron detectors effectively register the vibration signatures of all the components of interest within the reactor system.

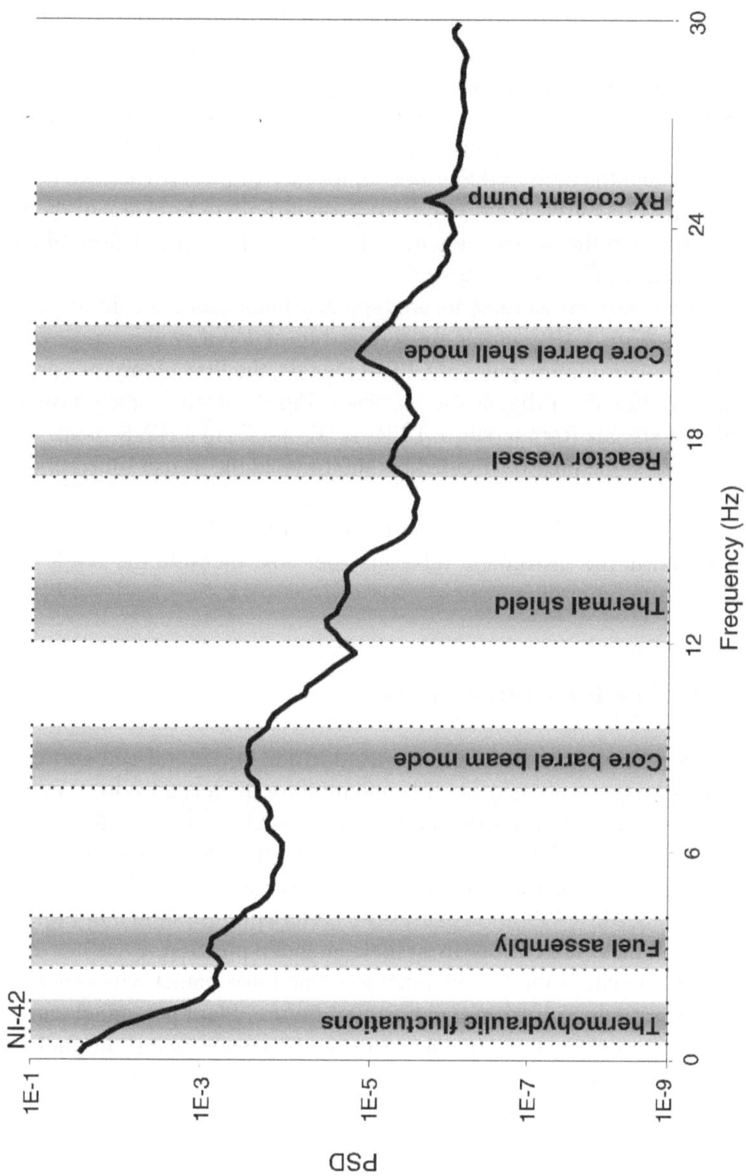

Fig. 1.8. PSD containing vibration signatures of reactor internals

1.6 Detecting Core Flow Anomalies

In Fig. 1.1 we showed that there are a number of thermocouples on the top of the core in a PWR plant. These thermocouples, called *core-exit thermocouples*, are normally used to monitor the reactor coolant's temperature at the output of the core. They can also be used in conjunction with the ex-core neutron detectors to monitor for flow through the reactor system. More specifically, by cross correlating signals from the ex-core neutron detectors and core-exit thermocouples, it is possible to identify the time it takes for the reactor coolant to travel between the physical location of the neutron detectors and the core-exit thermocouples (see Fig. 1.9). The result, referred to as *transit time* (τ), can be used with core geometric data to evaluate the reactor coolant's flow through the system, identify flow anomalies, detect flow blockages, and perform a variety of other diagnostics.

In BWR plants, flux measurements are typically made using a column of in-core neutron detectors (Fig. 1.10), which are referred to as *local power range monitors (LPRMs)*. By cross-correlating pairs of LPRM signals, the flow along the core can be baselined and monitored for diagnostic purposes. Fig. 1.10 shows the phase-versus-frequency plot of signals from a pair of LPRMs (B and C) in a BWR plant. This is a straight line whose slope may be divided by 360 to yield the transit time between the two LPRMs.

LPRMs can be used in BWRs not only to monitor flow through the core, but also to detect vibration in the instrument tube and fuel box, measure the BWR stability margin, and perform other diagnostics.

1.7 CANDU Reactor Applications

In CANDU reactors, neutron detectors are used inside horizontal and vertical tubes that extend into the reactor to measure flux and monitor the reactor power. In addition to measuring flux, these neutron detectors can be used to measure the vibrational signatures of the reactor's internals. For example, some old CANDU reactors have experienced sagging in the fuel channels, as illustrated in Fig. 1.11. This sagging apparently occurs because vibration causes the garter springs (shown in Fig. 1.11) to become loose, and they move away from their intended position.

This sagging can cause the fuel channel to come into contact with other components in the core, creating problems such as fuel failure. Plant personnel can use the signal from the neutron detector shown in Fig. 1.11 to determine if the fuel channel has sagged, especially if baseline vibration signatures are available for comparison purposes. The neutron detectors in CANDU reactors can also be used to measure the vibration of other components within the reactor, such as the horizontal and vertical detector tubes that contain the neutron sensors.

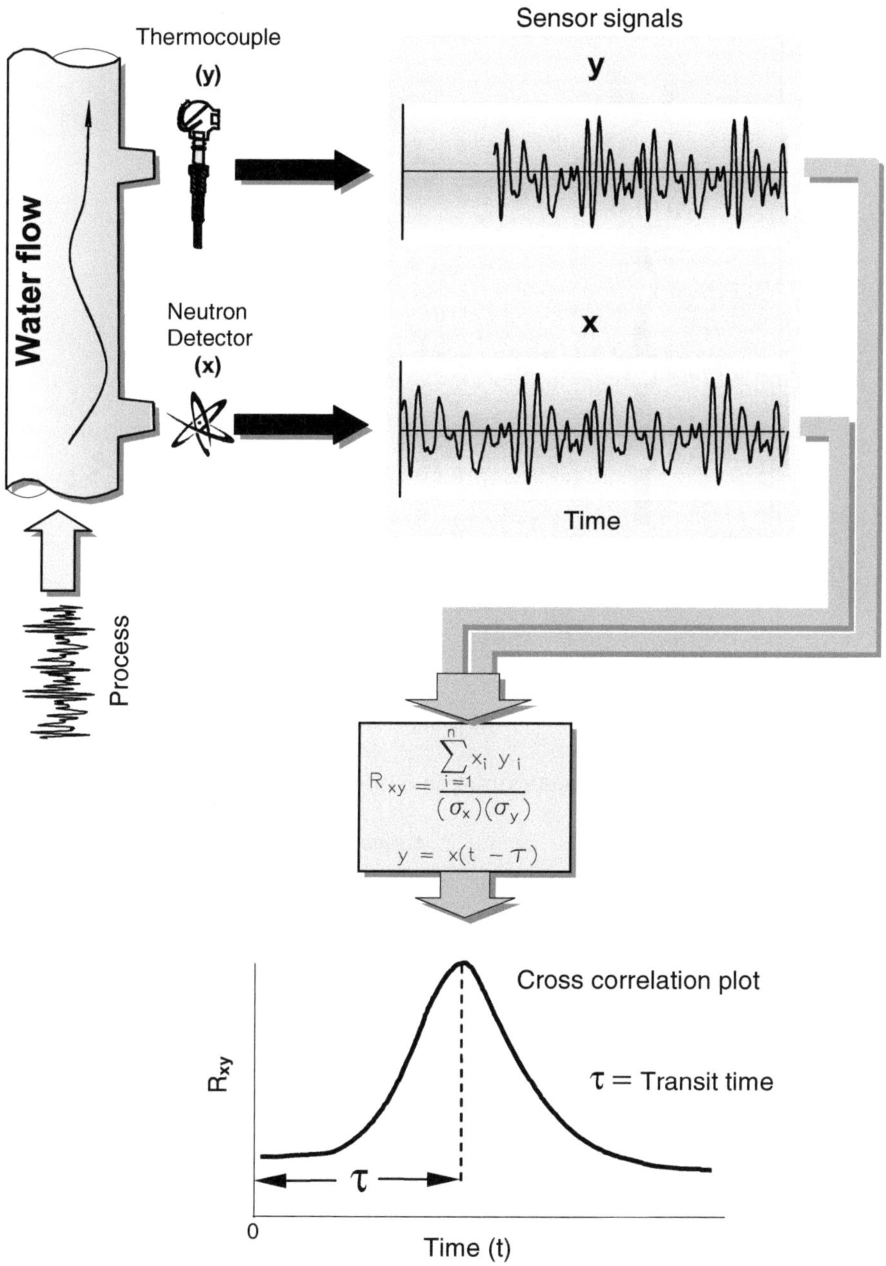

Fig. 1.9. Illustration of cross-correlation principle involving a neutron detector and a core-exit thermocouple to determine transit time (τ)

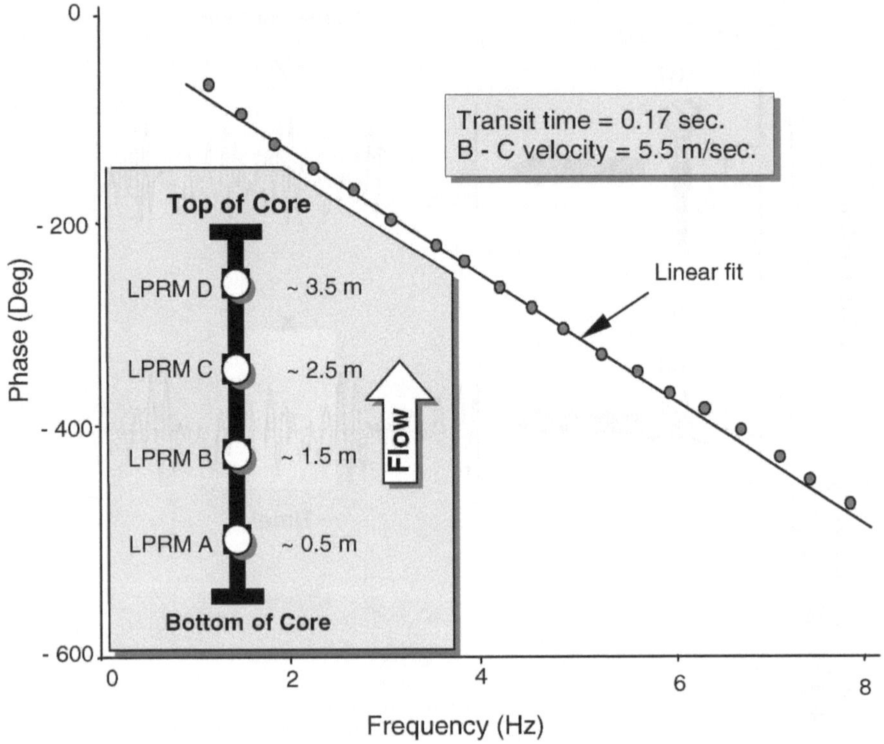

Fig. 1.10. BWR core flow diagnostics using an existing column of in-core neutron detectors

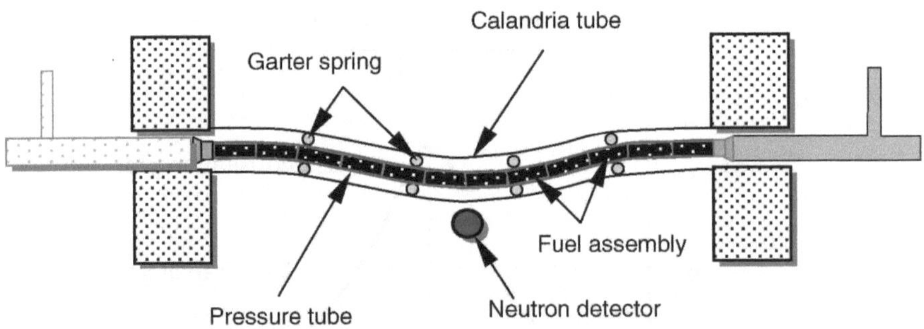

Fig. 1.11. Sagging of a fuel channel in a CANDU reactor

1.8 In-Situ Response-Time Testing of Temperature Sensors

Passive diagnostics based on readily available signals from sensors are not the only form of test signal in nuclear power plants. This book will also describe in-situ test methods that use externally applied active test signals for measuring equipment performance or for providing diagnostics and anomaly detection capabilities. For example,

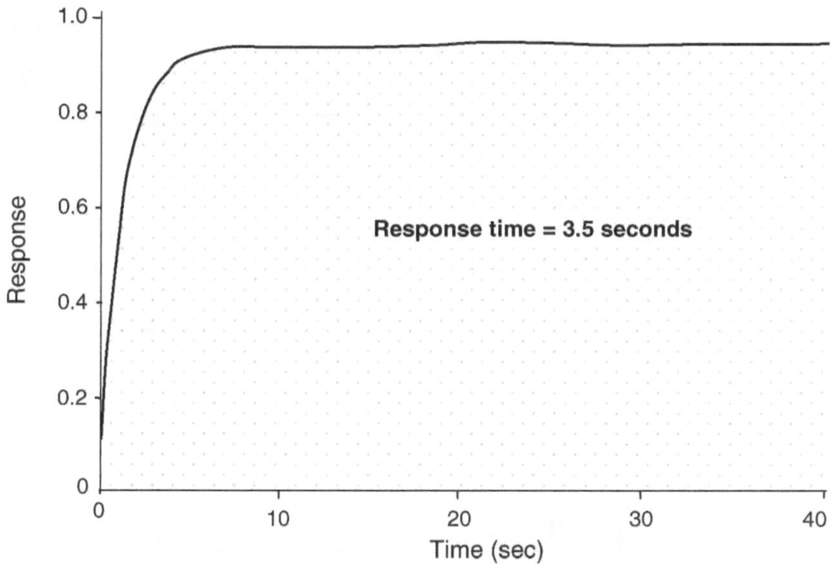

Fig. 1.12. Typical LCSR transient for a nuclear plant RTD

the response time of RTDs, thermocouples, and neutron detectors can be measured by sending a test signal to the sensor through these sensors' normal extension leads. These tests can be performed remotely from the process instrumentation cabinets in the control room area. Moreover, because these tests can be performed while the plant is operating, they make it possible to test the actual in-service response time of the sensors.

Specifically, the response times of primary coolant RTDs in nuclear power plants are sensitive to the flow rate, temperature, and pressure that they are exposed to. Their response times must therefore be measured at or near normal operating conditions. For this purpose, a method referred to as the *loop current step response* test was developed in the mid-1970s. This method involves sending a step change in current to the RTD sensing element which causes the sensor to heat internally. The test is performed by connecting the RTD to a Wheatstone bridge. The bridge includes a switch that allows the electrical current through the RTD to be switched from 1 or 2 mA to 30 to 50 mA for the LCSR test. This internal heating causes a transient increase in the RTD resistance that manifests itself as an exponential transit at the Wheatstone bridge's output. A typical LCSR transient for a nuclear plant RTD is shown in Fig. 1.12. This transient is recorded and analyzed to identify the RTD's response time.

1.9 Testing Cables In-Situ

In nuclear power plants, cables (including connectors, splices, and other components) are tested by evaluating the impedance relationships along the cable. Specifically, a

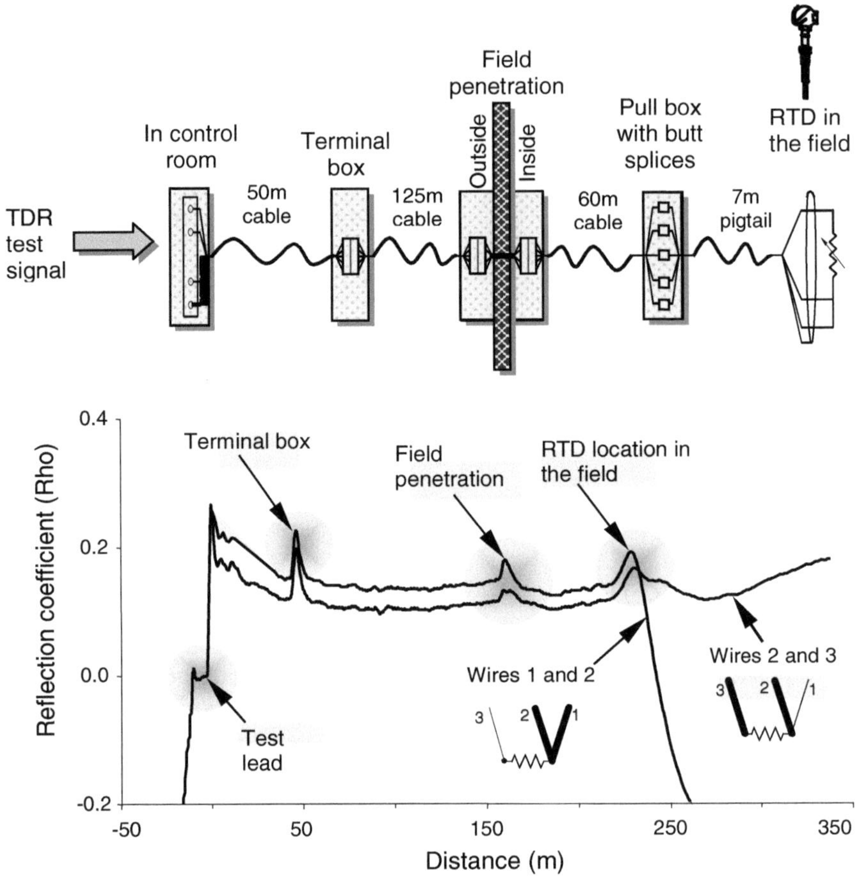

Fig. 1.13. Nuclear plant RTD circuit and corresponding TDR signatures

method called *time domain reflectometry* is used to test and troubleshoot cables in nuclear power plants. This involves sending an electrical signal through the cable and plotting its reflection as a function of time or distance along the cable (Fig. 1.13). The plot corresponds to the cable's impedance signature and is useful for locating such anomalies as an open, a short, or a shunt either along a cable or in the device at the end of the cable (e.g., an RTD, a thermocouple, or a neutron detector).

The TDR test is useful for performing cable diagnostics in nuclear power plants, especially if baseline TDR signatures are available for comparison purposes. For example, as soon as a nuclear power plant receives an anomalous signal from a sensor such as an RTD, a thermocouple, or a neutron detector, a question typically arises: is the problem inside or outside the reactor containment? If the problem is found to be inside the reactor containment, a second question usually arises: is the problem in the cables or in the end device (i.e., the sensor or detector)?

The TDR technique, when used with other electrical measurements such as resistance (R), capacitance (C), and inductance (L), can often help to answer these questions. The R, C, and L can all be measured using the same equipment referred to as an *LCR meter*.

The combination of TDR, LCR, and LCSR tests has proved very effective in separating cable problems from sensor problems in RTDs, thermocouples, and strain gauges. As for other nuclear plant sensors such as neutron detectors, the combination of TDR, LCR and the noise analysis technique are used to verify the integrity of the cables and performance of the end device, in this case, the neutron detector.

1.10 Automated Maintenance

In recent years, computer-aided maintenance has become popular in nuclear power plants. For example, in PWR plants, a significant number of control and shutdown rods are normally kept above the reactor core during normal plant operation at full power. When an event occurs that requires the reactor to be scrammed, these rods are suddenly released. They drop by force of gravity into the core and shut the plant down as quickly as possible. For this reason, the time it takes for the rods to drop from the top to the bottom of the core is often critical. It is therefore mandatory for most PWR plants to measure the drop time of their rods after each refueling outage and after they perform any maintenance work that involves removing the reactor head assembly.

Traditionally, measuring rod drop time has been done by dropping one rod at a time and recording the output of the corresponding rod position indicator on a strip chart recorder. With computer-aided data acquisition and data analysis, however, all the rods can now be dropped simultaneously and their drop time measured automatically. Typically, one bank of rods (comprising up to nine individual rods) is dropped simultaneously for this measurement. Fig. 1.14 shows the results of a rod drop-time measurement for a bank of rods in a PWR plant. This data represents the output of rod position indication coils as a function of time as the rods drop from the top to the bottom of the reactor and settle in their dashpots (a dashpot is a shock-absorbing section located at the bottom of the guide tubes through which the rods move). The plot is used to measure the rod drop times and also to detect any problems with rod movement (such as sticking or inadequate rod insertion).

Since rod drop time is typically measured during critical path at startup, using automated testing to test multiple rods saves hours of critical path time and yields great economic benefit to the plant.

To start the reactor or manipulate reactor power, the rods are moved in and out of the core using an electromechanical system called the *control rod drive mechanism*. In Westinghouse PWRs, a CRDM consists of three coils that operate arms that hold and/or move the rods. These coils are referred to as *stationary gripper coil* and *lift coil*. The stationary gripper coil holds the rod in place until the moveable coil latches onto it. The lift coil then moves the whole assembly. The operation of the three coils must occur with correct timing and sequencing or a rod can inadvertently fall into the core. To ensure the correct timing and sequencing of the CRDM system, the electrical

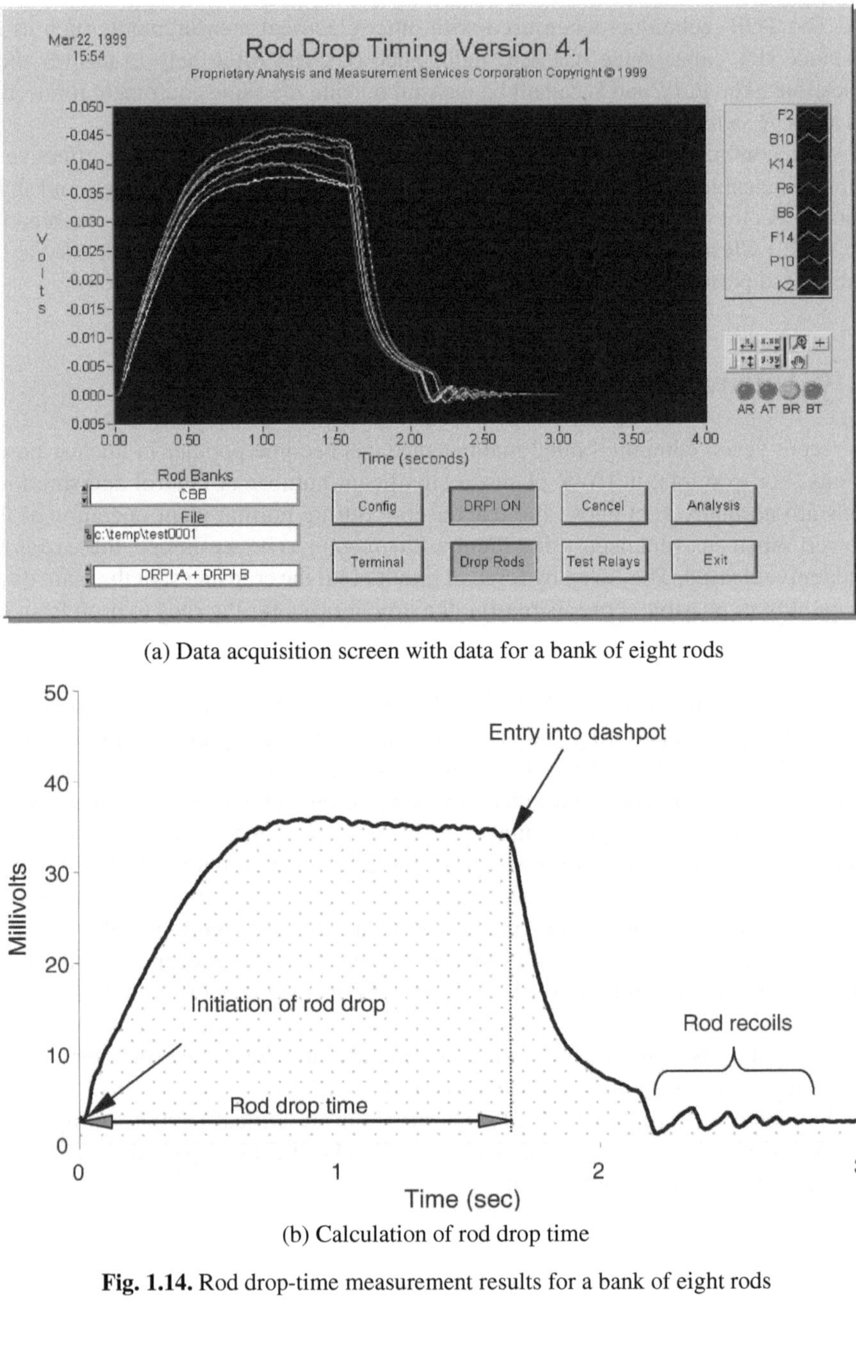

(a) Data acquisition screen with data for a bank of eight rods

(b) Calculation of rod drop time

Fig. 1.14. Rod drop-time measurement results for a bank of eight rods

currents that activate the coils are monitored and their timing and sequencing measured after each refueling outage or maintenance activity that involves the CRDMs. In the past, CRDM timing and sequencing tests have been performed on one rod at a time

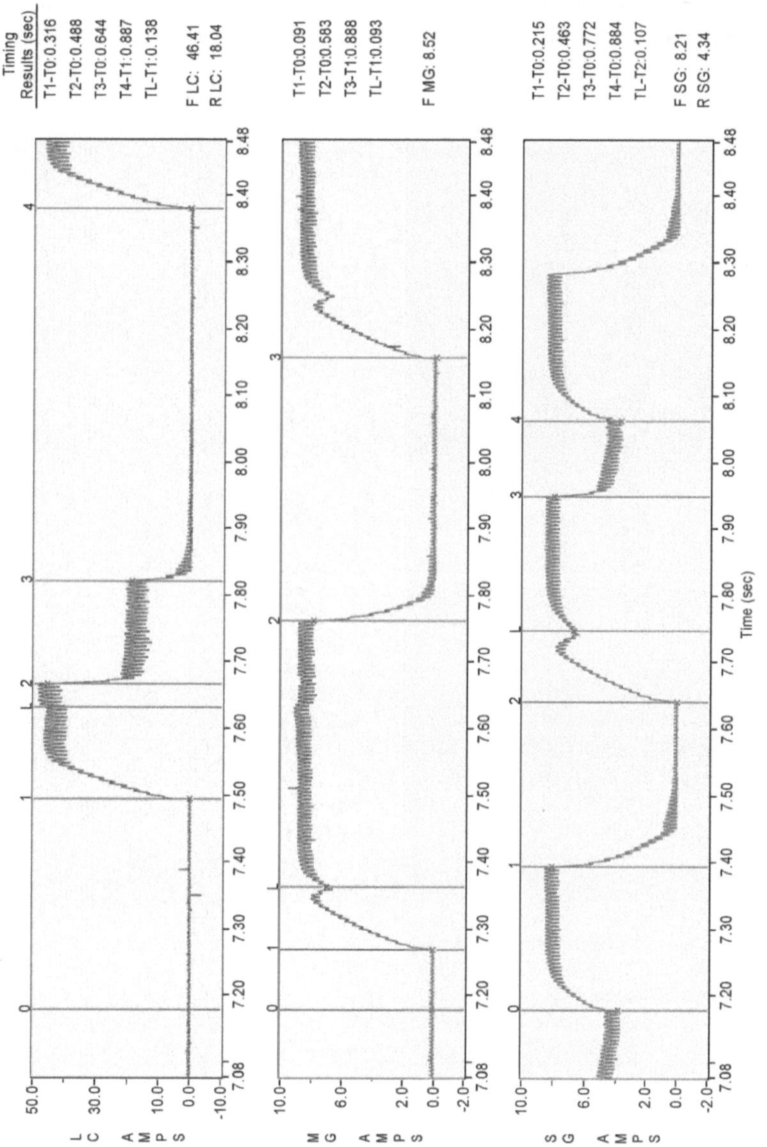

Fig. 1.15. Results of automated testing of CRDMs and calculation of timing events

and the data displayed on a strip chart recorder and visually examined to verify proper CRDM operation. Furthermore, the timing events were calculated manually. Obviously, this was a time-consuming exercise that was eventually automated. As a result, with computer-aided testing, multiple CRDMs are now tested simultaneously and their timing and sequencing are characterized automatically. Fig. 1.15 shows the results of an automated testing of a CRDM and the calculation of the CRDM timing events.

2

Origins of This Book

The material in this book stems from research and development (R&D) activities as well as measurements and diagnostics performed by the author and his associates at the Analysis and Measurements Services Corporation (AMS) from 1975 through 2006. This book complements the author's previous book, *Sensor Performance and Reliability*, published by the Instrumentation, Systems, and Automation Society (ISA) in 2005.[1] That earlier work presented the fundamentals of process instrumentation. This book will focus on process instrumentation testing and diagnostics, using actual examples and practical data from testing and diagnostic measurements performed in the process industries, aerospace applications, nuclear power plants, and simulated process conditions at the AMS laboratories.

The activities from which the material in this book are drawn have been performed in association with the Oak Ridge National Laboratory (ORNL), the University of Tennessee in Knoxville (UT), the Electric Power Research Institute (EPRI) and its Nuclear Maintenance Assistance Center (NMAC), Electricité de France (EdF), the Saclay laboratories of the Commissariat à l'Energie Atomique (CEA) of France, the NRC, the National Aeronautics and Space Administration (NASA), the U.S. Air Force, and utilities around the world that operate nuclear power plants. Moreover, the author's association with the International Atomic Energy Agency (IAEA), the International Electrotechnical Commission (IEC), and ISA has enabled him to help develop several national and international standards and guidelines for testing the instrumentation and control (I&C) systems of nuclear power plants. This book also draws from these activities.

A bibliography is provided in Appendix A that lists numerous technical papers, magazine and journal articles, reports, books and book chapters on the activities just mentioned.

2.1 Collaborative R&D

Back in the early 1970s, the Instrumentation and Control Division of ORNL was involved in several projects to develop new equipment and techniques for testing and performing diagnostics in nuclear power plants. For example, at that time an LMFBR called the *Clinch River Breeder Reactor (CRBR)* was being built in the United States, and ORNL played a supporting role in its development. Typically, the temperature of the liquid sodium used in the reactor coolant system of LMFBRs is measured using thermocouples. The dynamic response of these thermocouples is supposed to be fast so timely temperature measurements can be made if an unusual transient occurs in the reactor. For this reason, ORNL was tasked with developing an in-situ technique for measuring the dynamic response of thermocouples installed in liquid metal. ORNL engineers identified the LCSR method, originally conceived at NASA, as the best candidate for this application and began developing it at ORNL.

In the meantime, the NRC issued Regulatory Guide 1.118, which recommended that PWR plants measure the response time of their safety-related RTDs. This recommendation stimulated EPRI to fund an R&D effort at UT to adapt the LCSR method for RTDs. The author, then a graduate student at UT, worked on the EPRI project and, with the help of others, developed prototype equipment including hardware, software, and procedures for LCSR testing of RTDs in nuclear power plants. During these projects, the author worked on the LCSR technology not only at ORNL and UT but also in France, in collaboration with both EdF and CEA. Specifically, the work with EdF was carried out at the Les Renardieres laboratory near Paris, and the work with CEA was performed at the Saclay laboratory, also near Paris.

The Les Renardieres laboratory had a test loop for simulating PWR operating conditions in which EdF had installed a test section to accommodate the testing of RTDs at temperatures of up to 300° C (572° F), pressures of up to 150 bars (about 2,250 psi), and flow rates of up to 10 meters per second (about 30 feet/second). This loop was used to validate the LCSR method for response-time testing of RTDs at PWR operating conditions. Before this validation effort, almost all work on LCSR development had been conducted under laboratory conditions, with the exception of a limited number of tests at ORNL's High Flux Isotope Reactor (HFIR). The tests at the EdF loop in Les Renardieres provided data that demonstrated the validity and established the accuracy of the LCSR method for measuring the in-service response times of RTDs at PWR plants.

At the Saclay laboratory, where a flow loop had been developed to test sensors, additional LCSR validation tests were performed to supplement the work performed at Les Renardieres. The noise analysis technique was also examined as a way of testing the response time of RTDs and thermocouples. This technique was found to provide reasonable results, although not generally as accurate as those provided by the LCSR method. Some work on validating the noise analysis technique had also been performed earlier at the EdF loop in the Les Renardieres laboratory, and the same conclusion had been reached: the noise analysis technique has the potential to provide an in-situ means of measuring the response time of RTDs and thermocouples as installed in operating processes.

The first in-plant demonstration of the LCSR test was performed at the Millstone nuclear power station Unit , where 16 RTDs were tested for response time. The results of this and earlier R&D efforts on the LCSR method were then documented in a topical report on the Millstone plant.[2] This report was written by AMS under a contract with the Northeast Utilities Company, which operated the Millstone plant. Northeast Utilities submitted the topical report to the NRC with a request to approve the LCSR method for RTD response-time measurements in nuclear power plants. After about two years of debate, meetings, and question-and-answer sessions with the NRC, in 1980 the NRC approved the LCSR method as an acceptable method for meeting the Regulatory Guide 1.118 recommendations and complying with nuclear power plants' technical specification requirements for RTD response-time verification.

This is just one example of an R&D effort jointly undertaken by ORNL, UT, EPRI, EdF, CEA and a utility in support of the nuclear power industry. Some of these organizations have also been involved with AMS and others in developing testing and diagnostics techniques for a variety of other nuclear power plant applications. These applications include the in-situ response-time testing of pressure, level, and flow transmitters; the on-line detection of blockages, voids, leaks, and standing waves in pressure sensing lines; the measurement of vibration in reactor vessels and their internals; the measurement of stability margins in BWRs; applications for monitoring loose parts; and the on-line detection of core flow anomalies, flow blockages, and coolant transmission path. Aside from research in America and France, development work in these areas has also been carried out in Germany, Hungary, Japan, the Netherlands, Russia, South Korea, and other countries since 1975. Nuclear industry experts from around the world have published numerous papers on these efforts. The author has used updated summaries of these developments as much as possible in writing this book.

2.2 Government R&D

R&D efforts supporting nuclear energy that are funded by national governments and international government organizations are usually carried out at the major national and international laboratories and by their contractors. The ORNL in the United States and Saclay of CEA in France are just two examples. Internationally, the Halden Reactor Project (HRP) in Norway is an example of a laboratory that has the international funding to perform R&D work supporting nuclear energy and related technologies.

In the United States, a government R&D program was established in the early 1980s to stimulate innovation by individuals and small companies (defined as firms with up to 500 employees and annual revenues of less than $25 million in 2006). This program, referred to as *Small Business Innovation Research (SBIR)*, provides funding of up to about $1 million over three years to subsidize R&D and commercialization efforts in selected technical topics. These topics are identified by the government as those that meet the government's R&D needs and at the same time foster innovation in the private sector and the commercialization of government-funded work.

Under the SBIR program, AMS has conducted R&D work for the U.S. Department of Energy (DOE), the U.S. Department of Defense (DOD) for the U.S. Air Force, for NASA, and for the NRC. The results of these projects have been documented in several government reports, such as the NUREG/CR series of reports published by the NRC as well as NASA and Air Force reports. A representative list of these reports includes the following:

- NRC reports on the performance and aging characteristics of nuclear plant RTDs
  - NUREG/CR-4928
  - NUREG/CR-5560
- NRC reports on the performance and aging characteristics of nuclear plant pressure transmitters
  - NUREG/CR-5383
  - NUREG/CR-5851
- NRC reports on the development or assessment of advanced I&C maintenance technologies for nuclear power plants
  - NUREG/CR-6343
  - NUREG/CR-5501
- NRC reports on the development or assessment of new sensors for nuclear facilities
  - NUREG/CR-6312
  - NUREG/CR-6334
- Air Force reports on transient temperature measurements in jet engine test facilities
  - AEDC-86-46
  - AEDC-TR-91-26
- NASA reports on the development of in-situ methods for assessing the bonding quality of sensors to solid material
  - NASA-Phase I Report (not published by the government)
  - NASA/CR-4744

The material in these reports, as well as other R&D work—such as a noise analysis technique for diagnostics in nuclear power plants developed under an SBIR-funded DOE contract—provides the basis for some of the material covered in this book. Most of these reports are available publically from sources identified in Appendix A.

2.3 Utility R&D

The bulk of the R&D work funded by U.S. utilities is performed under the direction of EPRI. EPRI is headquartered in Palo Alto, California, but has affiliated offices such as the NMAC, which is located in Charlotte, North Carolina. EPRI is funded by utilities to support the power generation industry, and it sponsors R&D, develops reports and guidelines, and organizes technical meetings, seminars, and training courses. It also represents the interests of its member utilities in technical interactions with government agencies and others. Many utilities in the United States as well as some utilities in other countries are members of EPRI. These members contribute funding and advice to EPRI in return for access to its products and resources, some of which are available to nonmembers for a fee.

A recent example of an EPRI undertaking involving nuclear power plant instrumentation is the compilation and, to a lesser extent, development of on-line monitoring techniques for verifying instrument calibration. Not only did EPRI contribute to the development and compilation of these techniques, but it also helped to obtain the NRC's approval to use these techniques in nuclear power plants. In particular, EPRI wrote a topical report in the late 1990s on the subject of on-line monitoring of the calibration of pressure transmitters. This report was submitted to the NRC on behalf of the nuclear industry to obtain that agency's approval to implement an on-line monitoring approach that would extend the calibration interval of pressure, level, and flow transmitters in nuclear power plants. After a few years of discussion and debate, the NRC accepted the on-line calibration monitoring approach by publishing, in July 2000, a Safety Evaluation Report (SER) on this issue. The SER has cleared the way for the nuclear industry to proceed toward performance-based calibration of pressure, level, and flow transmitters.[3] To date, the Sizewell nuclear power plant in the UK has successfully implemented the on-line monitoring approach to verify the calibration of pressure, level, and flow transmitters in the primary and secondary systems of the plant, and the V.C. Summer nuclear power plant in the USA has applied to the NRC for a change in the plant technical specification requirements to switch from time-based calibration of process instruments to condition-based calibrations.

An example of EPRI's NMAC work is the development of a guideline document so nuclear power plant personnel can understand how to operate and maintain rod control systems in PWR plants. In addition to tutorials on the principles of operation of rod control systems, this guideline also presents automated techniques for measuring the drop times of control and shutdown rods and for testing the timing and sequencing of CRDMs and their associated slave cyclers in PWR plants.[4]

The R&D needs of utilities are met not only through EPRI but also through national and international laboratories as well as in-house R&D facilities and outside vendors. For example, the Duke Power Company and the Tennessee Valley Authority (TVA) have their own means for developing new equipment and techniques that support their plants or for conducting R&D to find the answers they need. Generally, in cases that are specific to a particular plant or situation, utilities may fund their own R&D work or contract a specialist or a vendor to find the answer.

As an example, the Arkansas Nuclear One (ANO) power station, which consists of two PWR units, contracted a research project to AMS to support its Unit 2 reactor (ANO-2). In the late 1970s, it was revealed that some of ANO-2's primary coolant RTDs could not readily meet the plant's technical specification requirement for a response time of 6.0 seconds. At that time, no other nuclear-qualified RTDs were available that could easily meet that requirement. For that reason, a thermal coupling compound called *"NEVER-SEEZ"* had to be used at the tip of the RTD's thermowell to improve heat transfer and reduce the response time. The response time of ANO-2's RTDs was reduced from an average of about 6.0 seconds to an average of about 4.0 seconds. This allowed the plant to continue to operate until an alternative RTD or thermowell could be found to meet the plant's requirement for better response time.

NEVER-SEEZ is normally used to lubricate threads when fitting metallic parts. It has good lubricating and thermal properties and works well at temperatures of up to about 200° C. At ANO-2, as at most PWRs, the operating temperature in the primary coolant system is 300° C or greater. NEVER-SEEZ could not therefore be used as a long-term solution since its thermal properties degraded at the plant's operating temperatures. AMS was tasked with identifying an alternative to NEVER-SEEZ. The research that ensued identified other thermal coupling compounds, which were tested in the AMS laboratory. None of them showed better long-term characteristics than NEVER-SEEZ. AMS then initiated new research on applying gold or silver plating to the sensing tip of RTDs to improve their response time. This effort showed good results, and the approach has since been used to improve the response time of several RTD models in nuclear power plants. In the meantime, RTD manufacturers developed new sensors and thermowells with better response-time characteristics, resolving the industry's struggle to meet response-time requirements.

2.4 IAEA Guidelines

The IAEA produces technical reports (referred to as *TECDOCs*) and guideline documents to disseminate information (existing as well as new), on a variety of subjects in support of the nuclear industry. Each IAEA report or guideline document is typically prepared by international experts whom IAEA invites to periodically share their information with IAEA member countries at IAEA headquarters in Vienna, Austria, or elsewhere. These experts, generally a committee of five authors, agree on the document's content and write the material on a consensus basis over a period of one to two years. The IAEA then invites advisors from other member countries to review the document, contribute to its accuracy and completion, and develop consensus. The documents are then published and provided to IAEA member countries at little or no charge, and meetings, conferences, and workshops are organized to explain the documents' content and make the nuclear industry aware of the information available through the IAEA.

A few examples of IAEA documents on subjects related to this book include the following:

- IAEA-TECDOC-1147, "Management of Aging of I&C Equipment in Nuclear Power Plants," June 2000.
- IAEA-TECDOC-1327, "Harmonization of the Licensing Process for Digital Instrumentation and Control Systems in Nuclear Power Plants," December 2002.
- IAEA-TECDOC-1402, "Management of Life Cycle and Aging at Nuclear Power Plants: Improved I&C Maintenance," August 2004.
- New IAEA document, "On-line Monitoring for Nuclear Power Plants, Part 1: Instrument Channel Performance Monitoring," to be published in 2006.
- New IAEA document, "On-line Monitoring for Nuclear Power Plants Part 2: Process and Component Condition Monitoring and Diagnostics," to be published in 2007.

The author has served on the IAEA committees that wrote these and other international documents, guidelines, and reports. This book draws on this experience.

2.5 ISA and IEC Standards

Consensus standards are developed by ISA, IEC, and other organizations, such as the Institute of Electrical and Electronics Engineers (IEEE) and the American Society for Testing and Materials (ASTM), to establish requirements for the nuclear and other industries' process operations and maintenance and hardware and software development efforts. Sometimes, these standards are endorsed by national standard organizations such as the American National Standards Institute (ANSI). A few examples of such standards include:

- ANSI/ISA Standard 67.06.01-2002, "Performance Monitoring for Nuclear Safety-Related Instrument Channels in Nuclear Power Plants," 2002.
- ANSI/ISA Standard 67.04.01-2000, "Setpoints for Nuclear Safety-Related Instrumentation," 2000.
- IEC Standard 62342, "Nuclear Power Plants – I&C Systems Important to Safety – Management of Aging," due for publication in 2007.
- IEC Standard 62385, "Nuclear Power Plants – I&C Methods for Assessing the Performance of Safety System Instrumentation Channels," due for publication in 2007.
- IEEE Standard 323, "IEEE Standard for Qualifying Class 1E Equipment for Nuclear Power Generating Stations," 2004.
- IEEE Standard 338, "IEEE Standard Criteria for the Periodic Surveillance Testing of Nuclear Power Generating Station Safety Systems," 1988.
- ASTM Standard E644, "Standard Test Methods for Testing Industrial Resistance Thermometers," 2004.
- ASTM Standard E230, "Specification and Temperature-Electromotive Force (EMF) Tables for Standardized Thermocouples," 2003.

The author has served as a writer, task leader, or contributor to some of these and other standards, and his reflections on this experience contributed to the writing of this book. Some of the more relevant standards can be briefly summarized as follows:

- ANSI/ISA Standard 67.06.01. This standard was originally written in the early 1980s to describe the methods for measuring the response times of temperature and pressure sensors in nuclear power plants. It was revised in the late 1990s to include on-line monitoring techniques for verifying the calibration of process sensors during plant operation. The title of the original 67.06 standard, published by ISA in 1984, is "Response Time Testing of Nuclear Safety-Related Instrument Channels in Nuclear Power Plants." The new revision was published in 2002. The title of the new revision is "Performance Monitoring for Nuclear Safety-Related Instrument Channels in Nuclear Power Plants."
- ASTM Standard E644-04. This standard is concerned with industrial RTDs, from construction and specification to testing requirements. For example, the standard describes the methods that sensor manufacturers and others shall use to measure RTD response times in reference laboratory conditions. The final version of this standard is dated 2004, and its title is "Standard Test Methods for Testing Industrial Resistance Thermometers."

- IEC Standard 62385 (to be published in 2007). This standard covers requirements for testing the performance of nuclear plant sensors and includes the LCSR and noise analysis methods for testing sensor response times. This standard supersedes the IEC Standard 61224, which was issued in 1993 to provide requirements for RTD response time testing using the LCSR and noise analysis techniques.
- IEC Standard 62342 (to be published in 2007). This standard provides general guidelines as to the steps that shall be taken in nuclear power plants to ensure that normal aging of safety-related instrumentation does not pose a threat to the plant safety. This standard was developed based on guidelines in IAEA TECDOC-1147 on "Management of Aging of I&C Equipment in Nuclear Power Plants."

Although standard-writing individuals, organizations, and committees work hard to produce up-to-date and accurate national and international standards, there is no guarantee that a standard takes into account all the necessary points or establishes all the relevant requirements. In some instances, satisfying a standard's requirements may not be enough to ensure proper operation and safety, no matter how up to date the standard.

Standards are typically written by a small group of volunteers who have expertise and interest in the subject as well as the means to participate in the standard's preparation. Often, vendor organizations get involved in standard-writing activities to help establish requirements for the industry. Sometimes, in contributing to a standard, the vendors also promote the use of their products and ideas. Neither single vendors or individuals nor communities of vendors or special interest groups are supposed to influence or dominate a standard, and consensus is normally achieved in all the subjects covered in the standard. However, the potential for conflict of interest always remains. As such, the users of a standard document should not rely on a standard or even a group of standards as the sole source of information, guidelines, or requirements on ensuring proper operation and safety.

3

Maintenance of Nuclear Plant Instrumentation

Maintenance of nuclear plant instrumentation should typically involve the following tests:[5]

1. Verifying calibration;
2. Measuring response time;
3. Testing cables; and
4. Performing noise diagnostics.

These tests are performed for a variety of reasons. For example, in almost all plants, calibration verification is mandatory for important instrumentation. In other plants, in addition to calibration verification, the response time of process instrumentation sensors must be measured to ensure compliance with plant technical specifications and/or regulatory requirements. Cable testing and noise diagnostics are not typically mandatory, but they are usually performed for troubleshooting purposes or to identify the root cause of signal anomalies and other I&C problems. Trending of calibration and response-time testing results, together with cable testing and noise diagnostics, are often recommended in nuclear industry standards and guidelines as a means for performing predictive maintenance and managing the aging of I&C equipment.

The term *maintenance* in this book means verifying equipment performance not fixing anything. The maintenance may involve active or passive tasks. Examples of active tasks are: measuring, monitoring, calibrating, or analyzing. Examples of passive tasks are: looking, listening, feeling, or smelling. For example, the dynamic performance of a sensor is verified by measuring its response time (an active maintenance task), and the condition of its cable insulation may be assessed by visually examining its color, texture, and integrity (a passive maintenance task).

4

Nuclear Plant Temperature Instrumentation

Most critical process temperatures in nuclear power plants are measured using RTDs and thermocouples. For example, in a PWR plant, the primary coolant temperature and feedwater temperature are measured using RTDs, and the temperature of the water that exits the reactor core is measured using thermocouples. These thermocouples, called *core-exit thermocouples*, are mainly used for temperature monitoring purposes and are therefore not generally subject to very stringent requirements for accuracy and response-time performance. In contrast, primary coolant RTDs typically feed the plant's control and safety systems and must, therefore, be very accurate and have good dynamic performance. Because of the importance of RTDs and the stringent requirements for verifying their performance in nuclear power plants, the remainder of the material on temperature measurement in this book will focus mostly on how to verify that RTDs are working properly.

4.1 History of RTDs

Used since the nineteenth century, RTDs have had sensing elements made out of platinum, copper, nickel, and other metals or alloys in which electrical resistance is proportional to temperature. Today, the sensing element of industrial RTDs are almost always made from platinum wire. Early RTDs were often fragile and unstable because the platinum sensing element became contaminated. Today, industrial RTDs are very rugged and reliable, and can be used in applications as extreme as the measurement of: (1) brake temperatures of over 1,000° C in high-speed aircraft; (2) primary coolant temperature in PWRs at temperatures up to about 350° C, flow rates of over 10 meters per second, and pressures of about 150 bar; and (3) oceanographic temperature under very high hydraulic pressures where high accuracy and quick response (less than 0.5 sec.) are important.

In 1821, Sir Humphry Davy observed that the conductivity of various metals decreased with changes in temperature. The first attempt to use this property to measure temperature, however, was made around 1871 by C.W. Siemens. Siemens constructed RTDs made of platinum wire that were wrapped around a clay mandrel and inserted into an iron sheath. Siemens' original RTDs suffered from large changes in RTD resistivity at high temperatures and could not be used for precision thermometry.

Sir Humphry Davy
1778-1829
England

C.W. Siemens
1823-1883
Germany

H.L. Callendar
1863-1930
England

Around 1891, H.L. Callendar recognized that the problem with Siemens' RTDs was caused by contamination of the platinum wire by the clay and the iron. He therefore produced a new design that consisted of a 0.15 mm platinum wire wound on an approximately 9 cm mica strip with thin mica disks spread along its sheath to minimize convection effects. It used either copper or silver leads and was built with highly pure, strain-free platinum. Callendar's work lead to the development of classical RTDs by C.L. Meyers in 1932. The RTDs used today originated from Meyers' work and have been ruggedized for industrial applications using such material as a ceramic mandrel to secure the sensing element and magnesium oxide (or equivalent) for insulation.

4.2 Nuclear-Grade RTDs

There are nearly 100 suppliers of RTDs around the world, but fewer than 10 of them manufacture RTDs for safety-related applications in nuclear power plants. This is because the size of the market for this application is relatively small and the performance and reliability required of these RTDs are very stringent. For example, nuclear safety-related RTDs must pass environmental and seismic testing per IEEE standards to demonstrate that they can survive a loss-of-coolant accident (LOCA), withstand a seismic event, and continue to provide reliable service under postaccident conditions.

The ability of a set of nuclear-grade RTDs to meet these requirements was demonstrated after the accident at the Three Mile Island Unit 2 (TMI-2) nuclear power plant in the United States in 1979. In particular, the primary coolant RTDs (Rosemount Model 177) at TMI-2 were tested after the accident and found to have maintained their calibration, dynamic response, and integrity. In contrast, the core-exit thermocouples at TMI-2 largely failed because of radiation damage. Some of the thermocouples were indicating erroneously high temperatures, and others had very low and implausible temperature indications (including negative values) at nearly the same locations. An evaluation of these thermocouples indicated that the inconsistent readings were due to inhomogeneity, which developed in the thermocouple wires after they were exposed to high radiation.

RTDs with nuclear safety-related qualifications are mostly used in the primary coolant system of PWR plants. Depending on the plant, there are normally between 16 and 32 RTD elements in a plant. Some nuclear power plants such as ANO-2 in the United States and the Sizewell B nuclear power plant in Great Britain have more RTDs for specific reasons. For example, ANO-2 has more RTDs because of a temperature streaming problem in its hot-leg pipes. The temperature streaming is inherent in the hot-leg pipes of PWR plants, and its severity depends on the plant. At ANO-2, the problem happens to be more severe than other plants, but it does not pose a risk to the plant's operation or safety. The problem is referred to at ANO-2 as the "hot-leg anomaly." The additional RTDs are used to help offset the impact of this "hot-leg anomaly." At Sizewell B, there are 60 primary coolant RTDs simply because this plant has a digital I&C system with a complete analog backup system and therefore twice as many sensors as most other PWR plants.

A simple calculation based on the number of RTDs per PWR plant and the total number of PWR plants around the world shows that there are fewer than 10,000 nuclear safety-related RTDs in the worldwide inventory of plants, including spares. Considering that the average life of these RTDs is about 20 years, the overall market is rather small (i.e., fewer than 1,000 new RTDs per year). Therefore, the cost of nuclear safety-related RTDs is generally very high compared with their commercial counterparts, and only a handful of manufacturers are in the business of producing these RTDs. In fact, most manufacturers of nuclear safety-related RTDs are small companies (fewer than 500 employees) because large companies cannot normally justify the overhead, liability, and stringent quality assurance (QA) required to supply equipment to nuclear power plants. As for non-safety-related RTDs and other temperature sensors for nuclear power plants, such as thermocouples and thermistors, many suppliers and a variety of options are available. Table 4.1 lists some of the suppliers of nuclear-grade RTDs.

In addition to excellent reliability and accident survivability, nuclear safety-related RTDs are expected to have good calibration and fast dynamic response time, as these characteristics are important to plant safety and economy. Fig. 4.1 shows a simplified diagram of a primary coolant loop for a PWR plant. In principle, the reactor power (P) is the product of the temperature difference (ΔT) across the core and the mass flow rate ($\dot{m}$) in the primary coolant system (i.e., $P \approx \dot{m}\Delta Tk$). The ΔT is typically about 30° C; thus, a one-degree error in measuring the ΔT corresponds to 3.33 percent in power output. Therefore, calibrating RTDs accurately is very important to the plant's economy. For that reason, the primary coolant RTDs in PWR plants are typically calibrated to an accuracy of 0.3° C or better before installation. Furthermore, this accuracy is verified periodically while the RTD remains installed in the plant using the RTD cross-calibration technique described in Chap. 5.

Fig. 4.2 illustrates a scenario in which the primary coolant temperature in a PWR plant experiences a step change. In such a scenario, the RTDs are expected to react in a

Table 4.1. Partial listing of suppliers of nuclear-grade RTDs

RTD Manufacturer	Model Number	RTD Type
Conax	7N10	Thermowell-mounted
	7RB4	Direct-immersion
	7N13	Thermowell-mounted
RdF	21204	Direct-immersion
	21297	Direct-immersion
	21232	Thermowell-mounted
	21458	Thermowell-mounted
	21459	Thermowell-mounted
	21465	Thermowell-mounted
Rosemount	104AFC	Thermowell-mounted
	176KF	Direct-immersion
	177HW	Thermowell-mounted
	177GY	Direct-immersion
Sensycon	1703	Thermowell-mounted
	1717	Thermowell-mounted
Weed	N9004	Thermowell-mounted
	N9007	Direct-immersion
	N9019	Direct-immersion

timely manner and trigger a mitigating action, including a reactor scram if needed, to ensure safety. For this reason, there are stringent requirements for the response time of primary coolant RTDs in PWR plants. These requirements differ from plant to plant. For example, in plants where RTDs are installed in thermowells in the primary coolant pipes, typical response-time requirements have a range of 4.0 to 8.0 seconds. This is in contrast with the 1.0 to 3.0 seconds that are required of the direct-immersion RTDs that are installed in bypass loops. As we will see later in Chap. 6, some plants use bypass lines to help sample the reactor water from all coolant loops and mix it before it is used to measure the primary coolant temperature. Because of this, the RTDs in bypass loops normally must be fast to make up for the time delay resulting from diverting the water from the primary coolant pipe to the location where its temperature is measured.

Fig. 4.3 shows a photograph of three nuclear-grade direct-immersion RTDs: one from Rosemount, one from RdF Corporation, and one from Weed Instrument Company. The fastest (<0.5 second response time), the Rosemount RTD, is built with its element attached to the sheath, as shown in the two x-rays and the cross-sectional drawing of Fig. 4.4.

In some PWR plants, direct-immersion RTDs are used in primary coolant pipes as opposed to bypass loops. Fig. 4.5 shows photographs and x-rays of two direct-immersion Rosemount Model 177GY RTDs. These RTDs are installed directly in the primary coolant pipes of PWR plans by Babcock and Wilcox (B&W). Fig. 4.6 shows the thermowell-mounted counterpart of this RTD, which is the Rosemount Model

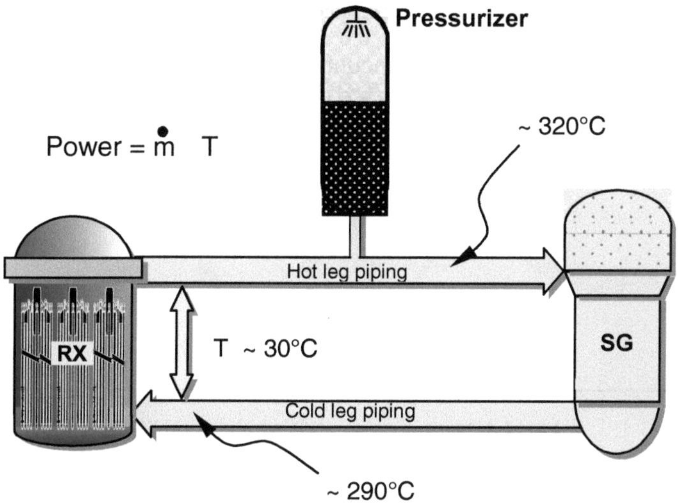

Fig. 4.1. Simplified diagram of a primary coolant loop of a PWR

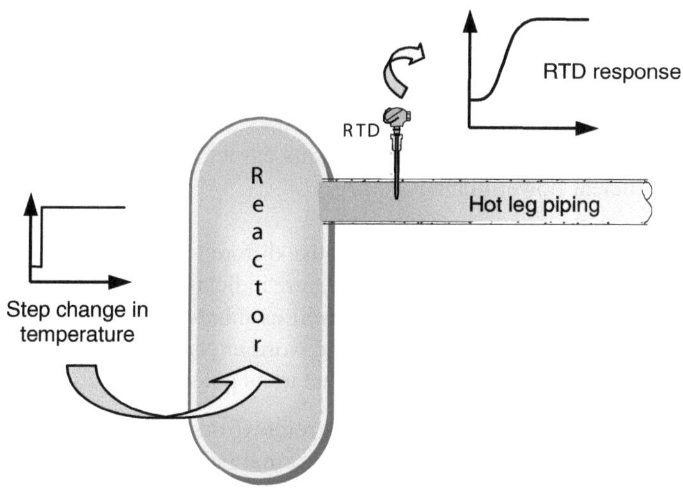

Fig. 4.2. Illustration of RTD response to a step change in temperature in the reactor

177HW, and is also used predominately in B&W plants. This RTD is silver brazed at the tip for improved response time. Other RTDs, such as those from RdF Corporation, also use silver plating or silver brazing on the RTD tip to improve response time. Fig. 4.7 shows a photograph of a silver-plated RdF RTD of the type used in nuclear power plants.

A complete RTD assembly is shown in Fig. 4.8. This RTD, a thermowell-mounted Rosemount Model 104 RTD, is a single-element sensor with a dummy compensation loop. The internals of the RTD are shown in Fig. 4.9. Both the RTD and its thermowell

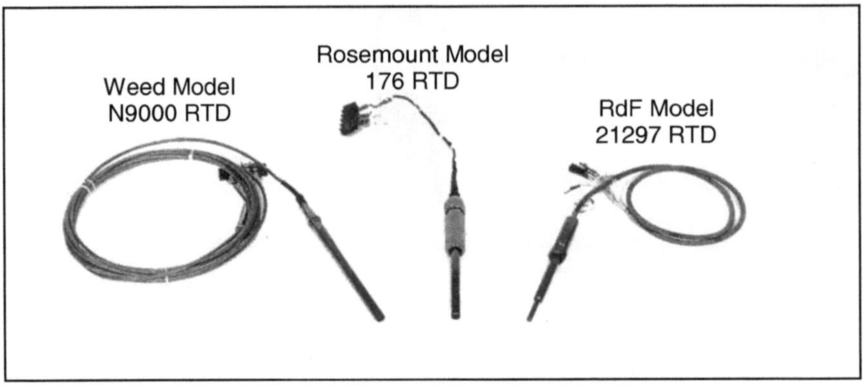

Fig. 4.3. Nuclear-grade direct-immersion RTDs

are tapered at the sensing tip for improved RTD/thermowell mating and better response time. Tapered-tip thermowells come in many varieties (see Fig. 4.10) and are common in nuclear power plants.

4.3 Nuclear Plant Temperature Measurement Terminology

The following are examples of common terms associated with temperature instrumentation in nuclear power plants:

- **Accuracy.** The maximum positive or negative difference that may exist between the actual process temperature and the temperature indicated by the temperature sensor. The term includes calibration errors as well as inherent RTD errors such as hysteresis, repeatability, and self-heating. The word *uncertainty* is a more appropriate term than *accuracy* but it is rarely used due to its potential negative connotation.
- **Aging.** The term *aging* as used in this book refers to decalibration or the degradation of a sensor's response time over time, in normal environments and under normal operating conditions. This definition is based on the NRC's Nuclear Plant Aging Research (NPAR) Program definition of aging, which is, "the cumulative degradation that occurs with the passage of time in a component, system, or structure which can, if unchecked, lead to loss of function and impairment of safety."

  Since the performance of nuclear plant temperature sensors such as RTDs is tested periodically, the degradation does not accumulate. Therefore, the word *cumulative* was deleted in the definition of RTD aging given above.
- **Calibration.** The relationship between sensor output and temperature. A chart that lists an RTD's resistance as a function of temperature is called a *calibration chart* or *calibration table*. A plot of resistance-versus-temperature is called a *calibration curve*. For the levels of accuracy required in the nuclear industry, calibration must

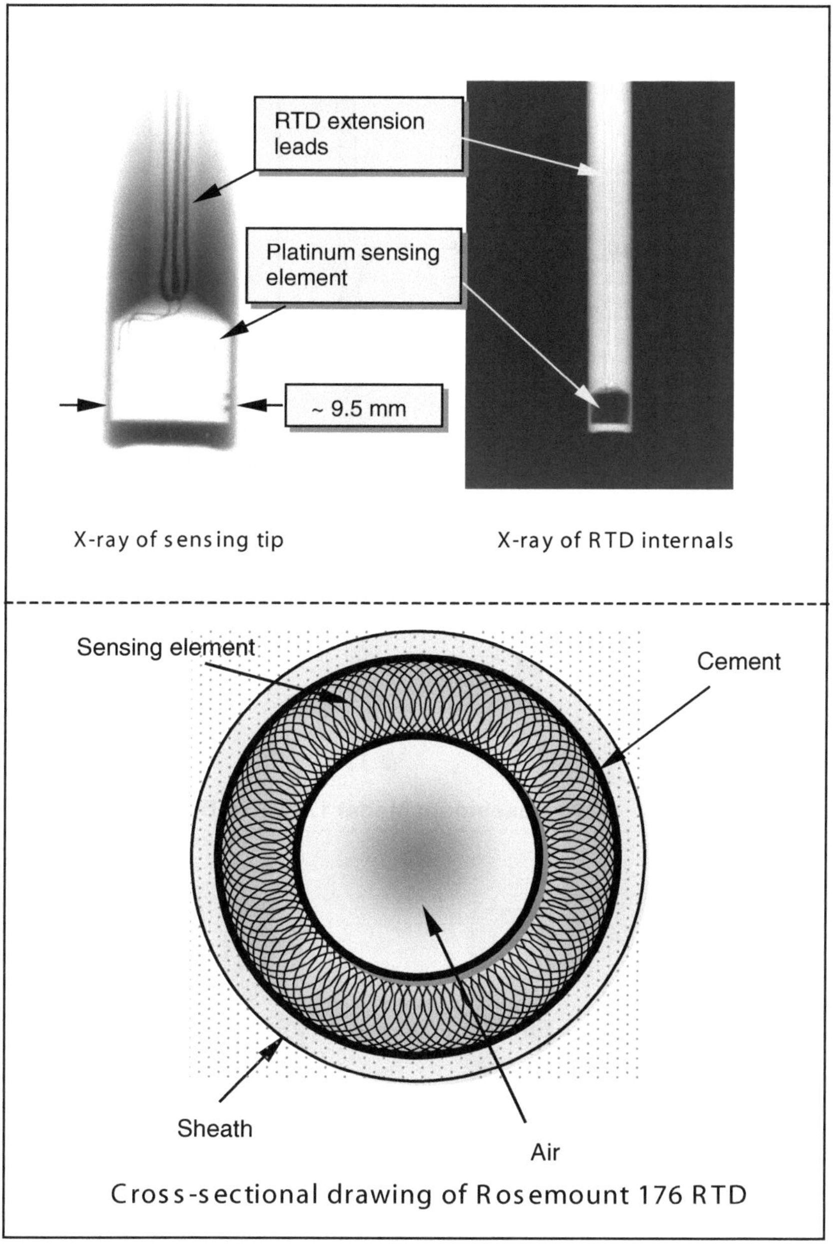

Fig. 4.4. X-rays and cross-sectional drawing of Rosemount Model 176 RTD

be uniquely determined for each RTD. Therefore, nuclear plant RTDs are individually calibrated in a laboratory before being installed in a plant. Thermocouples are not usually calibrated individually. Rather, a representative sample of thermocou-

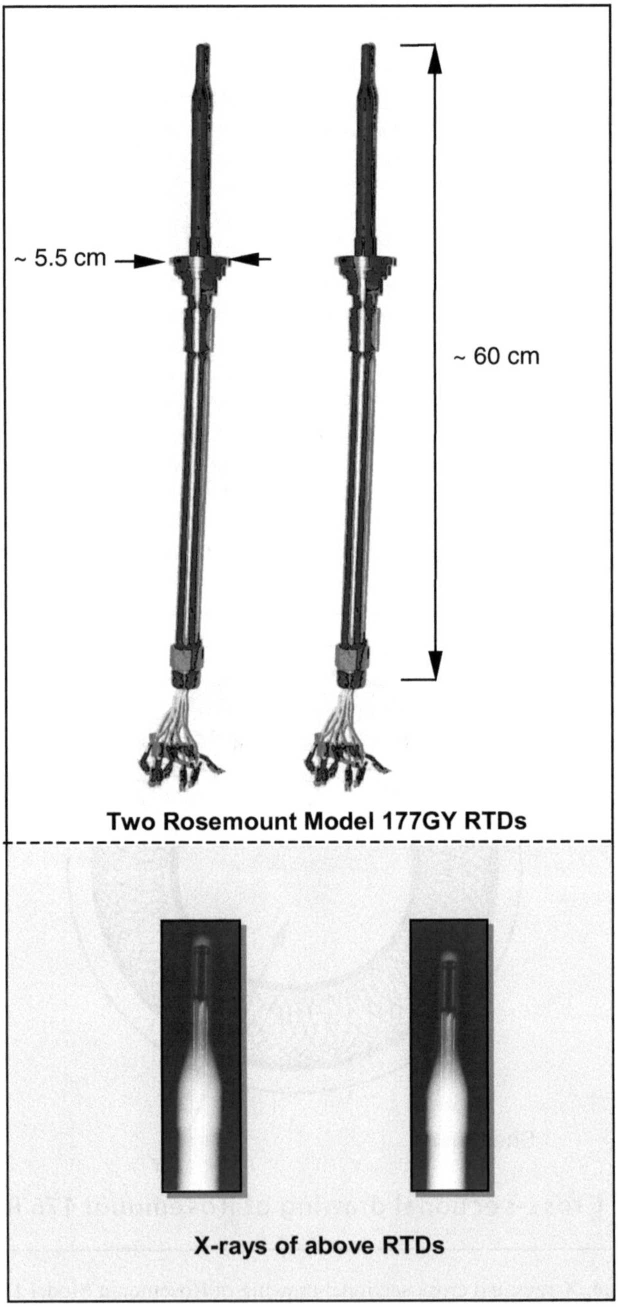

Fig. 4.5. Photograph and x-rays of direct-immersion Rosemount Model 177GY RTDs

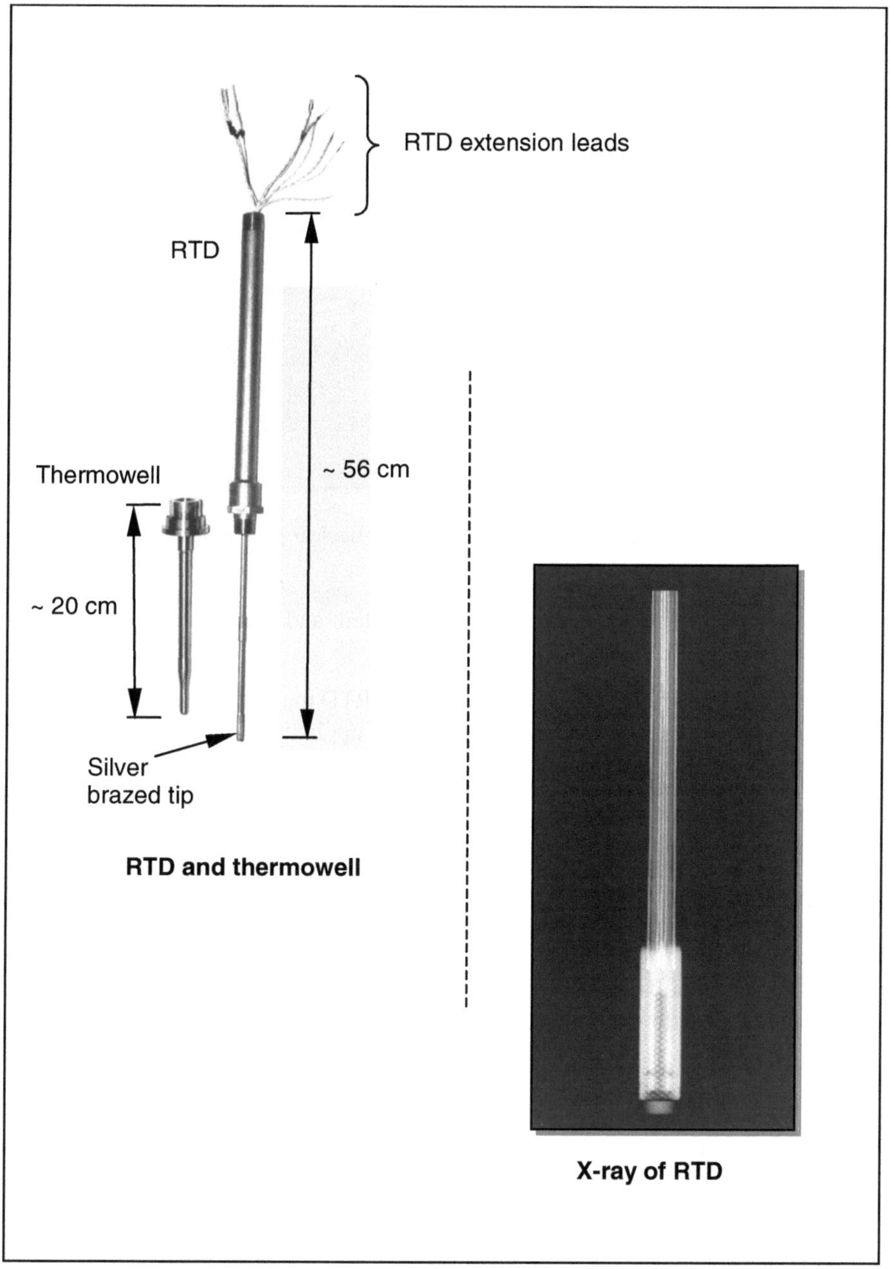

Fig. 4.6. Photograph and x-ray of Rosemount Model 177HW RTD

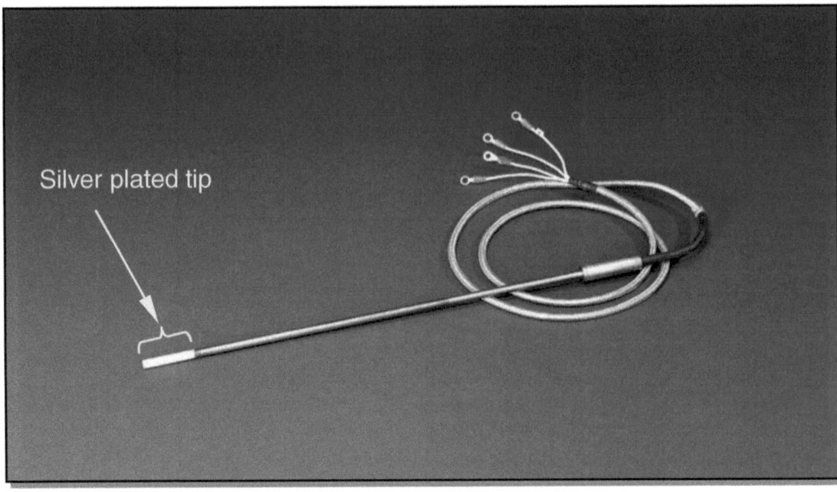

Fig. 4.7. Silver-plated RdF RTD for nuclear power plants

ples made to the same specification is calibrated, and the results are used for the remaining thermocouples in the same batch.

- **Commercial-grade RTD.** A general-purpose RTD made for general industrial applications as opposed to nuclear safety-related applications.
- **Common mode drift.** Unidirectional drift of a group of redundant sensors. If the group drifts all in a positive direction or negative direction, they are said to have "common mode" drift.
- **Cross calibration.** Comparison of the average indication of redundant RTDs with each individual indication in order to check for consistency and identify outliers. Cross-calibration is a method for verifying on-line that the calibration of redundant RTDs have not suffered a significant change. It is based on the assumption that redundant RTDs do not suffer common mode (unidirectional) drift.

  Thermocouples are also cross-calibrated in some plants. Usually, thermocouples are cross-calibrated against the average reading of redundant RTDs.
- **Degradation.** Changes in the calibration or response time of a temperature sensor. Response-time changes are usually called *degradation*, and calibration changes are called *drift* or *shift*.
- **Direct immersion sensors.** See definition of *wet-type sensors*.
- **Drift.** Changes in accuracy over time. Also called *calibration drift, calibration shift, stability,* or *instability.*
- **EMF.** Refers to voltage output of a thermocouple. In RTDs, EMF is an undesirable effect that may arise due to poor design and use of dissimilar metals.

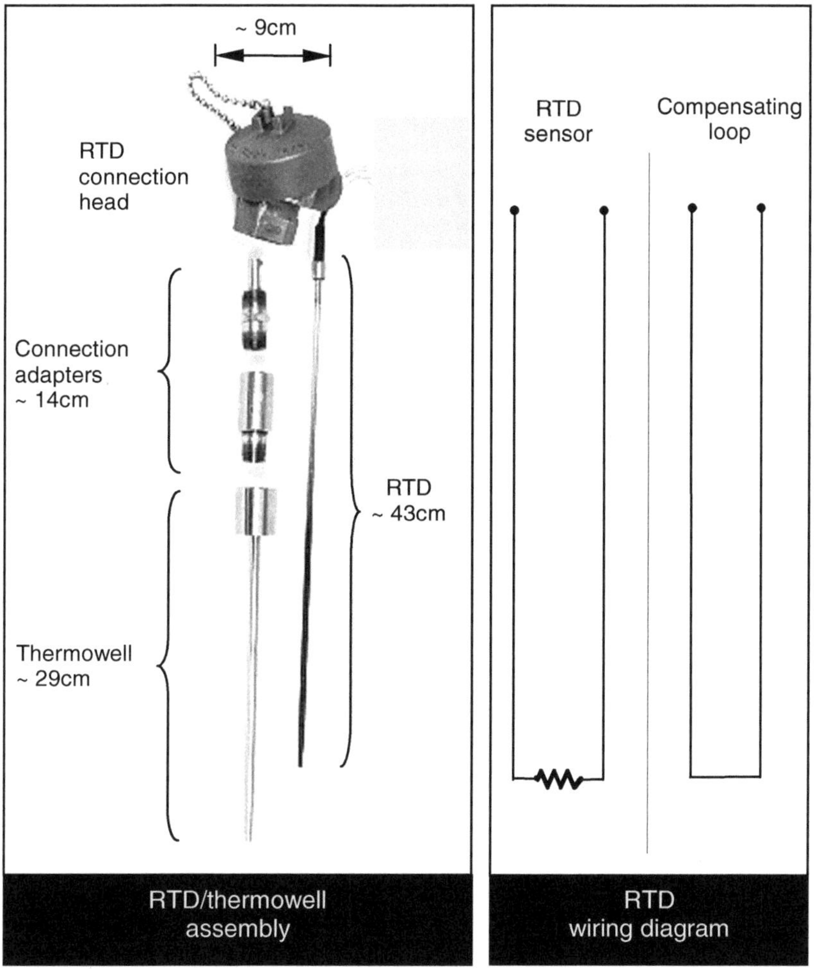

Fig. 4.8. Components of a complete RTD/thermowell assembly (Rosemount Model 104) (Note: New Weed RTDs that replace this RTD have approximately the same configuration and same dimensions)

- **Error.** Synonymous with uncertainty, inaccuracy, or accuracy.
- **In-situ.** See definition of *on-line testing*.
- **Insulation resistance.** The electrical resistance between any extension lead that exits the sensor and earth ground.
- **LER.** A compilation of reportable failures of certain components in nuclear power plants.
- **NIST.** Formerly known as the National Bureau of Standards, or NBS.

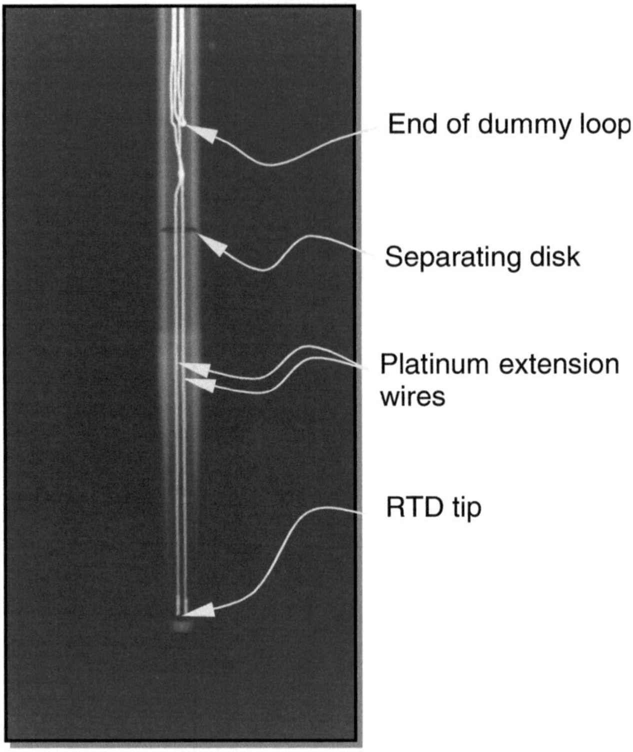

Fig. 4.9. Internal wiring of Rosemount Model 104 RTD of the type used in PWR plants (four-wire RTD including a dummy loop for lead-wire compensation)

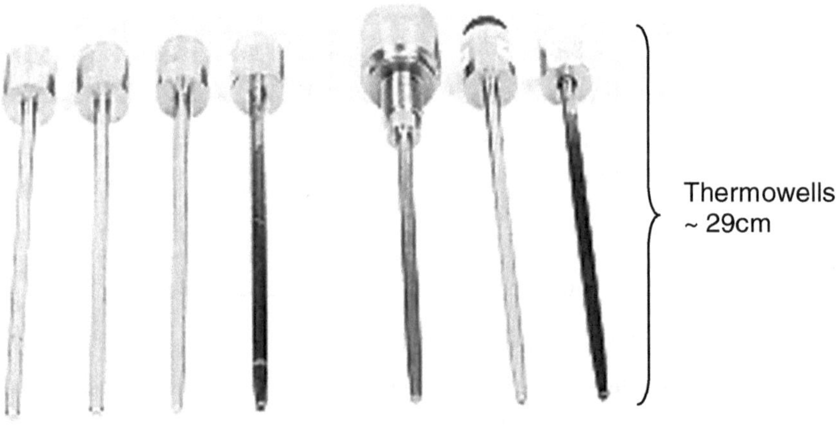

Fig. 4.10. Examples of RTD thermowells of the type used in nuclear power plants

The calibration of temperature sensors and associated test and measurement equipment for safety-related applications in nuclear power plants must be traceable to NIST using transfer standards such as standard platinum resistance thermometers (SPRTs) as well as resistance and voltage standards. The traceability to NIST is typically established by ensuring that the standard that is used for calibration has been calibrated at NIST.

- **Normal aging.** The natural degradation of a sensor's performance as it is subjected to normal environments and typical operating envelopes.
- **NPAR (Nuclear Plant Aging Research Program).** A program initiated by the NRC in the early 1980s to understand how components, systems, or structures in nuclear power plants age.
- **NPRDS (Nuclear Plant Reliability Data System).** A compilation of failure reports for certain nuclear power plant components that nuclear utilities voluntarily file with the Institute of Nuclear Power Operations (INPO).
- **Nuclear-grade RTD.** A platinum RTD designed for use in safety-related applications in nuclear power plants. A representative sample of a nuclear-grade RTD must have been qualified by IEEE standards. The extent of the qualification depends on where the RTD is used in a plant.
- **On-line testing.** Remote testing of installed sensors while the plant is operating. Also called *in-situ testing*.
- **Performance.** A general term used to refer to the static (calibration or accuracy) and dynamic (response time) characteristics of sensors.
- **Precision.** See definition of *repeatability*.
- **R-vs.-T curve.** The resistance-versus-temperature relationship, curve, table, or chart of an RTD.
- **Random error.** Errors whose value can be positive or negative with respect to the actual temperature. Random errors are sometimes called *accidental errors*.
- **Repeatability.** The ability to obtain the same output using the same sensor in the same conditions. Repeatability is the maximum difference between the results of repeated readings of the same sensor using the same equipment and procedure at given conditions. Also called *precision*.
- **Response time.** The time required for the output of a sensor to reach 63.2 percent of its final value following a step change in temperature. Also called a *time constant*. (Note that the term *time constant* is meaningful only for a first-order system. Although temperature sensors are not necessarily first-order, the term *time constant* is often used to quantify the speed of their dynamic response.)
- **RTD.** A term used to refer to industrial-resistance thermometers. If the RTD's sensing element is made of platinum wire, the RTD is called *platinum-resistance thermometer, PRT,* or *platinum RTD*.

- **Self-heating.** The phenomenon in which the electric current used to measure an RTD's resistance generates heat in the RTD.
- **Sensing element.** The wire (usually platinum) inside the RTD whose resistance changes with temperature. For thermocouples, the sensing element is the measuring junction where the two thermocouple wires come together at the tip of the sensor.
- **Shift.** Changes in an RTD's resistance-versus-temperature relationship, also called *drift*. Shift implies a sudden change occurring at the end of a period or a test, while drift implies gradual changes.
- **SPRT.** Also called *standard RTD* or *PRT*. SPRTs are typically calibrated at NIST and used as a transfer standard for calibrating industrial-temperature sensors in the laboratory.
- **Stability.** The ability of the temperature sensor to maintain its accuracy. Stability is quantified by drift or drift rate (° C/year). The term *stability* (or its opposite, *instability*) is also used to refer to the level of temperature fluctuations in a process. A relatively calm process is referred to as *stable* and a fluctuating process is referred to as *unstable* (or noisy).
- **Systematic error.** Additive errors. A constant error or bias.
- **Thermowell** A protective jacket (tube) that is used to protect the sensor from the process fluid and allow it to be easily replaced.
- **Time constant.** See definition of *response time*.
- **Uncertainty.** Potential difference between the true process temperature and the output of temperature instrumentation. (Also see definition of *accuracy*.)
- **Well-type sensors.** Sensors that are designed to be installed in a thermowell. Also referred to as *thermowell-mounted sensors*.
- **Wet-type sensors.** Sensors that are installed directly into the process fluid as opposed to being installed in a thermowell (also called *direct-immersion sensors*).

4.4 Problems with Nuclear-Grade RTDs

Nuclear-grade RTDs, like their commercial-grade counterparts, can suffer from calibration drift, response-time degradation, reduced insulation resistance, erratic output, wiring problems, and the like. Of course, these problems occur less often in nuclear-grade RTDs than in commercial-grade RTDs because the former are of much higher quality. A test in the late 1980s of nearly 100 nuclear-grade RTDs against comparable commercial-grade RTDs under both normal and harsh conditions showed that nuclear-grade RTDs are generally twice as resilient and immune from performance problems as commercial-grade RTDs.

In the early 1970s, at the height of nuclear power plant development in the United States, almost all nuclear safety-related RTDs for PWR plants were supplied by Rosemount. Over the years, other manufacturers have entered the market, and Rosemount has gradually reduced its presence. Rosemount RTDs are nevertheless still used in some PWR plants and have provided excellent service to the nuclear power industry. The nuclear industry's experience with nuclear-grade RTDs from other manufacturers is also very good. However, as Rosemount reduced its market presence, problems arose in the 1980s as new manufacturers of nuclear-grade RTDs entered the market. These problems eventually subsided as new manufacturers gained experience with design, development, and testing of nuclear grade RTDs. Nevertheless, the following list is prepared to enumerate the problems that have typically been encountered with RTDs in nuclear power plants:

1. Dynamic response problems
2. Failure of extension leads
3. Low-insulation resistance
4. Premature failure
5. Wrong calibration tables
6. Loose or bad connections
7. Large EMF effects
8. Open element
9. Thinning of platinum wire
10. Lead-wire imbalance
11. Seeping of chemicals from connection head into thermowell
12. Cracking of the thermowell
13. Erroneous indication

Let's review each of these problems below.

4.4.1 Dynamic Response

As mentioned earlier, the response time of primary coolant RTDs in PWR plants is measured periodically using an in-situ test technique that is described in Chap. 6 called the *LCSR test*. This test has revealed numerous cases in which nuclear plant RTDs failed to meet their response-time requirements.

Three examples of RTD response-time failures in nuclear power plants are shown in Table 4.2. Note that the RTDs involved are from three different manufacturers and that these problems occurred in three different plants. Almost all such cases have been caused by problems at the RTD/thermowell interface at the sensing tip of the assembly, specifically, dirty RTDs, dirty thermowells, residue left from using thermal coupling compounds in the thermowell, and dimensional tolerance issues involving the RTD and/or thermowell.

4.4.2 Failure of Extension Leads

This problem occurred during the early 1980s, as new manufacturers of nuclear-grade RTDs were emerging. Typically, extension leads failed because of defective silver soldering inside the RTD where the RTD leads were attached to extension wires that protruded from the sensor.

Table 4.2. Examples of problems encountered with response time of nuclear plant RTDs

		Response Time (Seconds)		
Plant	**Date of Problem**	**Expected**	**Measured**	**Manufacturer**
A	1978	5.4	21	X
B	1984	4.5	37	Y
C	1988	3.6	12	Z

The measured values are from in-situ response-time testing performed using the LCSR method while the plant was operating.

4.4.3 Low Insulation Resistance

It is generally accepted by most manufacturers that industrial RTDs are required to have an insulation resistance (IR) of at least 100 megohm at room temperature (20° C) when measured with an applied voltage of 100 VDC. Most nuclear-grade RTDs readily meet this requirement, and their IR often reaches the giga-ohm range or higher. However, if moisture enters the RTD, the IR value can drop to as low as a few kilo-ohms. Often, even a very large drop in the IR is not apparent unless IR is measured. Therefore, before installation into a plant, RTDs should be tested to ensure sufficient IR.

4.4.4 Premature Failure

In the early 1980s, when new manufacturers were emerging, a batch of nuclear-grade RTDs from one of the recognized manufacturers experienced a failure rate of about 50 percent early in their life. Since then, however, the failure rate of new nuclear-grade RTDs has been rather low.

4.4.5 Wrong Calibration Tables

Cases have occurred in which the calibration charts of different batches of RTDs were interchanged, leaving a nuclear plant with a batch of RTDs but calibration charts belonging to a different batch.

4.4.6 Loose or Bad Connections

There are a number of transition points in an RTD circuit from the field to the instrument cabinets in the control room area. Along this path are terminal blocks, weld/solder joints, or splices where loose or bad connections have frequently been found.

4.4.7 Large EMF Errors

EMF, which stands for *electromotive force*, is a voltage signal that may develop in an RTD circuit if there are dissimilar metals in the RTD that can fall in a temperature gradient within the RTD. If this occurs, the resistance of the RTD will depend on the measurement polarity. That is, if the resistance is measured with one polarity, then the result will be slightly different than when the resistance measurement is repeated with reverse polarity. Table 4.3 shows the results of a laboratory experiment that involved six RTDs from two manufacturers of nuclear-grade RTDs. These RTDs were placed in an oil bath along with a standard RTD that was used to measure the bath temperature. The output of each RTD was measured with normal and reversed polarity. At the same time, the open-circuit voltage at the RTD output was measured. For Manufacturer A, the three RTDs showed 80 microvolts of EMF voltage, and the temperature indication of the RTDs depended on measurement polarity. For Manufacturer B, there was no EMF effect and almost no difference between the temperatures indicated by each RTD in the normal or reverse polarity. That is, the EMF effect, if present, will cause temperature error. When this occurs, the true temperature can still be obtained by averaging the results of the two measurements. In precision thermometry, resistance measurements are made using AC bridges as opposed to DC bridges. This is because an AC bridge cancels any EMF effect in the circuit and yields the true resistance of the RTD. In essence, an AC bridge works as if it measures the RTD resistance with forward and reverse polarities and displays the average of the two.

Table 4.3. Example of EMF problems with nuclear plant RTDs

	Bath Temp (° C)			
RTD I.D.	**Standard RTD**	**Normal Polarity**	**Reversed Polarity**	**EMF (Microvolts)**
		Manufacturer A		
A-1	285.33	285.59	285.45	80
A-2	293.59	293.83	293.66	80
A-3	300.36	300.62	300.41	80
		Manufacturer B		
B-1	285.33	285.28	285.28	0
B-2	293.59	293.56	293.58	0
B-3	300.36	300.33	300.31	0

4.4.8 Open Element

The platinum element in RTDs is very fragile and can crack or open as a result of vibration, stress, and interaction with other material in the RTD. Fig. 4.11 shows an electron microscope photo of a nuclear-grade RTD element and Fig. 4.12 shows an

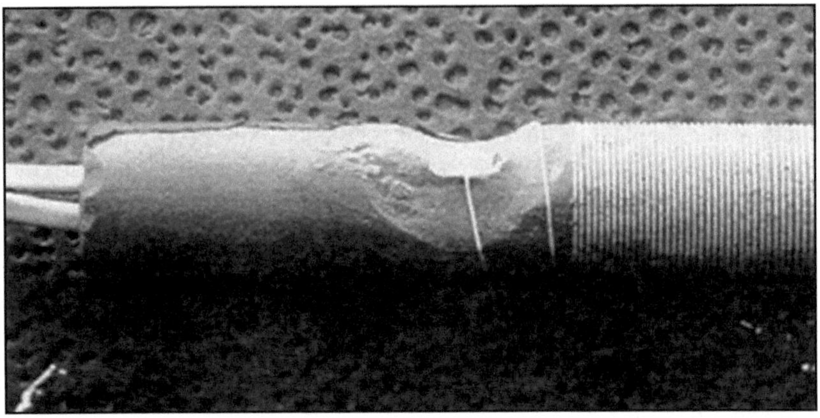

Fig. 4.11. Electron microscope photo of sensing element in a nuclear-grade RTD

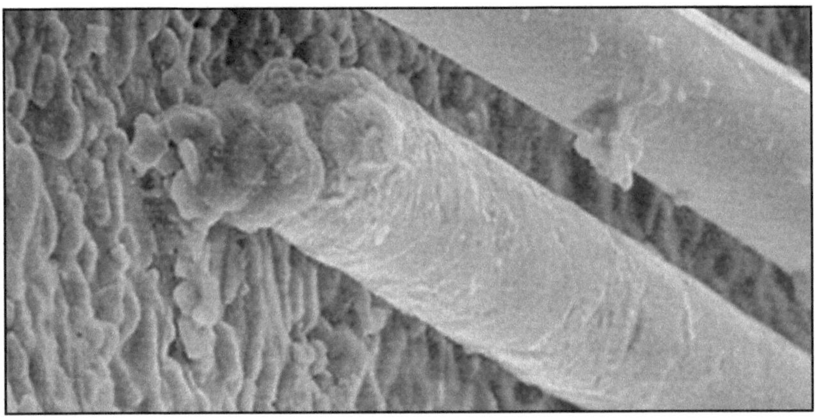

Fig. 4.12. Electron microscope photo of an open platinum wire in a nuclear-grade RTD

electron microscope photo of the platinum-sensing element of another nuclear-grade RTD that has failed open. Usually, the weak points where RTD elements fail are in weld points and places where the element is bent.

RTD failures due to open elements are sometimes preceded by erratic behavior whereby the RTD indication experiences large swings, spikes, and random shift. Fig. 4.13 shows on-line monitoring data of four hot-leg RTDs in a PWR plant. One of the four RTDs exhibits erratic behavior. A month or so after this observation, this RTD failed open and was replaced. It is interesting to point out that this behavior was not seen by the plant operators. In fact, during daily channel checks, the bad RTD continued to pass as its indication agreed well with the other three RTDs and met the plant's acceptance criteria.

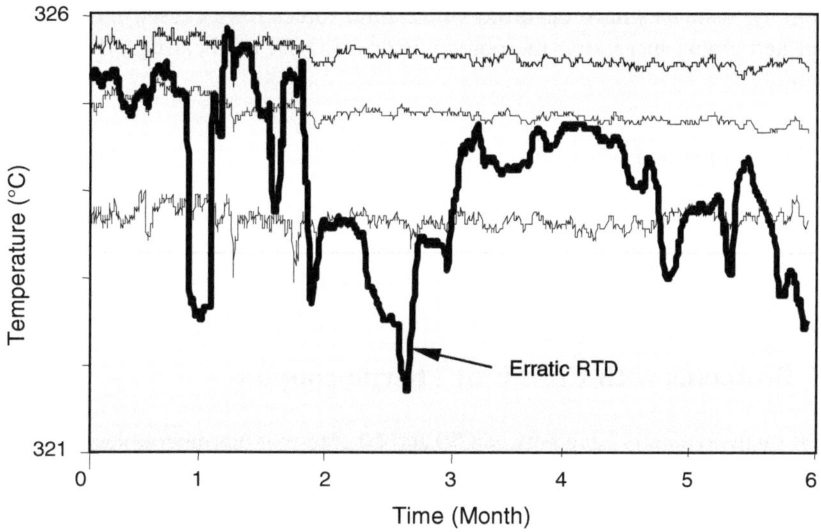

Fig. 4.13. rratic behavior preceding the failure of a primary coolant RTD at a PWR plant

4.4.9 Thinning of Platinum Wire

The sensing elements of nuclear-grade RTDs have experienced corrosive thinning caused by the chemicals that were used to clean the elements when they were manufactured or during the RTD's construction. This causes the cross section area of the sensing wire to decrease and its resistance to increase.

Thinning of the RTD element can also result from the chemical interaction between the element and the RTD insulation material.

4.4.10 Lead-Wire Imbalance

This is a potential problem in three-wire RTDs that are connected to three-wire Wheatstone bridges for measuring temperature. The two wires from across the RTD element that run to the two arms of the bridge must have equal resistances. Otherwise, the measurement of the RTD resistance can be erroneous. A similar problem can occur in RTDs with dummy compensating leads.

4.4.11 Seeping of Chemicals into Thermowell

In some RTDs, the connection head is filled with chemical foam to help it qualify for nuclear service. Nuclear power plants have experienced response time and other problems when the chemical seeps through into the RTD thermowell.

4.4.12 Cracking of Thermowell

The RTD should extend into the reactor coolant piping to a depth that is consistent with the fluid forces on the RTD and the mechanical strength of the RTD/thermowell assembly. Instances have occurred where fluid forces have caused the assembly to bend and crack, increasing the potential for a LOCA or shearing off of the entire assembly.

4.4.13 Erroneous Indication

Nuclear-grade RTDs have been found to have significant indication problems for a variety of reasons. Table 4.4 provides examples of some of the worst RTD indication problems observed by the author in U.S. nuclear power plants and their causes.

4.5 Problems with Core-Exit Thermocouples

PWR plants typically have between 50 and 60 core-exit thermocouples. An informal assessment of these thermocouples by the author in nearly 50 nuclear power plants has resulted in the following observations:

- Between 10 to 20 percent of core-exit thermocouples in PWR plants fail in the first 20 years of plant operation. The failures are in the form of large calibration shifts (e.g., 10 to 30° C errors at 300° C) erratic and noisy output, or saturated output.
- Some thermocouples develop cable problems while the thermocouple is still intact. Plants have been known to replace a core-exit thermocouple and later find out that the problem was not in the thermocouple.

Table 4.4. Examples of some of the worst problems encountered with indication of RTDs in nuclear plants

Indication Error (° C)	**Cause**
4° C	Calibration shift in two years
0.6° C	Error due to EMF
2.7° C	Difference between two elements of a dual RTD
0.6° C	Error due to wire-shielding problem
3.3° C	Dirty RTD contacts
1.1° C	Error due to low insulation resistance

Rather, the problem was in the thermocouple extension cables, connectors, or elsewhere in the circuit. Therefore, before replacing a thermo-couple, cable testing should be performed to distinguish cable/connector problems from thermocouple problems.

- Thermocouples can accidentally be reverse-connected, meaning that the positive and negative thermocouple wires may be crossed during installation or wiring of the thermocouple. In such cases, at room temperature, the thermocouple indication could appear to be normal, but as the temperature is increased, the thermocouple will show a negative reading.

In a U.S. nuclear fuel facility, engineers attempted to correct for a reverse-connected thermocouple by reversing its extension leads at the indicator. Originally, the thermocouple appeared to read correctly at room temperature, but its indication became negative when the process began to heat up. Thus, the thermocouple leads were interchanged at the indicator which made the reading positive and seemingly correct at low temperatures. However, as the temperature was increased, the thermocouple indication fell more and more below the actual process temperature to the point that at about 600° C, the thermocouple showed 450° C. This caused a fire at the nuclear fuel fabrication facility. The event is documented in an NRC Information Notice (IN-96-33).

- Thermocouples, even those that are properly connected, can have good indication at room temperature but diverge from true temperature as the plant heats up.

  Fig. 4.14 shows on-line monitoring results for a group of core-exit thermocouples in a PWR plant. It shows the thermocouple readings to be comparable at cold shutdown, but one deviates significantly from the others at plant operating temperature.

- Thermocouples can suffer response-time degradation as they age. Table 4.5 shows response-time results for core-exit thermocouples in four PWR plants. These are similar plants with nearly identical thermocouples. Note that the average response time of the thermocouples after 10 years of service is about 1 second and after 20 years of service is about 2 seconds, a 100 percent increase. This is not an operational or safety issue for the plant, but it indicates that as thermocouples age, their response time increases. This and other similar observations have motivated some nuclear power plants to perform response time testing on thermocouples as a means of aging management. In particular, it is important to establish the baseline response time of thermocouples when the plant is new or the thermocouples are first installed and then repeat the measurements periodically (e.g., once every five years) to determine if there is degradation from the nominal performance.

Table 4.5. Results of trending the performance of core-exit thermocouples in PWR plants

		Thermocouple Response Time (sec)		
Plant	**Years in Service**	**Average**	**High**	**Low**
A	30	2.01	2.9	0.6
B	30	1.96	2.8	0.6
C	10	0.97	1.5	0.5
D	10	1.10	1.5	0.6

- Thermocouples are not generally as accurate as RTDs. This is partly because thermocouples are not normally calibrated individually. Rather, thermocouple wires or

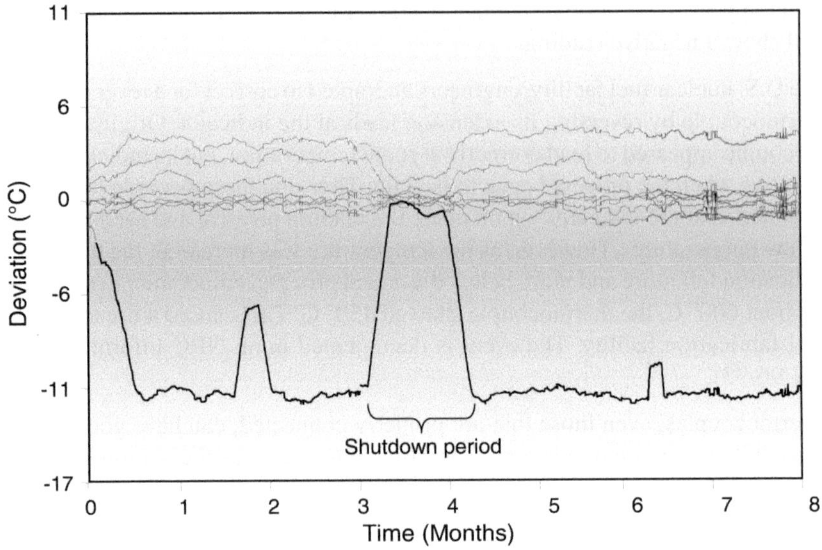

Fig. 4.14. On-line monitoring results for a group of core-exit thermocouples

a representative sample in a large batch of thermocouples is calibrated, and that calibration is used for all thermocouples in the batch. Table 4.6 shows estimated temperature measurement accuracies with industrial thermocouples over the range of 50° C to 500° C.

Table 4.6. Potential sources of error and their estimated values in industrial temperature measurements with thermocouples (for 50 to 500°C range)

Source	**Abbreviation**	**Range of Error**
Inherent thermocouple errors	(TC)	±0.5 to 5°C
Cold junction compensation error	(CJ)	±0.1 to 0.5°C
A/D errors	(AD)	±0.1 to 0.2°C
Precision errors	(Noise)	±0.1 to 0.5°C

Total Error = (TC)+(CJ)+(A/D)+(Noise)+(Other)
Other = Extension wire tolerance or mismatch, grounding, etc.

5

Cross-Calibration Technique

5.1 Background

In most PWR plants, the calibration of redundant sensors such as the primary coolant RTDs is verified periodically to ensure that any unacceptable drift or deviation is identified and corrected. For this purpose, the cross-calibration technique is used. This is a simple method that has received regulatory approval from the U.S. NRC, the British Nuclear Installation Inspectorate (NII), and others.

A more sophisticated and potentially more accurate technique also available for calibrating RTDs in-situ, is referred to as the "Johnson Noise" technique.[6] However, the Johnson Noise technique is still maturing and is, therefore, not currently used in nuclear power plants. ORNL and NIST in the United States and others in Australia, Germany, and elsewhere, have worked on the Johnson Noise technique for nearly three decades and continue to work to advance the technique for routine use in industrial applications. The Johnson Noise technique has application not only for calibrating RTDs in-situ, but also as the basis for developing a new and potentially very accurate sensor for high-temperature measurements.

A Johnson Noise thermometer is normally made out of an RTD that has sophisticated output electronics so it can measure a very small (in the nanovolt range) electrical signal that originates in the RTD element and varies with the temperature to which the RTD is exposed. The challenge posed by the Johnson Noise technique is in measuring the small signal in an industrial process at the end of long wires. This and other challenges posed by the Johnson Noise technique must be overcome before it will be ready for routine use in nuclear power plants or other processes.

In the meantime, the cross-calibration method has been improved and automated to better serve the nuclear industry. These improvements include: 1) analytical algorithms to correct the cross-calibration data for any significant instability in the plant temperature as cross-calibration data is collected; 2) corrections for temperature differences between primary coolant loops and hot legs and cold legs; 3) calculations of the uncertainties of cross-calibration results; 4) innovative techniques to handle sensors that do not meet the plant acceptance criteria for the cross-calibration test; 5) methods for retrieving data from the plant computer for the cross-calibration test; and

6) robust algorithms for handling data taken during temperature ramp conditions, at plant startup, or shutdown periods to provide cross-calibration results.

5.2 Test Principle

The cross-calibration technique is a means for verifying the calibration of a group of redundant sensors that measure the same process parameters. Its basic principle of operation is to record the reading of redundant sensors, average these readings, and calculate the deviation of each sensor from the average of the redundant sensors, less any outlier(s). The test may include the narrow-range and wide-range RTDs as well as the core-exit thermocouples. Normally, the acceptance criteria for the narrow-range RTDs are much tighter than those of wide-range RTDs, and the acceptance criteria for wide-range RTDs are much tighter than thermocouples.

Table 5.1 shows results from a typical cross-calibration test in a nuclear power plant. The data for this test was collected at isothermal conditions during plant heatup at approximately 280° C. This printout, referred to as a *cross-calibration run*, typically includes four data-collection passes for each sensor. After data for a run is collected, the results of the four measurements for each sensor are averaged and recorded under a column marked "Average Temp." In the example in Table 5.1, the average temperatures for all the narrow-range RTDs are averaged, and the result is subtracted from each individual average to identify the "deviation" of each sensor from the average of the narrow-range RTDs. The "Deviation" column is the result of the cross-calibration run. These results are referred to as the *preliminary results of the cross-calibration test*. Typically, the data is analyzed further, as described in Sect. 5.4 later in this chapter, to establish the final results of the cross-calibration test and to quantify the uncertainty of the deviation results. The final results include the necessary corrections for any significant instability and nonuniformity in the plant temperature when the cross-calibration data was collected.

The narrow-range RTDs are normally the most accurate temperature sensors in a PWR plant and are therefore used to provide the reference temperature for the cross-calibration test. In some plants, only the narrow-range RTDs are cross-calibrated; in others, the wide-range RTDs and/or core-exit thermocouples are also included in the cross-calibration test.

Any narrow-range RTD that deviates from the average by more than a predetermined criterion (e.g., 0.3° C) is excluded from the average. The criteria are different in different plants and usually depend on the plant's accuracy requirements for the primary coolant temperature. Table 5.2 provides the criteria for RTD cross-calibration used in seven nuclear plants and indicates both the deviation at which a narrow-range RTD is excluded from the average and the number of temperatures at which cross-calibration data is collected.[7]

The RTDs that are excluded (from the average) are referred to as *outliers*. An outlier is either replaced, or a new calibration table is developed for the outlier using the cross-calibration data. The procedure for recalibrating outliers is described later in this chapter in Sect. 5.13.

Table 5.1. Preliminary results of a typical cross-calibration run

RTD	Temperature (° C) Pass 1	Pass 2	Pass 3	Pass 4	Average Temp. (° C)	Dev. T (° C)
			Narrow-range RTDs			
1	280.3278	280.3274	280.3087	280.2956	280.315	−0.063
2	280.4091	280.3942	280.3853	280.3797	280.392	0.014
3	280.3616	280.3621	280.3426	280.3305	280.349	−0.029
4	280.3660	280.3655	280.344	280.3347	280.353	−0.026
5	280.4729	280.4599	280.4608	280.4571	280.463	0.084
6	280.3664	280.3329	280.3427	280.3274	280.342	−0.036
7	280.3392	280.3276	280.3230	280.3178	280.327	−0.051
8	280.4709	280.4574	280.4504	280.4453	280.456	0.078
9	280.3308	280.3312	280.3047	280.3029	280.317	−0.061
10	280.4369	280.4355	280.4081	280.4118	280.423	0.045
11	280.3765	280.3584	280.3477	280.3440	280.357	−0.022
12	280.4593	280.4584	280.4375	280.4296	280.446	0.068
			Wide-range RTDs			
13	280.0733	280.0612	280.0538	280.0352	280.056	−0.322
14	280.6964	280.6871	280.6741	280.6602	280.679	0.301
15	280.3290	280.3281	280.3067	280.3039	280.317	−0.061
16	280.4881	280.4899	280.4704	280.4686	280.479	0.101
			Core-Exit Thermocouples			
17	280.6723	280.6674	280.6261	280.6431	280.652	0.274
18	280.6301	280.6082	280.5928	280.6025	280.608	0.230
19	280.7786	280.7802	280.7640	280.7526	280.769	0.390
20	280.5482	280.5660	280.5474	280.5474	280.552	0.174
21	280.8232	280.8110	280.7940	280.7907	280.805	0.426
22	280.8978	280.8588	280.8483	280.8248	280.857	0.479
23	280.7680	280.7607	280.7445	280.7380	280.753	0.374
24	281.1411	281.1394	281.1394	281.1086	281.132	0.754
25	280.8037	280.7940	280.7510	280.7656	280.779	0.400

Average Temperature Indicated by the narrow-range RTDs = 280.378° C
The deviation column in this table (Dev.) is equal to the average temperature of each sensor minus the average temperature of the narrow-range RTDs.

5.3 Sources of Cross-Calibration Data

Cross-calibration data can be collected using a dedicated data acquisition system or retrieved from the plant computer. Fig. 5.1 shows the data flow diagram of cross-calibration for these two methods of data collection. The following two sections describe the details of each method.

5.3.1 Dedicated Data Acquisition System

To collect cross-calibration data using a dedicated data acquisition system, the sensors are usually accessed in the control room area in the process instrumentation cabinets. Fig. 5.2 shows a simplified diagram of a data acquisition system for cross-calibration. The sensors are disconnected from the plant and connected to the cross-calibration test equipment. The data acquisition procedure is as follows:

1. Sequence through all sensors, measuring their outputs, and convert into equivalent temperatures, if needed. This step yields one cross-calibration pass. To convert from resistance to temperature, if needed, the Callendar Equation (or equivalent) is used.
2. Repeat Step 1 to obtain four passes.
3. Average the four temperature measurements for each sensor.
4. Average the temperature indications from Step 3 for all the narrow-range RTDs.
5. Subtract the average temperature identified in Step 4 from the temperature indications of each sensor. The results are referred to as the deviation of each sensor and are denoted by ΔT.
6. If the deviation of any narrow-range RTD element exceeds a predetermined value (e.g., $\pm 0.3°$ C), remove the element's average measurement, obtained in Step 3, and repeat from Step 4. The RTD element that is removed from the average is referred to as an *outlier*.
7. Repeat Step 6 until all outliers have been eliminated from the average.

This procedure provides the preliminary results of the cross-calibration run. As we will see later in Sect. 5.4, additional analysis should be performed to improve the reliability of the results by correcting the data for any significant instability and nonuniformity in the plant temperature. The additional analysis will provide the final outcome of the cross-calibration test as well as the information needed to establish the uncertainty of the cross-calibration results.

This seven-step procedure is referred to as *traditional cross-calibration* and is illustrated in Fig. 5.3.

To convert RTD resistance into temperature if needed (see Step 1 above), the Callendar Equation is most often used. For temperatures above 0° C, the Callendar Equation is written as:

$$\frac{R(T)}{R(0)} = 1 + \alpha\left[T - \delta\left\{\left(\frac{T}{100°C}\right)^2 - \left(\frac{T}{100°C}\right)\right\}\right] \tag{5.1}$$

where:

$$
\begin{aligned}
T &= \text{Temperature } (^\circ \text{C}) \\
R(0) &= \text{Resistance at } 0^\circ \text{C } (\Omega) \\
Alpha(\alpha) &= \text{Calibration constant } (\Omega\ /\Omega\ /^\circ \text{C}) \\
Delta(\delta) &= \text{Calibration constant } (^\circ \text{C}) \\
R(T) &= \text{Resistance at any temperature } (\Omega)
\end{aligned}
$$

The terms R(0), α, and δ are referred to as the constants of the Callendar Equation. Alpha (α) is the average temperature coefficient of resistance over the 0 to 100° C in-

Table 5.2. RTD cross-calibration criteria in various PWRs

Nuclear Plant	Temperature Points	Outlier Criteria (° C)	Remarks
1	1	0.17	1
2	1	0.17	1
3	4	0.11	2
4	2	0.30	3
5	4	0.27	4
6	1	0.17	5
7	4	0.11	6

A. Plant stability criteria for RTD cross-calibration are typically about ± 0.15 to ± 0.3° C.
B. Temperature points: Number of temperatures at which cross-calibration data is collected.

Remarks:

1. *Cross-calibration data is taken for any number of plateaus. However, only the data for 292° C is used to meet acceptance criteria and adjust the temperature transmitters, as needed.*
2. *Data can be taken at a constant heatup rate. On 16 RTDs, data is taken as follows: RTD number 1 to 16, reverse current 16 to 1, reverse current 1 to 16, etc. This presumably corrects for both the ramping temperature and for EMF effects (reversing the current). Data is taken around 95° C, 170° C, 230° C, and 292° C.*
3. *Two plateaus: 170° C and 292° C. For deviations greater than 0.17° C, the deviations at 170° C and 292° C are used to determine the error offset and the slope and to apply the corrections to the temperature transmitter.*
4. *Data is taken at four temperatures on 16 RTDs sequentially, 1-16, 16-1, etc. The heatup rate is also measured.*
5. *The 16 RTDs in this plant are tested one channel (four RTDs) at a time. Data is taken for 25 minutes at five-minute intervals. This is repeated for all four channels. The plant stability requirement for the tests is 0.17° C (i.e., the temperature cannot change by more than 0.17° C from the beginning to the end of any test run).*
6. *Tests are performed at 120° C, 180° C, 230° C, and 275° C. Data is taken on 16 RTDs, 1 to 16, 16 to 1, 1 to 16, and 16 to 1. The four calibration points are used to determine a zero and a slope for the correction to temperature transmitters.*

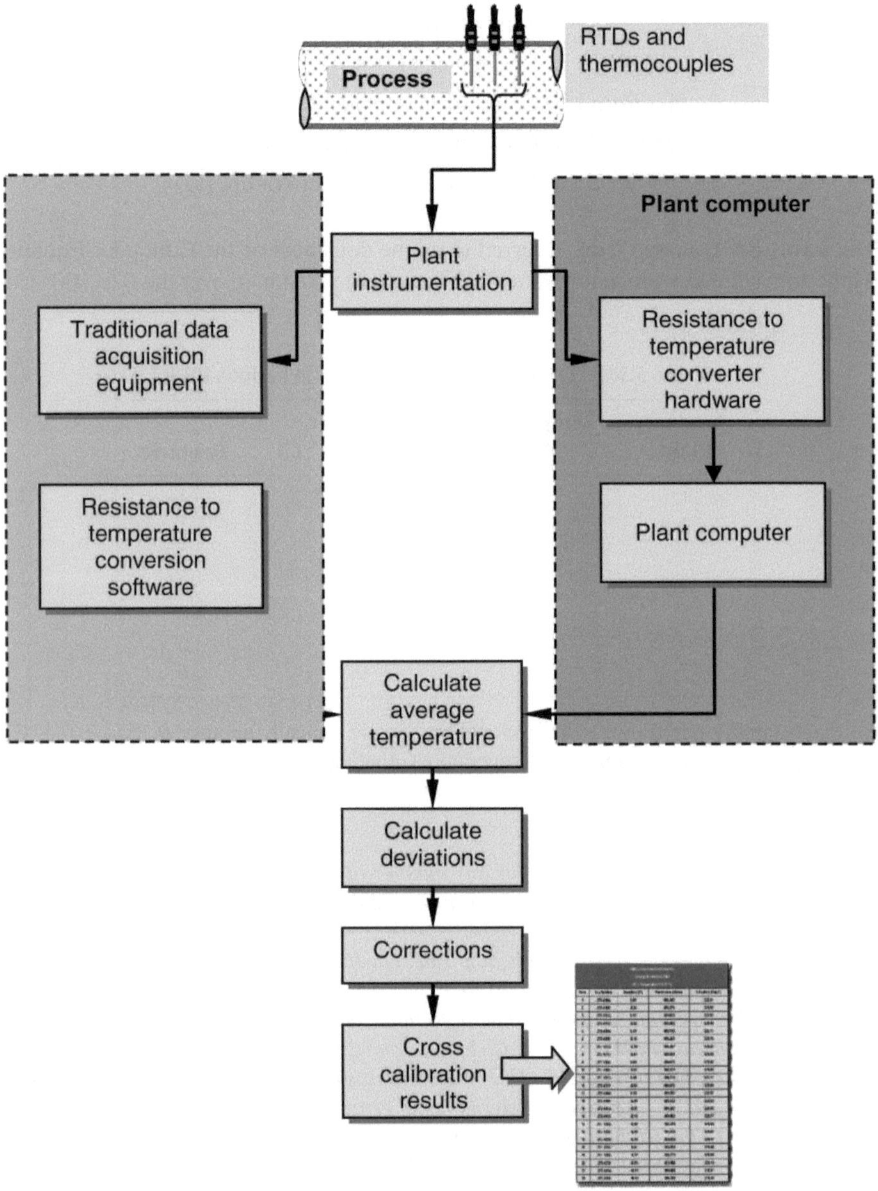

Fig. 5.1. Data acquisition options for cross-calibration

terval, and Delta (δ) is the index of the departure of the resistance-versus-temperature curve from a straight line. These two constants, as well as R(0), are normally identified for each RTD by calibrating the RTD in a constant temperature bath in a laboratory.

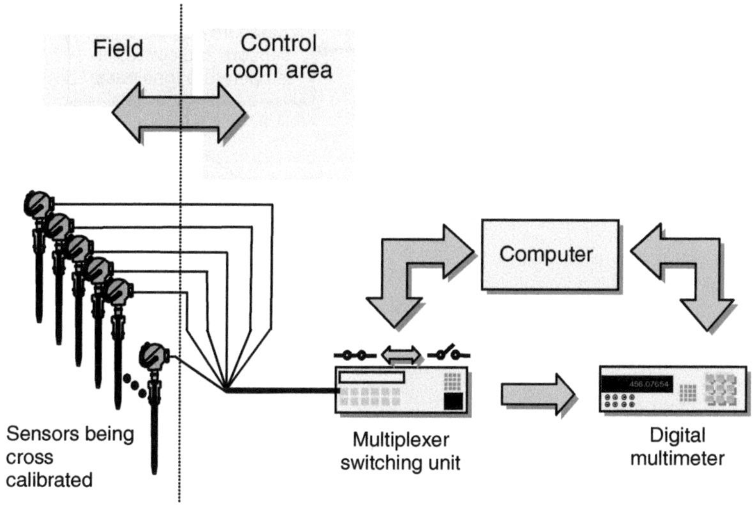

Fig. 5.2. Equipment setup for cross-calibration

Once the three constants are identified, they are substituted in Eq. 5.1 to provide a calibration table for the RTD.

In some Westinghouse PWRs, instead of the Callendar Equation, a second-order polynomial, called the "Westinghouse Reference Function," of the following form is used:

$$R(T) = \mathrm{Ref}(T) + \mathit{Offset} + (\mathit{Slope})\ (T - 525) \tag{5.2}$$

$$\mathit{Ref}(T) = 185.807 + (0.444693)(T) + 0.000036082(T^2) \tag{5.3}$$

where:

$R(T)$ = Resistance of the RTD in ohms as a function of temperature (T)
Ref(T) = Reference function
Offset and Slope = Constants of the Westinghouse Reference Function (obtained from RTD calibration)

The Westinghouse Reference Function has a fixed curvature with a linear adjustment to match the RTD calibration curve. The temperature (T) in Eqs. 5.2 and 5.3 is in °F.

5.3.2 Plant Computer Data

Nuclear power plants are often equipped with a means for collecting and storing the output of process sensors. This output can then be retrieved through commercial data management software packages. Two examples of such data management software

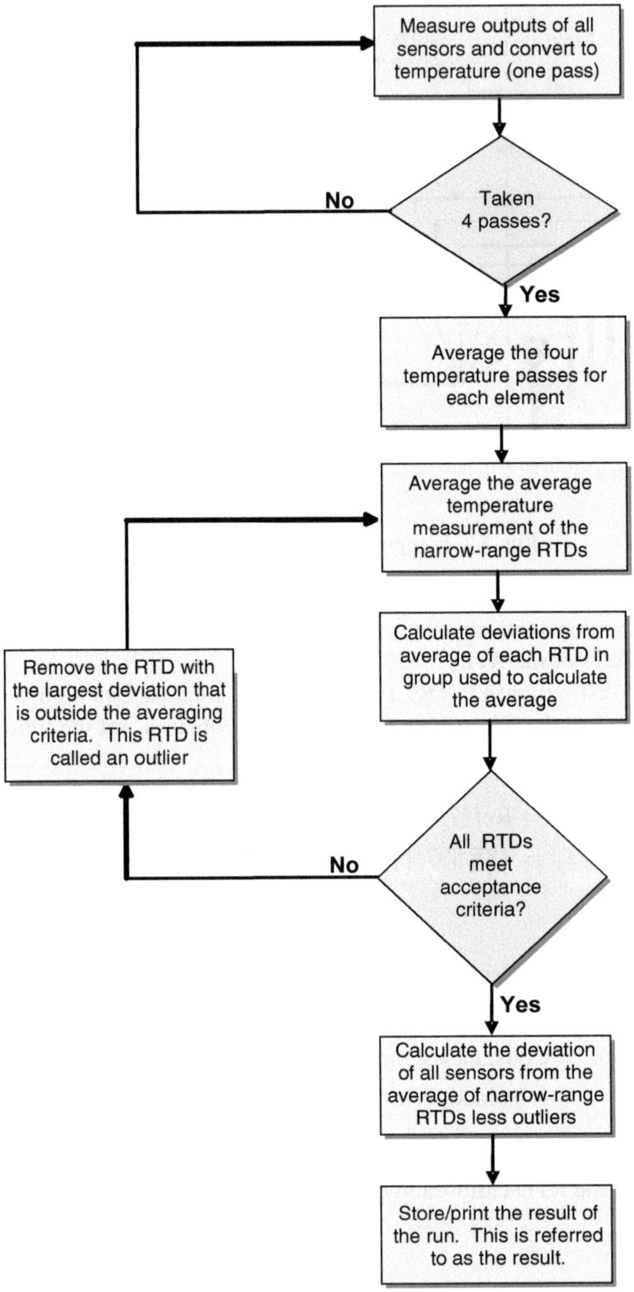

Fig. 5.3. Flowchart of cross-calibration procedure using a dedicated data acquisition system

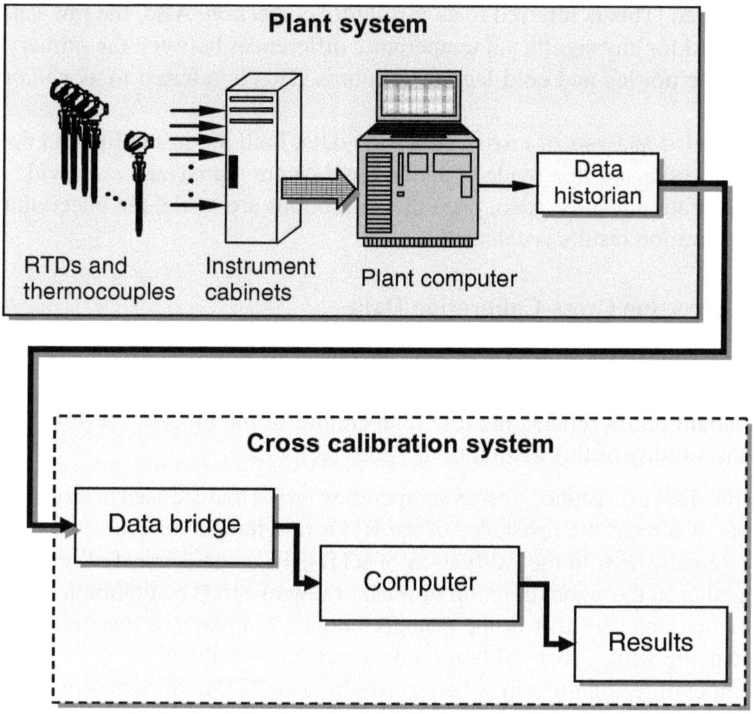

Fig. 5.4. Block diagram of cross-calibration data retrieval from the plant computer

packages are: 1) a product called PI Software© (from OSISoft Company); and 2) a product called eDNA Software© (from InStep Software Company).

To access the data stream from the plant computer, a software program typically referred to as a "Data Bridge" is used. The Data Bridge helps obtain the sensor data from the plant data management servers. The data are then analyzed on a local computer. Fig. 5.4 shows a block diagram of how cross-calibration data is acquired and analyzed using data from a plant computer. Typically, the outputs of plant sensors are sampled by the plant computer, converted into temperature, and stored. The sampling period depends on the plant and typically ranges from one to ten seconds for RTDs and ten to sixty seconds for core-exit thermocouples.

5.4 Detailed Analysis of Cross-Calibration Data

The cross-calibration process described results in a preliminary analysis of the raw cross-calibration data and provides the preliminary results of the cross-calibration test whether the data is collected using a dedicated data acquisition system or retrieved from the plant computer. For more accurate results, the raw data should be corrected for any significant temperature fluctuations that might have been present when data was collected. This is referred to as *instability correction*. Also, the raw data should be corrected for any significant temperature differences between the primary coolant loops or the hot-leg and cold-leg temperatures. This is referred to as *nonuniformity correction*.

In detailed analysis of cross-calibration data, both the instability and nonuniformity corrections are first made and then the data are reanalyzed to provide the final cross-calibration results. Also, once the corrections are made, the uncertainty of the cross-calibration results is calculated.

5.4.1 Correcting Cross-Calibration Data

The cross-calibration of RTDs in nuclear power plants is based on the assumption that, at isothermal plant conditions, the average temperature of a sufficient number of redundant RTDs reflects the true temperature of the process. Several factors can affect the validity of this assumption. These factors are:

1. Errors in the resistance-versus-temperature tables that are used in cross-calibration tests to convert the resistance of the RTDs into temperature.
2. Systematic drift in the calibration of RTDs. This can occur if all the RTDs drift together in the same direction upward or downward (i.e., common mode drift).
3. Fluctuations and drift in the primary coolant temperature that could have been occurring while cross-calibration data was taken at the plant.
4. Temperature nonuniformity between redundant RTDs. Since the cross-calibration method assumes that all RTDs are at the same temperature, any significant departure from this assumption can cause errors in the results of cross-calibration tests.

In the cross-calibration testing of a group of RTDs that have been used in a plant for one or more operating cycles, the first and second of the four possible factors just given may be accounted for by removing one or more of the RTDs from the plant and calibrating it in a laboratory. Another alternative is to replace one of the RTDs with a newly calibrated RTD and then repeat the cross-calibration tests at the end of the outage while the plant is heating up toward power operation. A more practical way to rule out the second possible factor is to rely on the experimental data published in NUREG/CR-5560.[7] The data in NUREG/CR-5560 indicates that the drift of a group of nuclear-grade RTDs is predominately random rather than systematic. If this is assumed to be the case, bias errors are unlikely to occur in the results of the cross-calibration tests and the second of the above four factors would be moot.

The third and fourth possible factors given above may be resolved by implementing numerical techniques as described in the following two sections, to correct the cross-calibration data for plant temperature instability and temperature nonuniformity.

5.4.2 Instability Correction

When temperature fluctuations or drift during cross-calibration tests occur, it is almost always because the plant temperature cannot be controlled perfectly at steady state.

The method used to correct temperature instability depends on the plant conditions under which the data was acquired. If the plant temperature is changing at a slow and constant rate, then ramp data acquisition is used because it automatically compensates for the changes that occur in plant temperature while the data is being acquired. If the plant is being maintained at stable isothermal conditions, then plateau data acquisition is used, and the plant temperature fluctuations are compensated for during the detailed analysis.

Ramp data acquisition accounts for constant temperature changes by sampling the RTDs in reverse order during the second and fourth passes of a cross-calibration run. For example, with 24 RTDs, the sampling order for the four passes would be 1 to 24, 24 to 1, 1 to 24, and 24 to 1. The reversal of the sampling order inverts the effect of the temperature ramp so any error is cancelled when all four passes are averaged together.

During plateau data acquisition, compensating for constant temperature ramping is not critical, so more emphasis is placed on short-term fluctuations. This is essential because of the changes in heat removal that are frequently required to keep the plant at a fixed temperature. In this case, the RTDs are sampled in the same sequence for each pass, making short-term fluctuations more apparent. Fig. 5.5 shows an example of data before and after it has been corrected for instability; it consists of all the data points of a complete run involving four passes. First, a straight line is fit (5.5a) to the pass averages. This line fit represents a linear regression performed on the average temperature of the RTDs in each of the four passes in the run. The straight line is subtracted from the data to remove any effect of process temperature drift. Figure 5.5b shows the data after it has been corrected by the fit to the pass averages. The data shown in Fig. 5.5b represent the deviations of individual RTDs from the average of all RTDs. If these deviations are subtracted out, any remaining process temperature fluctuations are evident in the data (Fig. 5.5c). These remaining fluctuations are referred to as *residual temperature fluctuations* whose standard deviation is used in calculating the overall uncertainty of the cross-calibration results as described later in Sect. 5.8.

For plants at which data is retrieved from the plant computer, fluctuation corrections are performed by calculating the standard deviation of each set of data and using the value of the standard deviation to reject the unacceptable data sets. The procedure is as follows:

1. Calculate the standard deviation of each individual data set at each temperature and call it STD.
2. Calculate the average and standard deviation of the STDs and call them AVE and σAVE.
3. Calculate the difference between the results of Steps 1 and 2, i.e., Δ = STD - AVE for each data set at each temperature.
4. The data set is accepted if $|\Delta|$ is less than "mσAVE". That is, we should have $|\Delta| \leq$ m (σAVE) where m is a multiplier typically equal to 1 or more. This multiplier is referred to as the *standard deviation criteria*. If $|\Delta|$ is not less than m(σAVE), the corresponding data set is excluded from analysis.

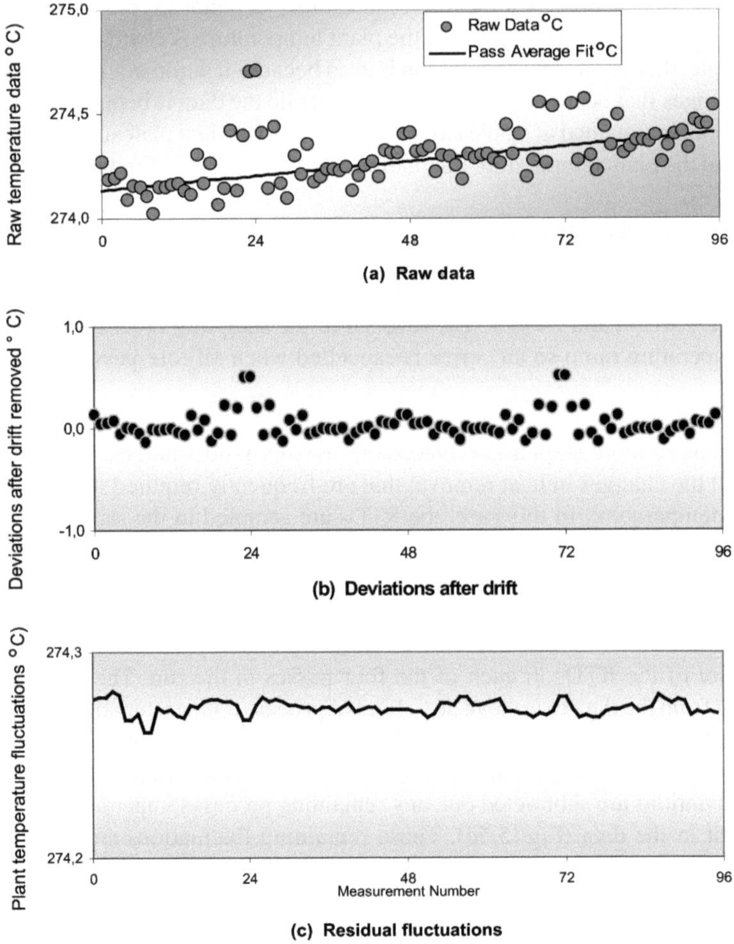

Fig. 5.5. Effect of instability correction on cross-calibration data

Table 5.3 shows an example involving 18 cross-calibration runs at three temperatures (171° C, 238° C, and 277°) on which this four-step procedure was applied. The runs that were rejected by the criteria just stated are identified in the table by an asterisk.

5.4.3 Nonuniformity Correction

The nonuniformity correction is made in order to account for any gross differences that could have existed (during cross-calibration data acquisition) between the hot-leg and cold-leg temperatures in each loop or between each loop across the reactor. These differences can occur as a result of incomplete mixing of the reactor coolant or differences in the heat removal of the steam generators.

Table 5.4 shows representative average temperatures calculated from the cross-calibration data after it has been corrected for plant temperature stability. Such data is used to determine if temperature uniformity problems exist. If temperature differences are very small, then nonuniformity corrections are not necessary. Otherwise, the data should be corrected for temperature differences between the hot-leg RTDs and the cold-leg RTDs, or temperature differences between the reactor coolant loops.

Table 5.3. Standard deviations of cross-calibration runs calculated for instability correction

	Standard Deviation (° C)		
Run Number	**171° C**	**238° C**	**277° C**
1	0.214	0.058	0.053
2	0.223*	0.059	0.053
3	0.231*	0.064	0.055*
4	0.209	0.071	0.052
5	0.218	0.075*	0.052
6	0.224*	0.074*	0.052
7	0.237*	0.074*	0.052
8	0.187	0.064	0.052
9	0.164	0.062	0.052
10	0.192	0.056	0.051
11	0.179	0.059	N/A
12	0.147	0.057	N/A
13	0.144	0.061	N/A
14	0.142	0.061	N/A
15	0.186	0.062	N/A
16	0.188	0.060	N/A
17	0.177	0.059	N/A
18	0.178	0.059	N/A

(*Runs rejected by STD criteria = 1.0. That is, if a run is out by any more than one standard deviation, it is rejected.)

5.5 Presenting Cross-Calibration Results

After the cross-calibration data is corrected for any plant temperature instability and nonuniformity, it is re-analyzed to provide the final (corrected) results. Table 5.5 shows typical cross-calibration results for raw data and corrected data for a set of narrow-range primary coolant RTDs in a PWR plant. The raw data results are labeled as *preliminary results* and the results from analysis of corrected data are labeled as *final results*.

Table 5.4. Representative averages of primary coolant temperatures calculated for evaluating temperature nonuniformity

	Temperature (° C)					
Description of Averages	**204**	**221**	**232**	**249**	**266**	**277**
Average of Whole Plant	204.06	219.24	233.76	248.89	265.35	276.41
Average of Hot-leg RTDs	204.08	219.26	233.77	248.91	265.35	276.41
Average of Cold-leg RTDs	204.02	219.22	233.74	248.88	265.36	276.42
Average of Loop 1 RTDs	204.06	219.26	233.84	249.06	265.47	276.55
Average of Loop 2 RTDs	204.05	219.21	233.69	248.81	265.29	276.35
Average of Loop 3 RTDs	204.03	219.19	233.66	248.76	265.24	276.30
Average of Loop 4 RTDs	204.08	219.32	233.83	248.96	265.39	276.45

Normally, corrections are made in cross-calibration data for only narrow-range RTDs and wide-range RTDs. That is, the data from core-exit thermocouples are not normally corrected.

5.6 Effect of Corrections on Cross-Calibration Results

Correcting the raw data for a plant's temperature instability and non-uniformity often makes a difference in the final results of a cross-calibration test. The degree to which the results are affected by the corrections depends on the plant. Sometimes the corrections make a huge differ- ence, and at other times, the corrections make only a small difference.

Fig. 5.6 shows cross-calibration results for 12 RTDs in a PWR plant before and after corrections are made for plant temperature instability and nonuniformity. It is apparent that before the corrections, all RTDs show positive deviations, while after corrections the deviations become random, as expected. Furthermore, in this example, the corrections reduced the absolute values of the RTD deviations.

5.7 Automated Software for Cross-Calibration

The cross-calibration process is simple but involves numerous calculations. As such, it is prudent to automate the process to facilitate the data collection and data analysis tasks. Fig. 5.7 shows raw cross-calibration data obtained from a plant computer during startup. The automated software reads and plots the raw data, performs the analysis including corrections for plant temperature instability and nonuniformity, and prints the results.

Table 5.5. Comparison of preliminary and final cross-calibration results

	Results (° C)	
RTD Tag Number	**Preliminary**	**Final**
1NCRD5420	0.133	0.087
1NCRD5421	0.044	−0.002
1NCRD5422	0.050	0.003
1NCRD5430	0.067	0.019
1NCRD5440	−0.061	−0.107
1NCRD5460	0.011	0.043
1NCRD5461	0.000	0.044
1NCRD5462	−0.050	−0.014
1NCRD5470	−0.128	−0.092
1NCRD5480	−0.006	0.029
1NCRD5500	−0.011	0.010
1NCRD5501	−0.006	0.020
1NCRD5502	0.000	0.022
1NCRD5510	−0.039	−0.014
1NCRD5520	−0.061	−0.038
1NCRD5540	0.128	0.121
1NCRD5541	−0.011	−0.020
1NCRD5542	0.083	0.076
1NCRD5550	−0.122	−0.129
1NCRD5560	−0.039	−0.048
1NCRD5850	0.217	0.171
1NCRD5860	−0.094	−0.138
1NCRD5870	0.189	0.216
1NCRD5880	0.500	0.531
1NCRD5900	0.289	0.310
1NCRD5910	−0.333	−0.313
1NCRD5920	0.639	0.632
1NCRD5930	0.256	0.249

The preliminary results are from analysis of raw data and final results are from analysis of corrected data.

5.8 Uncertainty of Cross-Calibration Results

The uncertainty of a cross-calibration test depends on whether the data is collected using a dedicated data acquisition system or the plant computer. The uncertainties associated with each of these two situations are discussed in the following two sections.

5.8.1 Uncertainty with Dedicated Data Acquisition System

Four types of uncertainty or error are involved when a dedicated data acquisition system is used for RTD cross-calibration. These are: test equipment uncertainty, precision error, instability error, and nonuniformity error. Table 5.6 shows typical uncertainty

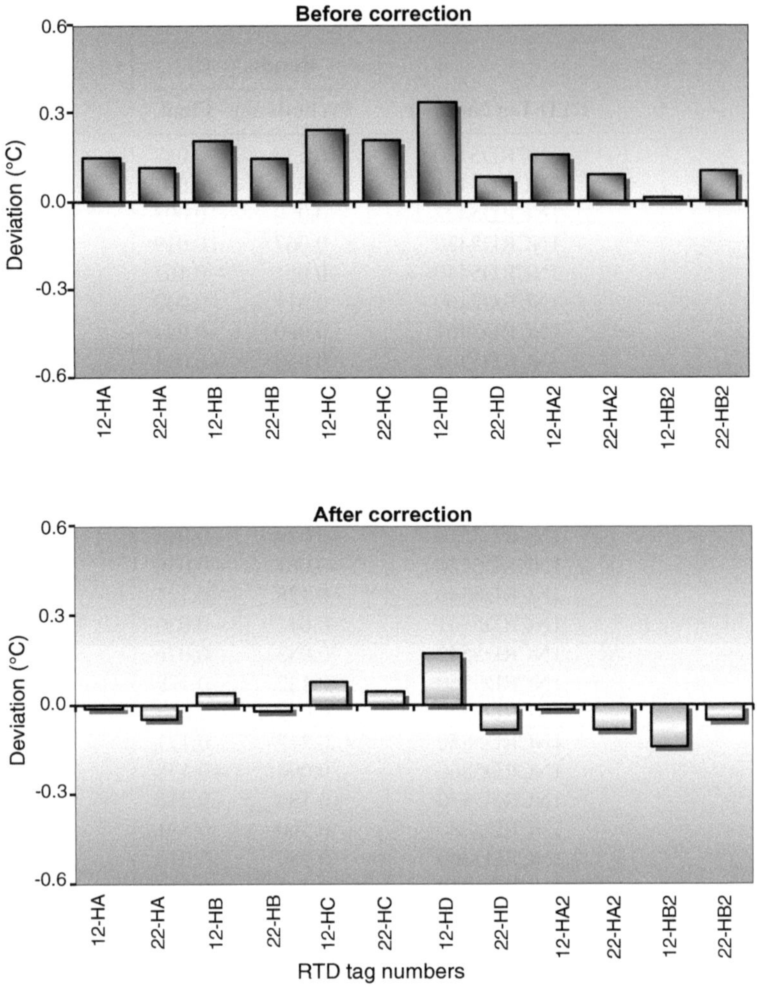

Fig. 5.6. Cross-calibration results before and after correcting for plant temperature instability and nonuniformity

estimates for a set of RTD cross- calibration tests performed at seven different temperatures in a PWR plant. Theses uncertainties were calculated by combining the errors that arise from the following four sources:

1. **Test Equipment Uncertainty.** With a dedicated data acquisition system, resistance measurements are made and converted into temperature. Therefore, the uncertainty of the results depends on the accuracy and drift of the resistance measurement equipment. Typically, the accuracy and short-term drift of resistance measurement equipment available today correspond to 0.01° C to 0.03° C

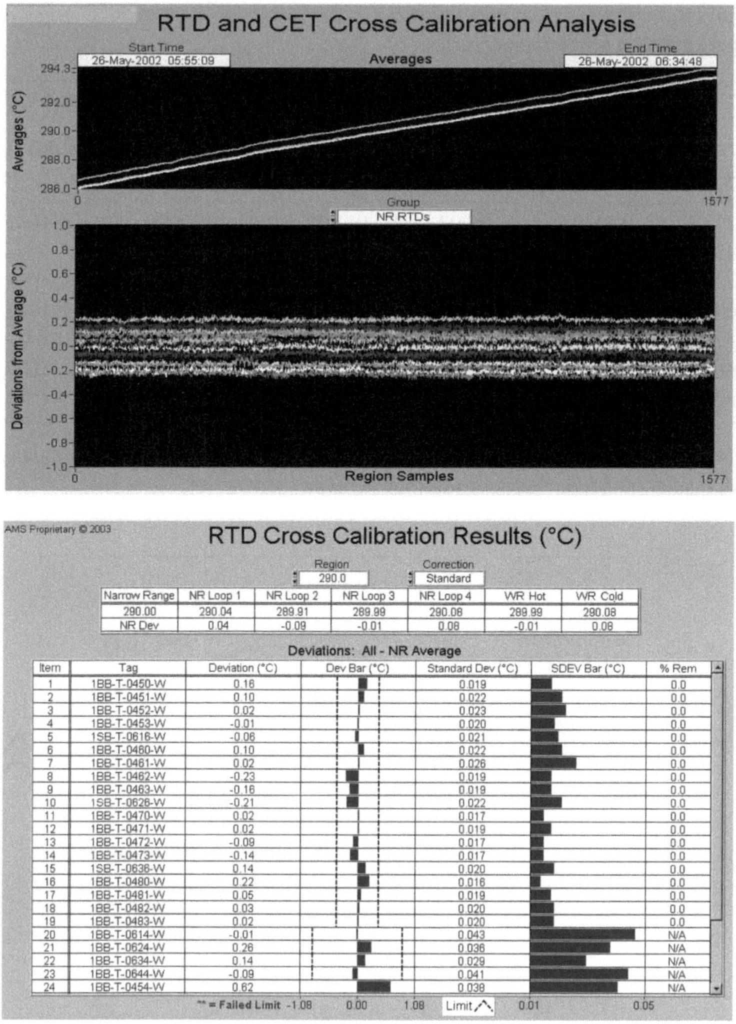

Fig. 5.7. Raw cross-calibration data and results of analysis from automated software for data retrieval and data analysis

in temperature, depending on the resistance values measured and the equipment used.

2. **<u>Precision Error.</u>** Precision error is identified by measuring the sample standard deviation of the measurement noise. Usually, a fixed resistor is connected to the data acquisition equipment to mimic RTDs as a means to identify the precision error. The resistor value is selected based on the temperatures at which the cross-calibration tests are performed, and the resistance of the RTDs being cross-calibrated.

Table 5.6. Example of typical uncertainties for the results of a set of RTD cross-calibration testing performed at seven temperatures

Temp (° C)	Individual Errors (° C) Test Equipment e_1	Precision e_2	Instability e_3	Non-uniformity e_4	Total Error (° C- RSS)
121	0.011	0.003	0.001	0.001	0.011
149	0.013	0.003	0.003	0.004	0.014
171	0.013	0.003	0.018	0.021	0.031
204	0.013	0.003	0.010	0.014	0.022
232	0.013	0.003	0.021	0.027	0.037
260	0.014	0.003	0.019	0.026	0.035
282	0.014	0.003	0.018	0.021	0.031

The numbers under "Total Error" column are the error bars (I) that should accompany the RTD deviation results (i.e., a deviation result may be expressed as $|\Delta| \pm$ RSS).

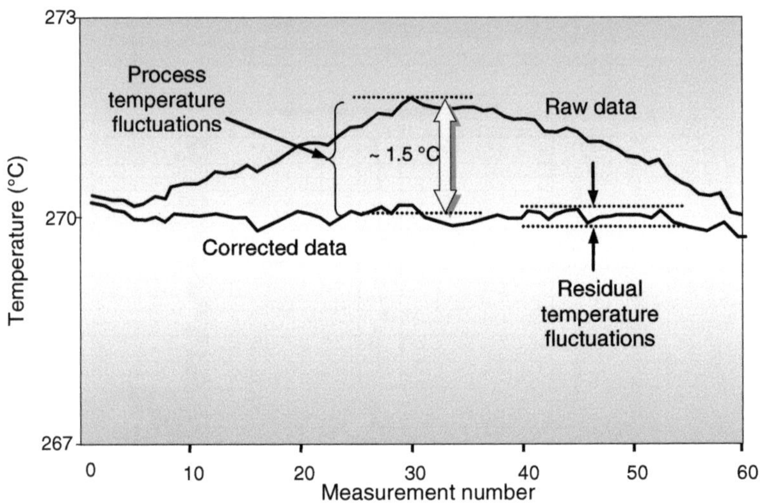

Fig. 5.8. Example of cross-calibration data before and after correcting for process temperature fluctuations

3. **Instability Error.** To illustrate the type of instability errors that are encountered in cross-calibration of nuclear plant RTDs, Fig. 5.8 is presented here from a set of cross-calibration tests performed in a PWR plant on 16 RTDs. The data includes 64 points which represent four data collection passes for the 16 RTDs. The plant's temperature instability is apparent in the raw data. A conceptual example of the corrected data is also shown. The corrected data no longer has a large swing but still has some fluctuations. These fluctuations are referred to as *residual temperature fluctuations*.

In this example, the instability correction was performed by a least-square fitting of the raw data to two straight-line segments. The straight lines were then subtracted from the raw data to remove the large temperature swings. Doing this reduced the effect of fluctuations from about 1.5° C to about 0.15° C.
To calculate the instability error, one must identify the standard deviation of the residual temperature fluctuations. These are the fluctuations that remain after the data is corrected for: (1) the deviations of individual RTDs; (2) the process temperature fluctuations and drift; and (3) gross nonuniformities between the hot-leg and cold-leg RTDs or the primary coolant loops. The instability error (standard deviation of residual fluctuations) for the corrected data shown in Fig. 5.8 is about 0.015° C.
Table 5.7 shows standard deviations of cross-calibration data at three temperatures, with and without corrections for instability. These are the standard deviations of 64 data points representing four sets of cross-calibration data on 16 RTDs. The table clearly shows that the instability correction reduced the average standard deviation by a factor of three, from 0.09° C to 0.03° C. This data, along with results from other similar measurements, show that the uncertainties in cross-calibration results because of a plant's temperature instability typically range from 0.001° C to 0.1° C.

4. **Nonuniformity Error.** Incomplete mixing and differences in loop-heat removal could cause the temperature of primary loops to differ by as much as 0.5° C between the hot-leg and cold-leg temperatures at isothermal conditions. Fig. 5.9 shows differences, at four temperatures, between the average temperature of the hot legs and cold legs in two loops of a PWR plant. These differences are referred to as *nonuniformity error*. This error should be identified and subtracted from the data to make the cross-calibration results more accurate.

Table 5.7. Effect of instability correction on standard deviation of raw and corrected cross-calibration data

		Standard Deviation (° C)	
Temperature (° C)	**Run #**	**Raw Data**	**Corrected Data**
280° C	1	0.10	0.03
	2	0.08	0.03
220° C	1	0.12	0.02
	2	0.06	0.02
170° C	1	0.09	0.03
	2	0.11	0.03
Average		0.09	0.03

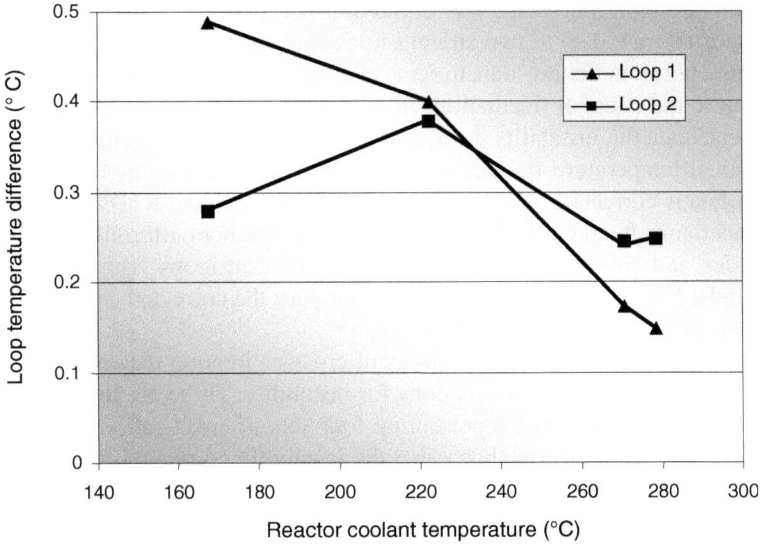

Fig. 5.9. Difference between the hot-leg and cold-leg temperatures in each loop of a two-loop PWR

The nonuniformity errors for each temperature are identified by measuring both the standard deviations of the RTDs in each loop and the standard deviations of the hot-leg and cold-leg RTDs, and then using the largest result.

Now, the total error for each temperature is calculated using the root sum squared (RSS) formula as follows:

$$RSS\ Error\,(^{\circ}C) = \sqrt{e_1{}^2 + e_2{}^2 + e_3{}^2 + e_4{}^2} \tag{5.4}$$

where e_1, e_2, e_3, and e_4 are the four errors just described.

The results of this uncertainty calculation were shown in Table 5.6. These results range from about 0.01° C to about 0.03° C. These are the error bars that should go with the deviation results from a cross-calibration test. That is, the cross-calibration results including the uncertainty of the results may be expressed as $|\Delta| \pm \mathrm{RSS}$, where RSS is the combined error calculated using Eq. 5.4 and $|\Delta|$ is the deviation result from the cross-calibration test.

5.8.2 Uncertainties with Plant Computer Data

When data is acquired from the plant's computer or a data historian, several other error sources must be included when calculating the uncertainty of the cross-calibration results, in addition to errors discussed in the previous section. For example, the resistance of RTDs in nuclear power plants is often converted into a voltage by a circuit card, which usually consists of a Wheatstone bridge or other similar circuit. These

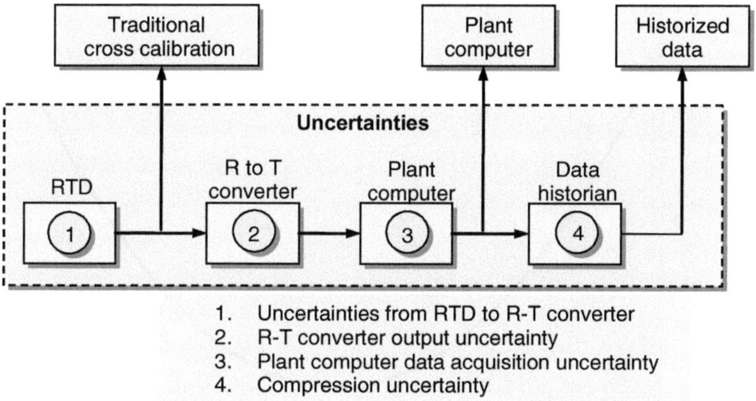

Fig. 5.10. Example of a temperature measurement channel and corresponding sources of uncertainties that may be involved in RTD cross-calibration using data from plant computer

circuit cards are typically calibrated so as to output a voltage that is proportional to the input resistance. This voltage is also proportional to the temperature that the RTD is reading. Fig. 5.10 shows a block diagram of typical components of a temperature measurement channel, which may be involved in a cross-calibration test.

Calibrating circuit cards generates an error that must be included when calculating the uncertainty of cross-calibration results. In addition, circuit cards typically perform a linear conversion from resistance to temperature. This linear fit results in another error at the output of the circuit card because the RTD curve is not actually linear; it is quadratic. Fig. 5.11 shows this error as a function of temperature for a narrow-range RTD. This is referred to as the *resistance-to-temperature (R-T) uncertainty* (see Item 2 in Fig. 5.10).

Another source of uncertainty must be considered in cross-calibration testing using data from a plant computer. The inputs to the plant computer are typically analog signals that must be converted into digital values be fore they can be stored in the computer. This can result in sampling uncertainty (see Item 3 in Fig. 5.10).

The data historian may also contribute to uncertainty. Typically, data historians such as PI© or eDNA© compress or "historize" the data from the plant computer so as to reduce storage space. This is normally accomplished by saving a data sample only if the sample differs from the previous sample by a certain amount. For example, a data historian may only save a temperature value if the current sample differs from the previous sample by 0.1° C. This results in a compression error that must be included when calculating the uncertainty of cross-calibration results using plant computer data (see Item 4 in Fig. 5.10).

5.9 Validating the Cross-Calibration Technique

The cross-calibration technique has been validated for both RTDs and thermocouples in a laboratory setting involving an oil bath, an SPRT, and a set of cross-calibration

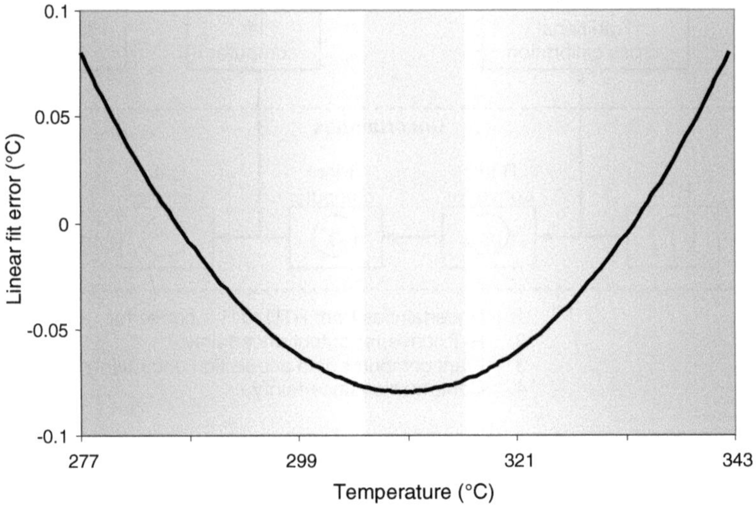

Fig. 5.11. Error between linear fit and quadratic equation over a narrow temperature range

test equipment.[7] The goal of the validation work was to demonstrate that the "true" temperature of the oil bath as measured with an SPRT is very close to the average temperature indicated by a group of RTDs or thermocouples. The results of the validation work are summarized in Tables 5.8 for RTDs and 5.9 for thermocouples.

It is clear in Table 5.8 that the average temperature of the bath as indicated by the 18 RTDs (less one outlier) being 300.786° C is very close to the true temperature of the bath as indicated by the two SPRTs, one reading 300.788° C and the other 300.762° C. The agreement is also very good for the thermocouples, as seen in Table 5.9. More specifically, the average temperature indicated by the thermocouples is 200.33° C compared with the "true" temperature of the batch, which is 200.38° C as measured by the SPRT.

The results in Table 5.8 are for four-wire RTDs. The same type of validation work has been performed on these same RTDs in three-wire configuration. The results are discussed in Sect. 5.10 below in terms of uncertainties that may be involved in three-wire RTDs.

5.10 Uncertainty in Cross-Calibrating Three-Wire RTDs

The uncertainties discussed in Sect. 5.8 excluded the errors caused by lead-wire imbalances in three-wire RTDs. This error arises from differences between the resistances of the wires that extend from the sensing element to the resistance measuring equipment. Fig. 5.12 shows a three-wire and a four-wire arrangement. In a four-wire arrangement, the lead-wire resistances are completely compensated, while in a three-wire arrangement, the resistance of wire 3 (R_3) must be equal to the resistance of wire 1 (R_1) or wire 2 (R_2), depending on which wire is used as the common wire in

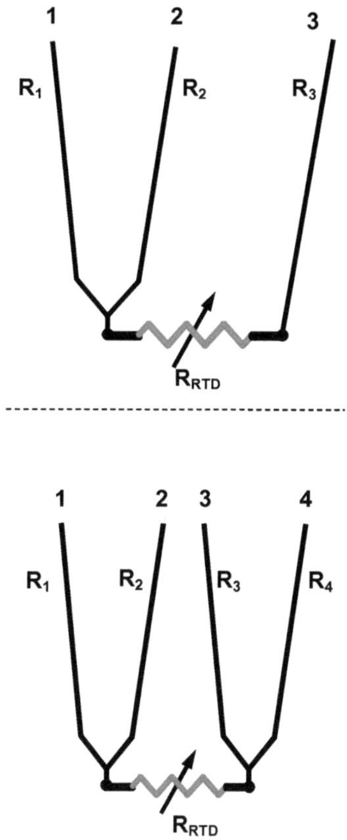

Fig. 5.12. Three-wire and four-wire RTD configurations

the three-wire bridge. If the resistance of wire 3 (R_3) is not equal to the resistance of one of the remaining two wires, then an error will arise in temperature measurement with the three-wire RTD. This error is referred to as *lead-wire imbalance error*. As will be seen later, the lead-wire imbalance can cause about 0.10° C error, on average, at a temperature of about 300° C.

5.10.1 Cross-Calibration Procedure for Three-Wire RTDs

The cross-calibration procedure for three-wire RTDs must include three separate four-wire resistance measurements in order to obtain the resistance of each RTD that is involved in the cross-calibration test:

$$Measurement\ 1:\ R_{13} = R_1 + R_{RTD} + R_3 \tag{5.5}$$

$$Measurement\ 2:\ R_{23} = R_2 + R_{RTD} + R_3 \tag{5.6}$$

Table 5.8. Results of laboratory validation of cross-calibration technique for four-wire RTDs

RTD Tag #	Resistance Measurements (Ohms) Pass 1	Pass 2	Pass 3	Pass 4	Avg. Res. (Ohms)	*Bath Temperature: 300° C* Temp. (° C)	Dev. (° C)
21	212.2644	212.2643	212.2741	212.2634	212.2666	*303.537	# 2.751
19	429.4928	429.4910	429.5060	429.4940	429.4960	300.759	-0.028
12A	429.1156	429.1152	429.1298	429.1134	429.1185	300.811	0.025
12C	429.0084	429.0088	429.0224	429.0102	429.0125	300.807	0.021
03	424.6416	424.6324	424.6384	424.6494	424.6405	300.785	−0.001
18	429.8286	429.8200	429.8228	429.8394	429.8277	300.819	0.033
15A	424.6800	424.6818	424.6862	424.6988	424.6867	300.774	−0.012
15C	424.2652	424.2698	424.2740	424.2818	424.2727	300.774	−0.012
13A	428.9974	428.9962	429.0018	429.0144	429.0025	300.762	−0.024
13C	428.9734	428.9702	428.9726	428.9840	428.9751	300.771	−0.015
9C	430.2376	430.2382	430.2478	430.2404	430.2410	300.787	0.001
9A	430.1488	430.1532	430.1658	430.1480	430.1540	300.780	−0.006
17A	424.3216	424.3148	424.3264	424.3152	424.3195	300.840	0.054
17C	424.2266	424.2230	424.2322	424.2186	424.2251	300.833	0.047
16A	424.7866	424.7882	424.7912	424.7800	424.7865	300.806	0.020
16C	424.5208	424.5294	424.5258	424.5228	424.5247	300.806	0.020
07	430.0636	430.0732	430.0620	430.0628	430.0654	300.724	−0.062
20	430.3424	430.3492	430.3344	430.3448	430.3427	300.749	−0.037
SPRT-1	54.7764	54.7778	54.7761	54.7767	54.7768	**300.788**	0.002
SPRT-2	54.7930	54.7954	54.7930	54.7935	54.7937	**300.762**	−0.024
					Average Temperatur:	**300.786° C**	

* *Not used in average* # Deviation limit exceeded

Table 5.9. Results of laboratory validation of cross-calibration technique for thermocouples

Sensor I.D.	Type	EMF (mv)					Temp (°C)	•T (°C)
		Pass 1	Pass 2	Pass 3	Pass 4	Avg.		
1	K	8.103	8.103	8.103	8.103	8.103	199.16	-1.17
2	K	8.116	8.115	8.115	8.115	8.115	199.46	-0.87
3	K	8.201	8.201	8.201	8.201	8.201	201.61	1.28
4	K	8.123	8.124	8.130	8.125	8.126	199.74	-0.59
5	E	13.471	13.471	13.471	13.470	13.471	200.71	0.38
6	E	13.500	13.500	13.499	13.499	13.500	201.10	0.77
7	E	13.513	13.514	13.510	13.512	13.512	201.26	0.93
8	E	13.433	13.442	13.430	13.420	13.431	200.17	-0.16
9	J	10.755	10.758	10.757	10.757	10.758	199.67	-0.66
10	J	10.831	10.834	10.834	10.833	10.834	201.04	0.71
11	J	10.725	10.725	10.725	10.725	10.725	199.07	-1.26
12	J	10.830	10.829	10.830	10.829	10.830	200.96	0.63
SPRT	N/A	45.334	45.334	45.334	45.334	45.334	200.38	0.05
		AVERAGE TEMPERATURE (°C):					**200.33°C**	

$$Measurement\ 3: \ R_{12} = R_1 + R_2 \tag{5.7}$$

For the cross-calibration results to be very accurate, the following must be true:

$$R_3 = \frac{R_1 + R_2}{2} \tag{5.8}$$

If this is true, then:

$$R_{RTD} = \frac{Measurement\ 1 + Measurement\ 2}{2} - Measurement\ 3 \tag{5.9}$$

The resistance of wire 3 cannot be measured in-situ during the cross- calibration test. Therefore, it is not possible to determine the exact impact of the lead-wire imbalances on cross-calibration results. What can be done is to identify the values of R_1 and R_2 and use the differences between these values to estimate the uncertainties caused by lead-wire imbalances. This was done for a set of cross-calibration tests performed in a PWR plant on 16 three-wire RTDs. The results are shown in Table 5.10. The average difference between R_1 and R_2 for this example is about 0.07 ohms. This corresponds to a temperature uncertainty of about 0.09° C for a 200 ohm RTD.

It should be pointed out that the lead-wire imbalance error can also be a problem in four-wire RTDs that include a dummy loop (see Chap. 4, Fig. 4.9).

Table 5.10. Lead-wire imbalance at 280° C plateau

RTD Tag Number	Resistance (ohm) R_1	R_2	$\Delta R\ (R_1 - R_2)$
1	2.228	2.476	0.248
2	2.968	2.955	0.013
3	2.257	2.303	0.047
4	2.310	2.417	0.108
5	2.302	2.311	0.009
6	2.239	2.284	0.045
7	2.916	3.018	0.102
8	2.243	2.251	0.008
9	2.567	2.671	0.095
10	2.530	2.540	0.010
11	2.304	2.373	0.069
12	2.088	2.072	0.016
13	2.552	2.662	0.070
14	2.248	2.218	0.030
15	2.697	2.636	0.061
16	2.336	2.098	0.238
Average of ΔR (Ω)			0.073
Temperature Error (° C) (for a 200 Ω RTD)			0.091

5.10.2 Cross-Calibration Validation for Three-Wire RTDs

The total cross-calibration error of a three-wire RTD shares the same components as those shown previously in Table 5.6, except for the uncertainties associated with lead-wire imbalances. To validate the cross-calibration technique for three-wire RTDs, laboratory cross-calibration tests were performed in an oil bath at 300° C on the same 18 RTDs as in Table 5.8. The three-wire tests were performed with the four-wire RTDs used in three-wire configuration.

The validation results for four-wire cross-calibration were shown in Table 5.8. The results for three-wire cross-calibration are shown in Table 5.11. Clearly, the RTD deviations for the three-wire configuration are larger than for the four-wire configuration. This is normally expected due to the effect of lead-wire imbalance on three-wire cross-calibration.

5.11 Validation of Dynamic Cross-Calibration

Cross-calibration data can be collected at temperature plateaus or when the plant temperature is undergoing a ramp change during startup or shutdown. If data is

Table 5.11. Results of laboratory validation of cross-calibration technique for three-wire RTDs

Test Temperature: 300° C

Tag	Resistance Measurements (Ohms) Pass 1	Pass 2	Pass 3	Pass 4	Avg. Res. (Ohms)	Temp (° C)	Dev.
21	212.2397	212.2376	212.2349	212.2344	212.2366	*303.452	#2.908
19	429.3999	429.4064	429.3937	429.3864	429.3966	300.622	0.078
12A	428.8234	428.8349	428.8227	428.8128	428.8234	300.403	−0.141
12C	428.7745	428.7909	428.7758	428.7656	428.7767	300.481	−0.063
03	424.7646	424.7887	424.7765	424.7661	424.7740	*300.973	#0.429
18	429.6882	429.7120	429.7032	429.6962	429.6999	300.643	0.099
15A	424.5699	424.5908	424.5827	424.5752	424.5797	300.623	0.079
15C	424.1568	424.1741	424.1683	424.1659	424.1663	300.625	0.081
13A	428.7102	428.7284	428.7178	428.7214	428.7195	300.371	#−0.173
13C	428.6318	428.6460	428.6365	428.6451	428.6399	300.308	#−0.236
9C	430.0284	430.0315	430.0301	430.0450	430.0338	300.500	−0.044
9A	429.9458	429.9449	429.9430	429.9646	429.9496	300.497	−0.047
17A	424.1572	424.1518	424.1428	424.1502	424.1505	300.602	0.059
17C	424.0595	424.0545	424.0462	424.0589	424.0548	300.593	0.049
16A	424.6559	424.6485	424.6553	424.6712	424.6577	300.625	0.081
16C	424.3978	424.3937	424.4001	424.4085	424.4000	300.631	0.087
07	429.8613	429.8435	429.8580	429.8836	429.8616	300.445	−0.099
20	430.2722	430.2506	430.2682	430.2763	430.2668	300.645	0.101
SPRT-1	54.7627	54.7614	54.7617	54.7634	54.7623	300.632	0.088
SPRT-2	54.8013	54.7996	54.8012	54.8046	54.8017	*300.848	#0.304

* Not used in average # Deviation limit exceeded Average Temperature: **300.544° C**

taken during temperature ramp conditions, the test is referred to as *dynamic cross-calibration*.[8,9]

In dynamic cross-calibration, the sensors are scanned from first to last and then from last to first, and the results are averaged. When performed correctly, dynamic cross-calibration should produce results that are comparable to the results obtained when data is taken at temperature plateaus. This has been verified using both laboratory and in-plant data. The results are shown in Tables 5.12 and 5.13. In the laboratory validation tests, the bath temperature was ramping down from 300° C to room temperature at a rate of about 60° C/hr.

The results of this work have shown that: (1) the ramp test results are typically within about 0.03° C of plateau test results; and (2) the differences between ramp and plateau results are within the normal range of repeatability and uncertainty of the cross-calibration technique. Plant test experience has shown that process temperature is typically more stable and uniform during ramp conditions than when the operators are attempting to maintain a steady temperature plateau.

5.12 Cross-Calibrating Core-Exit Thermocouples

In PWR plants, at isothermal conditions, the primary coolant RTDs and core-exit thermocouples are at essentially the same temperature. For that reason, the core-exit thermocouples are cross-calibrated against the primary coolant RTDs because the latter are usually more accurate than thermocouples.

To perform thermocouple cross-calibration, subtract the indication of each core-exit thermocouple from the average temperature indicated by the narrow-range RTDs. Then tabulate the results, as shown in Table 5.14. The data for this table was obtained from a plant computer and analyzed using automated cross-calibration software.

5.13 Recalibrating Outliers

If cross-calibration data is taken during plant heatup or cooldown at three or more widely spaced temperatures, then a new calibration table can be generated for any outlier RTD according to the procedure described in this section.

5.13.1 Recalibration

An outlier RTD may be recalibrated to avoid premature RTD replacement, provided the following conditions are met:

- There are only a few outliers.
- A new calibration chart is generated for an RTD no more than once or twice at the most.

Table 5.12. Results of laboratory validation of dynamic cross-calibration technique

Tag	Resistance Measurements (Ohms)				Avg. Res.	Temp	Dev.
	Pass 1	Pass 2	Pass 3	Pass 4	(Ohms)	(° C)	(° C)
21	212.1559	209.3864	209.3155	206.6431	209.3752	*295.338	#2.631
19	429.1486	423.7962	423.3428	418.2018	423.6224	292.681	−0.026
12A	428.7144	423.6962	422.9530	418.1374	423.3752	292.885	#0.177
12C	428.4792	423.7342	422.7188	418.1828	423.2788	292.892	#0.184
03	423.8728	419.4584	418.1750	413.9560	418.8656	292.686	−0.022
18	428.8116	424.5744	422.9724	418.9354	423.8235	292.575	−0.132
15A	423.7890	419.9490	418.1166	414.4610	419.0789	292.900	#0.193
15C	423.2386	419.6854	417.5794	414.1920	418.6739	292.905	#0.197
13A	427.7112	424.3816	421.9590	418.7786	423.2076	292.766	0.058
13C	427.5354	424.4894	421.7880	418.8812	423.1735	292.765	0.057
9C	428.5696	425.8194	422.8422	420.2084	424.3599	292.664	−0.043
9A	428.3434	425.8806	422.6212	420.2622	424.2769	292.663	−0.045
17A	422.3640	420.2154	416.7224	414.6750	418.4942	292.655	−0.052
17C	422.1288	420.2594	416.5016	414.7212	418.4028	292.646	−0.061
16A	422.5072	420.9188	416.8842	415.3698	418.9200	292.567	−0.141
16C	422.1058	420.8016	416.4910	415.2438	418.6606	292.565	−0.143
07	427.4872	426.4386	421.6986	420.6922	424.0792	292.526	#−0.182
20	427.6642	426.9180	421.9240	421.1984	424.4262	292.617	−0.090
SPRT-1 54.4276	54.3723	53.6961	53.6425	54.0346	292.793	0.085	
SPRT-2 54.4188	54.4000	53.6885	53.6699	54.0443	292.692	−0.016	

* Not used in average # Deviation limit exceeded Average Temperature: **292.707° C**

Table 5.13. Results of in-plant validation of dynamic cross-calibration technique

RTD Tag Number	Plateau Results (° C)	Ramp Results(° C)
2NCRD5420	−0.03	−0.03
2NCRD5421	−0.06	−0.06
2NCRD5422	0.08	0.09
2NCRD5430	0.01	0.00
2NCRD5440	0.01	0.00
2NCRD5460	0.04	0.00
2NCRD5461	0.13	0.12
2NCRD5462	−0.01	−0.04
2NCRD5470	−0.10	−0.06
2NCRD5480	−0.06	−0.02
2NCRD5500	0.01	−0.02
2NCRD5501	−0.01	−0.03
2NCRD5502	−0.01	−0.03
2NCRD5510	0.06	0.10
2NCRD5520	−0.05	−0.01
2NCRD5540	−0.10	−0.07
2NCRD5542	−0.07	−0.05
2NCRD5550	0.05	0.02
2NCRD5560	0.02	−0.03

Both columns of results are based on in-plant data collected in a PWR plant at a temperature of approximately 300° C.

Once these conditions are satisfied, the outlier may be recalibrated according to the following procedure:

1. Produce a table of two columns. The first column is for listing the temperatures at which cross-calibration data were collected. Each temperature registered in this column should represent the best estimate of the process temperature as determined by calculating the average of the narrow-range RTDs. The second column is for listing the corresponding resistances of the outlier RTD. To obtain the RTD resistance when performing cross-calibration using plant computer data, add the RTD's temperature deviation to the average temperature and insert the result into the Callendar Equation or the quadratic equation using constants taken from the existing RTD calibration. Then solve for the corresponding resistance.
2. Fit the resistance-versus-temperature data to the Callendar Equation or quadratic equation.
3. Identify the constants of the Callendar Equation or quadratic equation.
4. Use the new Callendar Equation or quadratic equation to produce a new calibration table for the outlier RTD.

The Callendar Equation was given by Eq. 5.1 in Sect. 5.3.1. The quadratic equation that may be used to generate a new calibration table for an outlier is as follows:

Table 5.14. Results of thermocouple cross-calibration

Item	Tag	Deviation
1	1BB-T-0325-W (H15)	5.33
2	1BB-T-0331-W (J12)	0.47
3	1BB-T-0337-W (L10)	-0.53
4	1BB-T-0338-W (L12)	1.02
5	1BB-T-0339-W (L14)	-1.25
6	1BB-T-0346-W (N14)	2.12
7	1BB-T-0349-W (R10)	3.58
8	1BB-T-0326-W (J2)	-0.71
9	1BB-T-0327-W (J4)	2.04
10	1BB-T-0329-W (J8)	2.83
11	1BB-T-0334-W (L4)	-0.65
12	1BB-T-0335-W (L6)	2.42
13	1BB-T-0340-W (N2)	0.91
14	1BB-T-0343-W (N8)	3.19
15	1BB-T-0347-W (R6)	3.02
16	1BB-T-0348-W (R8)	1.70
17	1BB-T-0300-W (A6)	1.68
18	1BB-T-0303-W (C2)	0.97
19	1BB-T-0304-W (C4)	0.37
20	1BB-T-0310-W (E2)	0.45
21	1BB-T-0311-W (E4)	4.28
22	1BB-T-0317-W (G2)	3.45
23	1BB-T-0318-W (G4)	-0.85
24	1BB-T-0301-W (A8)	5.87
25	1BB-T-0302-W (A10)	0.65
26	1BB-T-0306-W (C8)	-0.19
27	1BB-T-0308-W (C12)	2.75
28	1BB-T-0313-W (E8)	1.35

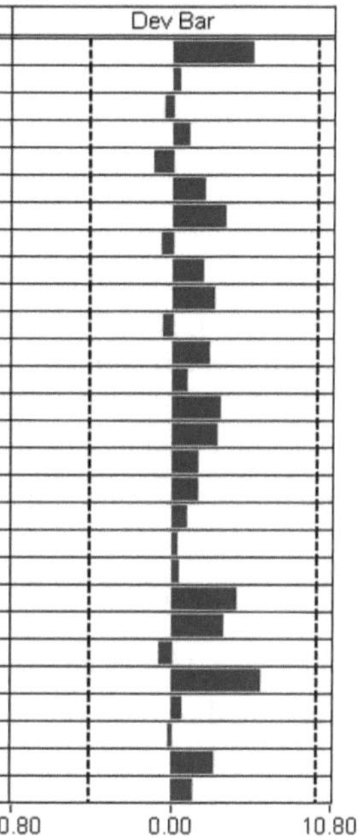

$$R(T) = R_0(1 + AT + BT^2) \tag{5.10}$$

where:

$R(T)$ = Resistance at any temperature T (Ω)
T = Temperature in ° C
R_0, A, and B = RTD calibration constants

Generally, creating a new calibration table for an outlier RTD is similar to performing a calibration without an ice point. The impact on RTD accuracy of calibration without an ice point has been studied by laboratory testing using nuclear-grade RTDs.[7] The study in question involved a four-point calibration performed on six RTDs at 0° C, 100° C, 200° C, and 300° C. The data was analyzed with and without the ice point. The results are shown in Table 5.15 in terms of the differences at 0° C, 200° C, 280° C, and 300° C.

As expected, the differences are large at 0° C and small at higher temperatures. This expectation was further verified by repeating the calibration at 12 temperatures in the range of 0° C to 300° C. The data was analyzed using the four normal calibration points (0° C, 100° C, 200° C, and 300° C) and four high-temperature calibration points (160° C, 200° C, 240° C, and 300° C). The differences are shown in Table 5.16 at four temperatures: 0° C, 200° C, 280° C, and 300° C. Again, the differences are small at high temperatures and large at 0° C.

5.13.2 New Calibration Table

After an outlier RTD is recalibrated per the procedure outlined in Sect. 5.13.1, a new calibration table is generated and used to adjust the corresponding plant instruments to bring the outlier in line with the other redundant RTDs in the plant. The calibration constants may be tabulated as shown in Fig. 5.13. A quadratic equation was used to generate these results.

Fig. 5.13 shows the data that was used to generate a new calibration table for the outlier, the constants of the quadratic equation, and a curve showing the difference between the old calibration of the outlier and its new calibration.

5.13.3 Uncertainty of Recalibration Results

A concern when recalibrating an outlier is that the new calibration curve may get extrapolated beyond the last temperature point used in the cross-calibration test. For example, when data is retrieved from the plant computer for narrow-range RTDs, the three cross-calibration temperature points may be at 270° C, 280° C, and 290° C. Because of this narrow temperature range, the uncertainties posed by a new calibration curve that is generated for an outlier could be large, especially at temperatures that are significantly above 290° C or below 270° C. Fig. 5.14 shows how potential extrapolation errors can increase at higher temperatures, depending on how the errors at the three temperature points are combined to calculate the extrapolation uncertainties. This data is based on using a quadratic or Callendar fit for extrapolation. If a linear fit is used, the extrapolation uncertainties decrease significantly, as shown in Fig. 5.15.

The extrapolation error for a 0.05° C uncertainty in each of the three calibration points may be calculated as follows. If the calibration points are 270 ± 0.05° C, 280 ± 0.05° C, and 290 ± 0.05° C, then all possible permutations are as listed in

Table 5.15. Calibration errors caused by a lack of ice point in a four-point calibration

RTD	Difference (° C)			
Tag Number	**0° C**	**200° C**	**280° C**	**300° C**
15A	0.09	0.013	0.002	0.006
15C	0.09	0.013	0.002	0.005
16A	0.09	0.014	0.003	0.004
16C	0.07	0.011	0.002	0.004
17A	0.08	0.013	0.002	0.005
17C	0.09	0.016	0.004	0.005

The differences are between the fitted results of a four-point calibration with and without the ice point.

Tag

1BB-T-0450-W

Recal Data

RCS Temp. (°C)	Res. (ohms)	Uncertainty (°C)
292.81	209.212	0.047
279.50	204.469	0.047
266.20	199.705	0.047

Quadratic Calibration

Constant	Original	New Calibration	Units
Ro	99.943	99.189	Ohms
A	3.902351E-3	3.993068E-3	1/°C
B	-5.822314E-7	-6.997264E-7	1/°C/°C

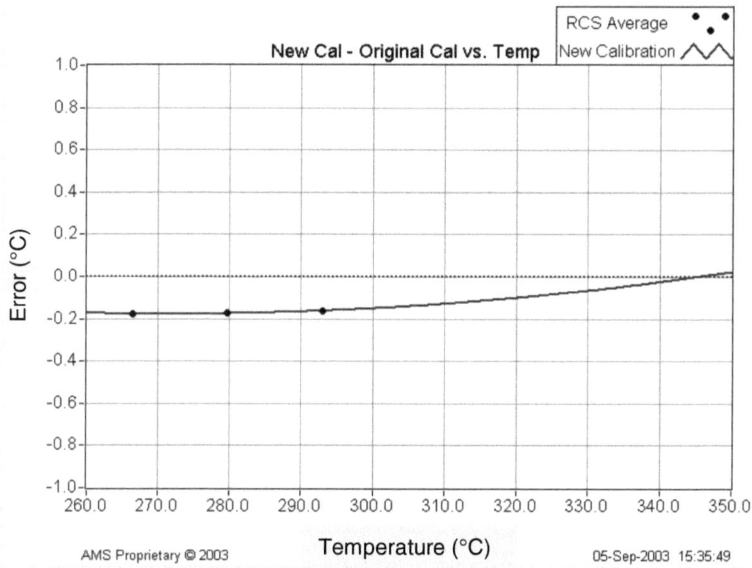

Fig. 5.13. Results of recalibration of an outlier using automated software

Table 5.17. From these permutations, new coefficients are calculated based on the type of calibration equation and fitting method used (i.e., Callendar second order, Callendar linear correction, etc.). Then, the calibration coefficients curve that is fit through 270° C, 280° C, and 290° C is subtracted from each permutation coefficient curve, and the result is plotted as was shown in Fig. 5.14. The extrapolated error is obviously smaller when a linear correlation is used, as seen in Fig. 5.15. This is because a first-order difference is always less than a second-order difference as the temperature is moved beyond the range of the data used in the calculations.

Fig. 5.14 represents the worst-case errors and assumes that the temperature could fluctuate within the uncertainty bounds at each temperature. Realistically, the uncertainty for data acquired over a short period will be a bias term of approximately the same amount within the calculated uncer- tainty. As such, the actual extrapolation errors will normally be much less than the worst case shown in Fig. 5.14.

Tag

1BB-T-0450-W

Recal Data

Temperature (°C)	Resistance (ohms)	Uncertainty (°C)
292.81	209.212	0.047
279.50	204.469	0.047
266.20	199.705	0.047

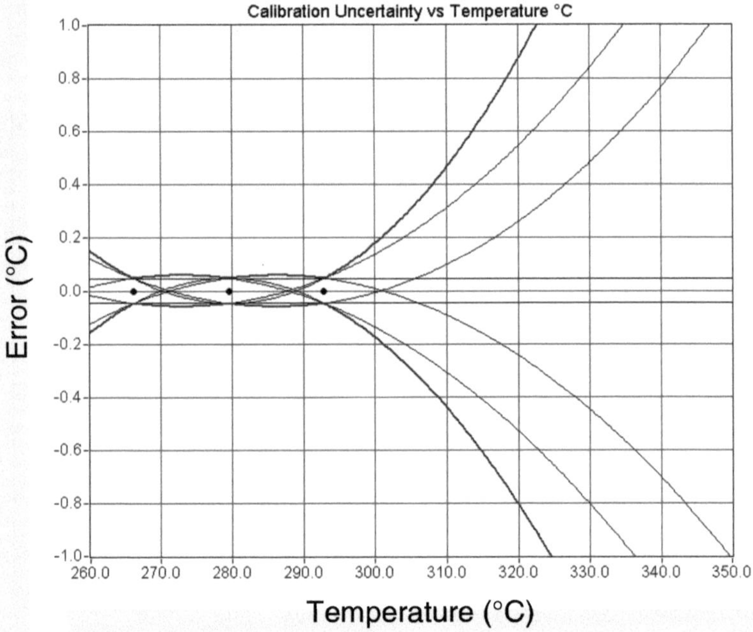

Fig. 5.14. Extrapolation errors when the Callendar or a quadratic equation is used

A complete recalibration table for an outlier RTD is shown in Table 5.18 for the temperature range of 0 to 400° C in one-degree increments. This table lists the resistance of the RTD at each temperature and is based on a quadratic equation.

5.14 NRC Position on RTD Cross-Calibration

The NRC's position on RTD cross-calibration is presented in NUREG-0800.[10] More specifically, Appendix 13 of Chapter 7 of NUREG-0800, which is referred to as *Branch Technical Position 13* or BTP-13, presents the NRC's view of RTD performance testing, including both response-time testing and cross-calibration. The following summarizes some of the key points of BTP-13:

Tag

1BB-T-0450-W

Recal Data

Temperature (°C)	Resistance (ohms)	Uncertainty (°C)
292.81	209.212	0.047
279.50	204.469	0.047
266.20	199.705	0.047

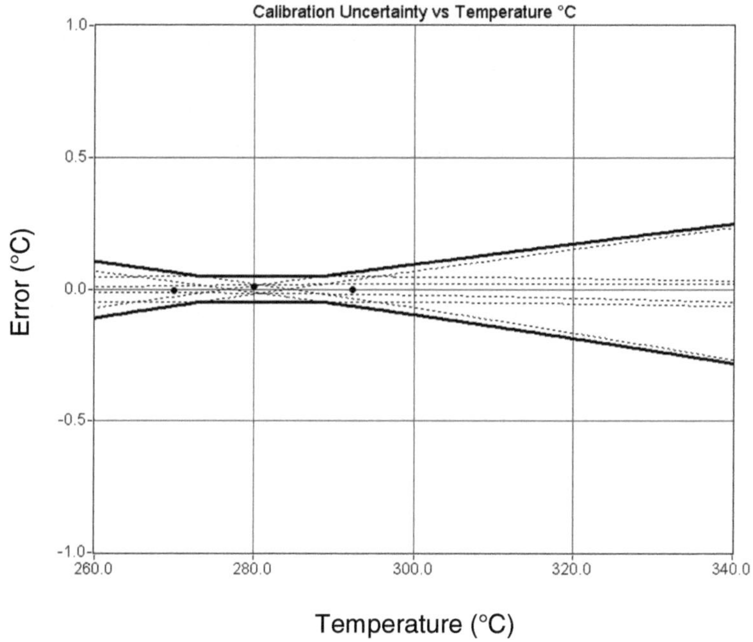

Fig. 5.15. Extrapolation errors when a linear fit is used

- The performance of an RTD is evaluated by its accuracy and response time. To ensure adequate performance of the RTD, its accuracy and response time should be verified.
- The cross-calibration test data should be corrected for process temperature fluctuations and drift, which may occur during the test.
- The cross-calibration method and calibration and response-time data should be examined to identify calibration inaccuracies, uncertainties, and errors.

A copy of the BTP-13 document is provided in Appendix B of this book.

Table 5.16. Calibration errors caused by a lack of ice point in a twelve-point calibration

RTD	Difference (° C)			
Tag Number	**(0° C)**	**(200° C)**	**(280° C)**	**(300° C)**
	Calibration Points 300° C, 260° C, 200° C			
15A	0.099	0.018	0.004	0.006
15C	0.072	0.017	0.003	0.006
16A	0.112	0.015	0.004	0.005
16C	0.121	0.013	0.004	0.004
17A	0.065	0.014	0.002	0.005
17C	0.129	0.040	0.004	0.005
	Calibration Points 300° C, 260° C, 200° C, 160° C			
15A	0.117	0.017	0.004	0.006
15C	0.109	0.016	0.003	0.006
16A	0.110	0.015	0.004	0.005
16C	0.107	0.013	0.004	0.004
17A	0.065	0.014	0.002	0.005
17C	0.071	0.016	0.004	0.004

The differences are between the fitted results of a four-point calibration with and without the ice point.

Table 5.17. Temperature permutations for calculating extrapolation errors

Permutation	270 ± 0.05° **C**	280 ± 0.05° **C**	290 ± 0.05° **C**
1	269.95	279.95	289.95
2	270.05	279.95	289.95
3	269.95	280.05	289.95
4	270.05	280.05	289.95
5	269.95	279.95	290.05
6	270.05	279.95	290.05
7	269.95	280.05	290.05
8	270.05	280.05	290.05

Table 5.18. RTD recalibration table

Constant	Value	Units
Ro	99.189	Ohms
A	3.993068E-3	1/°C
B	-6.997264E-7	1/°C/°C

Temp (°C)	+0	+1	+2	+3	+4	+5	+6	+7	+8	+9
0	99.19	99.59	99.98	100.38	100.77	101.17	101.56	101.96	102.35	102.75
10	103.14	103.54	103.93	104.33	104.72	105.11	105.51	105.90	106.30	106.69
20	107.08	107.48	107.87	108.26	108.66	109.05	109.44	109.83	110.22	110.62
30	111.01	111.40	111.79	112.18	112.58	112.97	113.36	113.75	114.14	114.53
40	114.92	115.31	115.70	116.09	116.48	116.87	117.26	117.65	118.04	118.43
50	118.82	119.21	119.60	119.99	120.37	120.76	121.15	121.54	121.93	122.32
60	122.70	123.09	123.48	123.87	124.25	124.64	125.03	125.41	125.80	126.19
70	126.57	126.96	127.35	127.73	128.12	128.50	128.89	129.28	129.66	130.05
80	130.43	130.82	131.20	131.59	131.97	132.35	132.74	133.12	133.51	133.89
90	134.27	134.66	135.04	135.42	135.81	136.19	136.57	136.96	137.34	137.72
100	138.10	138.48	138.87	139.25	139.63	140.01	140.39	140.77	141.16	141.54
110	141.92	142.30	142.68	143.06	143.44	143.82	144.20	144.58	144.96	145.34
120	145.72	146.10	146.48	146.86	147.23	147.61	147.99	148.37	148.75	149.13
130	149.51	149.88	150.26	150.64	151.02	151.39	151.77	152.15	152.53	152.90
140	153.28	153.66	154.03	154.41	154.78	155.16	155.54	155.91	156.29	156.66
150	157.04	157.41	157.79	158.16	158.54	158.91	159.29	159.66	160.04	160.41
160	160.78	161.16	161.53	161.90	162.28	162.65	163.02	163.40	163.77	164.14
170	164.52	164.89	165.26	165.63	166.00	166.38	166.75	167.12	167.49	167.86
180	168.23	168.60	168.98	169.35	169.72	170.09	170.46	170.83	171.20	171.57
190	171.94	172.31	172.68	173.05	173.41	173.78	174.15	174.52	174.89	175.26
200	175.63	176.00	176.36	176.73	177.10	177.47	177.83	178.20	178.57	178.94
210	179.30	179.67	180.04	180.40	180.77	181.14	181.50	181.87	182.23	182.60
220	182.97	183.33	183.70	184.06	184.43	184.79	185.16	185.52	185.89	186.25
230	186.61	186.98	187.34	187.71	188.07	188.43	188.80	189.16	189.52	189.89
240	190.25	190.61	190.97	191.34	191.70	192.06	192.42	192.78	193.15	193.51
250	193.87	194.23	194.59	194.95	195.31	195.67	196.03	196.40	196.76	197.12
260	197.48	197.84	198.20	198.56	198.91	199.27	199.63	199.99	200.35	200.71
270	201.07	201.43	201.79	202.14	202.50	202.86	203.22	203.58	203.93	204.29
280	204.65	205.00	205.36	205.72	206.08	206.43	206.79	207.14	207.50	207.86
290	208.21	208.57	208.92	209.28	209.63	209.99	210.35	210.70	211.05	211.41
300	211.76	212.12	212.47	212.83	213.18	213.53	213.89	214.24	214.59	214.95
310	215.30	215.65	216.01	216.36	216.71	217.06	217.42	217.77	218.12	218.47
320	218.82	219.18	219.53	219.88	220.23	220.58	220.93	221.28	221.63	221.98
330	222.33	222.68	223.03	223.38	223.73	224.08	224.43	224.78	225.13	225.48
340	225.83	226.18	226.53	226.88	227.22	227.57	227.92	228.27	228.62	228.96
350	229.31	229.66	230.01	230.35	230.70	231.05	231.39	231.74	232.09	232.43
360	232.78	233.13	233.47	233.82	234.16	234.51	234.85	235.20	235.54	235.89
370	236.23	236.58	236.92	237.27	237.61	237.96	238.30	238.64	238.99	239.33
380	239.67	240.02	240.36	240.70	241.05	241.39	241.73	242.07	242.42	242.76
390	243.10	243.44	243.78	244.13	244.47	244.81	245.15	245.49	245.83	246.17

6

Response-Time Testing of RTDs and Thermocouples

6.1 Reasons for Test

In-situ response-time testing is performed on RTDs and thermocouples in nuclear power plants for one or more of the following reasons:

1. To measure the sensor's "in-service" response time to meet technical specification requirements, regulatory regulations, or both.
2. To verify that plant sensors bottom out in their thermowells and to test for air gaps, dirt, and foreign objects in the thermowell.
3. To provide for predictive maintenance, incipient failure detection, and aging management and to establish objective schedules for replacing sensors.
4. To distinguish between sensor problems and cable or connector problems.
5. To diagnose sensor or process anomalies.

In almost all U.S.-made PWR plants, testing RTD response times is required and is performed on one or more RTD channels once every operating cycle. For thermocouples, response-time testing is not mandatory, but some plants perform the tests for one or more of the reasons just noted.

6.2 Historical Practices

Historically, the response time of RTDs and thermocouples has been characterized by a single parameter called the *plunge time constant* (τ). This is defined as the time it takes the sensor output to achieve 63.2 percent of its final value after a step change in temperature is impressed on its surface. This step change is typically achieved by suddenly immersing the sensor in a rotating tank of water at 1 meter per second. The water must be at either a higher or lower temperature than the RTD. Measuring τ in this way is referred to as *plunge testing*.

Until 1977, testing temperature sensors' response times in nuclear power plants was almost always performed using the plunge test. In nuclear reactors, however, plunge testing is inconvenient because the sensor must be removed from the reactor coolant piping and taken to a laboratory for testing. Nuclear reactor service conditions of 150 bar (2,250 psig) and 300°C (572°F) are difficult to reproduce in the laboratory. Therefore, all laboratory tests are performed at much milder conditions, and the results are extrapolated to service conditions. The combination of manipulating the sensor and extrapolating the results to service conditions leads to significant errors in the measurement of sensor response times, sometimes by as much as a factor of three.[11] These drawbacks of the plunge test motivated the industry to find a better way to test the response time of nuclear plant temperature sensors. As a result, the following methods were identified, developed, and implemented in nuclear power plants:

- **LCSR Test.** In the LCSR method, the sensing element is heated by an electric current, and the temperature transient in the element is recorded. From this transient, the response time of the sensor to changes in external temperature is identified. The method is useful for both RTDs and thermocouples.
- **Measurement of Self-Heating Index.** This method is applicable only to RTDs and is used to identify changes in response time as opposed to measuring the response time. In this method, as in the LCSR test, the sensing element is heated by an electric current. After the RTD output settles, the steady-state increase in RTD resistance is measured as a function of the electric power applied to the sensor. The result is referred to as the *self-heating index* (SHI). Any significant change in SHI indicates a change in RTD response time. Therefore, the SHI can be tracked to determine the degradation of RTD response time.
- **Noise Analysis Technique.** In this method, the natural fluctuations (noise) that normally occur at the sensor output during plant operation are recorded and analyzed to determine the sensor's response time. This method is useful for testing the response time of RTDs, thermocouples, and other sensors.

These three methods are compared in Table 6.1, and this chapter explains them. However, the emphasis in this chapter is on the LCSR method and RTDs because: 1) the LCSR method is the most commonly used means for measuring temperature sensors' response time; and 2) RTDs are subject to more stringent response-time testing requirements in nuclear power plants than are thermocouples.

6.3 LCSR

The LCSR method was developed to measure remotely the response time of RTDs and thermocouples while the sensor is installed in an operating process. The test involves injecting the sensor with an electrical current applied at the end of the sensor's extension leads. The current causes Joule heating in the sensor and results in a temperature transient inside the sensor. The time plot, of either the heating while the current is applied, or the cooling after the current is discontinued, is recorded during the LCSR test. From this plot, the sensor response time is obtained by means of the LCSR transformation.

The LCSR test accounts for all the effects of installation and process conditions on response time and thereby provides a sensor's actual "in-service" response time.

6.3.1 Test Equipment

The LCSR test equipment for RTDs and thermocouples are quite different. We will begin by describing LCSR test equipment for RTDs.

For RTDs, a Wheatstone bridge is used to perform the LCSR test (Fig. 6.1). First, the bridge is balanced with 1 to 2 mA of DC current running through the RTD. Then, the current is switched "high" to about 30 to 50 mA. This causes the RTD sensing element to heat up gradually and settle at a few degrees above the ambient temperature. The amount by which the temperature rises in the RTD depends on the magnitude of the heating current used and on the rate of heat transfer between the RTD and its surrounding medium. Typically, the RTD heats up about 5 to 15°C during the LCSR test.

Fig. 6.2 shows two LCSR test transients: one for a direct-immersion RTD and another for a thermowell-mounted RTD. These transients are from the LCSR testing of the RTDs in nuclear power plants using a heating current of about 40 ma. Clearly,

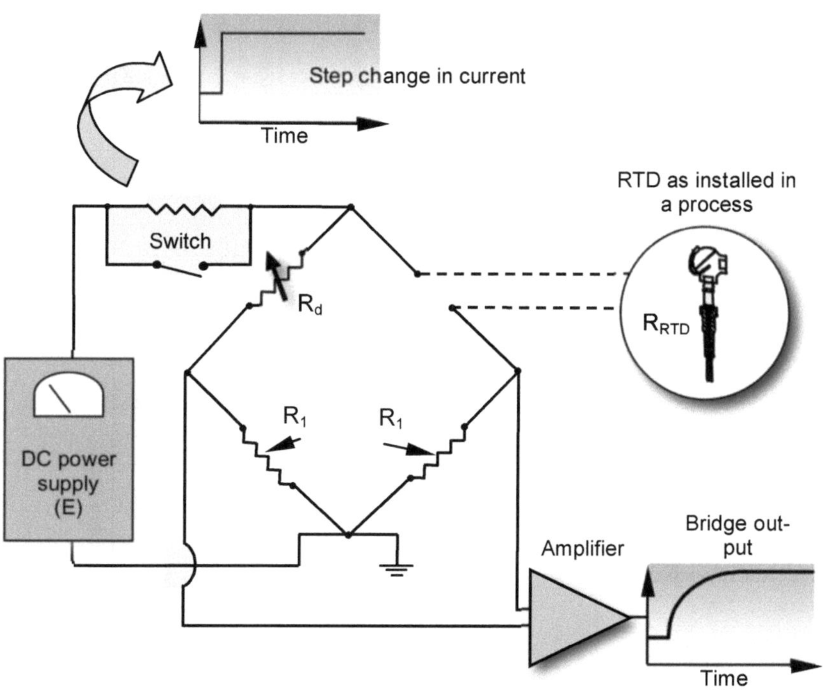

Fig. 6.1. Wheatstone bridge for LCSR testing of RTDs

Table 6.1. Characteristics of methods for response-time testing of nuclear plant RTDs and thermocouples

Test Method	**Application**	**Where Performed**	**Necessary to Take Sensor Out of Service?**	**Complexity of Measurement**	**Quality of Measurement**
Plunge Test	RTDs and thermocouples	Laboratory	Yes	Need to remove RTD and test in laboratory.	Plunge test measures τ directly, but measurement has poor quality for three reasons: (1) manipulating RTD may change its time response; (2) service conditions are usually not reproduced in the laboratory, and results must therefore be extrapolated to service conditions; and 3) for thermowell-mounted RTDs, the plunge test is only valid if the sensor is tested in the same thermowell as is used in the plant. The combined effect of these three elements can result in very large errors.
LCSR Test	RTDs and thermocouples	In-situ	Yes	Special test equipment needed. Data analysis is complex and requires special expertise.	- LCSR is the most accurate means of measuring the response time in RTDs and thermocouples in-situ. - LCSR provides τ of RTD or thermocouple without having to remove the sensor from the plant. - Results are generally accurate to within 10 percent. - Results account for the effect of process conditions and installation on response time.
Self-Heating Test	RTDs	In-situ	Yes	Test simple. Uses simple standard electronic test equipment. Data analysis is simple.	- SHI can be measured quite accurately. - Gross changes in RTD response time can be detected from changes in SHI. - An exact correlation does not exist between τ and SHI.
Noise Analysis Technique	RTDs and thermocouples	In-situ	No	Special test equipment needed. Data analysis requires special expertise.	A useful tool for detecting degradation of response time in RTDs. Not as accurate as LCSR but passive and can be performed on many sensors at the same time.

Source: NUREG-0809, Reference [11].

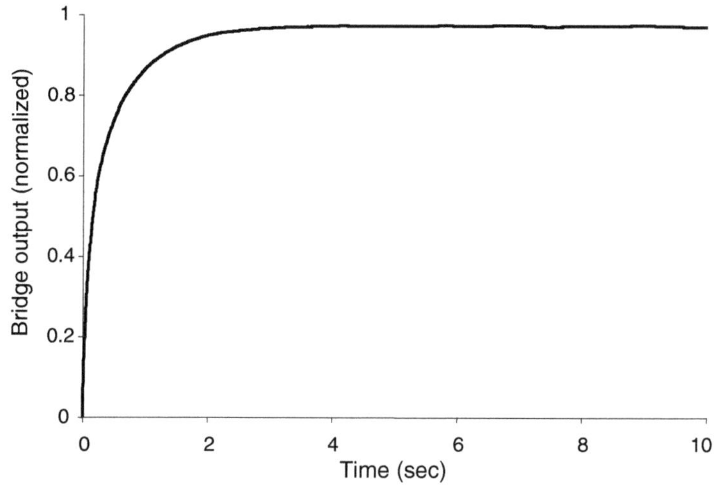

(a) Direct-immersion RTD

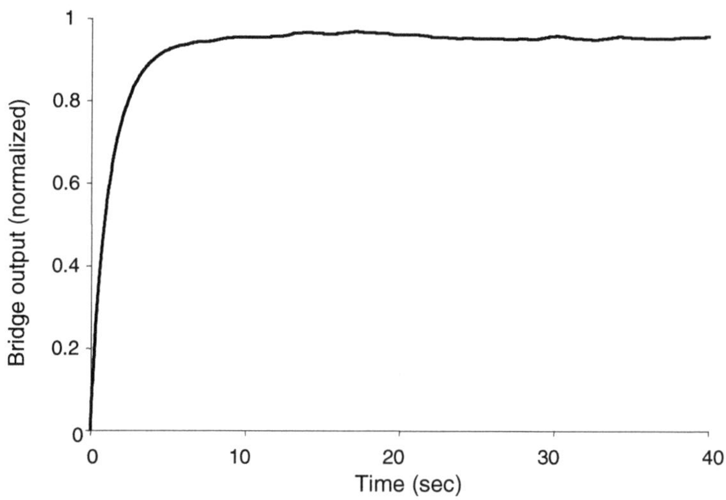

(b) Thermowell-mounted RTD

Fig. 6.2. Field data from LCSR testing a direct-immersion and a thermowell-mounted RTD

the direct-immersion RTD, which has a faster response time, heats up faster than the thermowell-mounted RTD. Of course, these transients are caused by internal heating and cannot provide the RTD's' response time until after the data is transformed.

To provide the necessary current for the LCSR test, the power supply in the Wheatstone bridge is adjusted so the high current is between about 30 and 50 mA, depending on the RTD and the process in which it is installed. If the RTD is in a

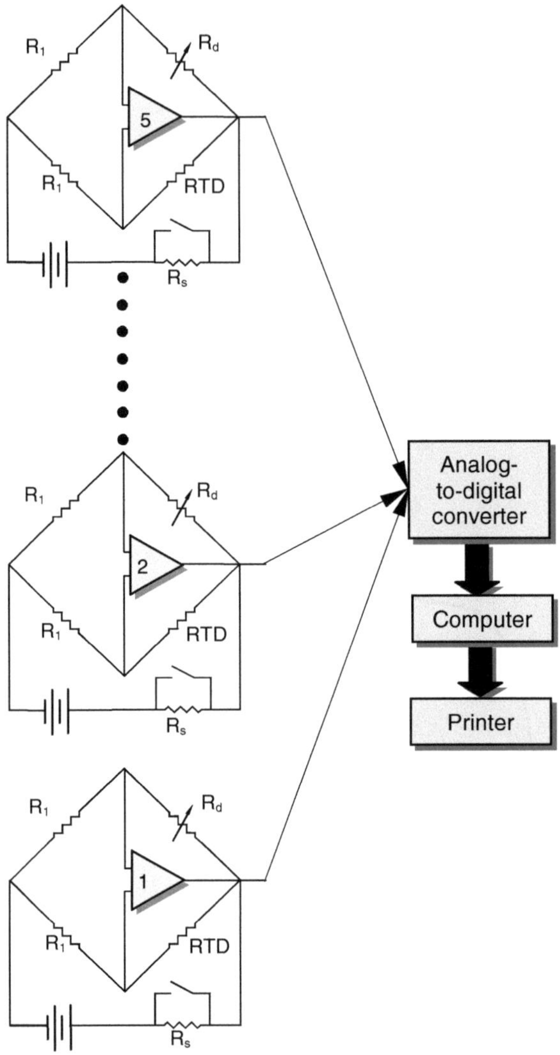

Fig. 6.3. Diagram for a multichannel LCSR test unit

stable process, then 30 mA is usually enough. On the other hand, if the RTD is in a process that has large temperature fluctuations, a higher current (e.g., 50 mA) is needed to improve the signal-to-noise ratio. Of course, an amplifier can also be placed at the output of the Wheatstone bridge to adjust the amplitude of the LCSR signal, as necessary.

The Wheatstone bridge's output voltage (V) changes almost linearly with changes in RTD resistance (δR) during the LCSR test. A simple circuit analysis shows that the bridge output is given by (Fig. 6.1):

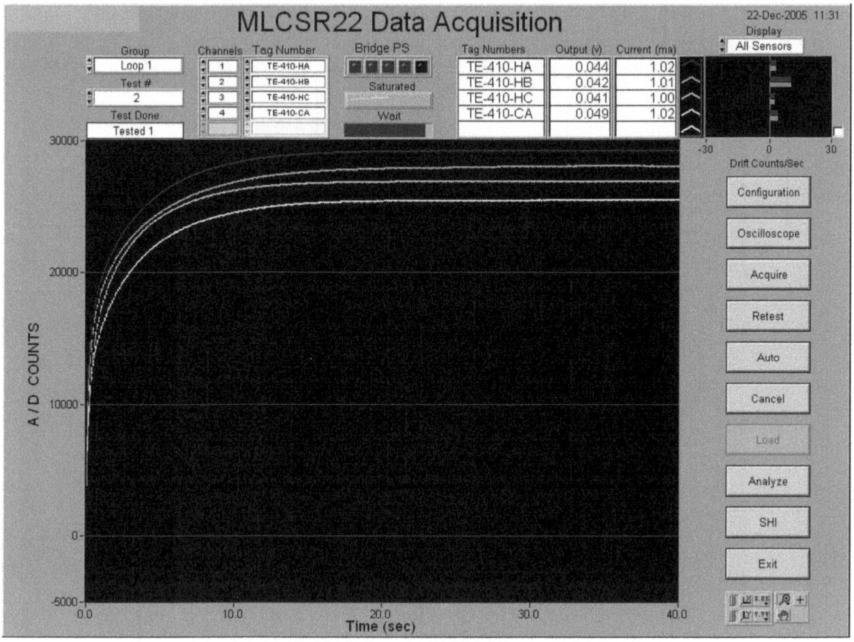

Fig. 6.4. LCSR data acquisition software screen

$$V = \frac{R_1(R_{RTD} - R_d)}{(R_1 + R_d)(R_1 + R_{RTD})} E\,. \tag{6.1}$$

where E is the bridge power supply voltage.

When the bridge is balanced in preparation for LCSR testing, $R_{RTD} = R_d$, and $V = 0$. As soon as the current is stepped up to begin the LCSR test, the bridge output rises exponentially, while the RTD resistance increases to $R_{RTD} + \delta R$. The bridge output eventually settles at a steady-state value. With these points in mind, the bridge output voltage can be written as:

$$V = \left(\frac{R_1}{R_1 + R_d}\right)\left(\frac{\delta R}{R_1 + R_d + \delta R}\right) E \tag{6.2}$$

If we assume that $R_1 + R_d$ is much greater than δR, then we can write:

$$V = C\,\delta R\,E \tag{6.3}$$

where C is a constant.

Equation 6.3 shows that the bridge output changes linearly with δR as long as $R_1 + R_d$ is much greater than δR. Typically, δR is less than 10 ohms, and $R_1 + R_d$ is in the range of 300 to 600 ohm, depending on the RTD and the temperature to which it

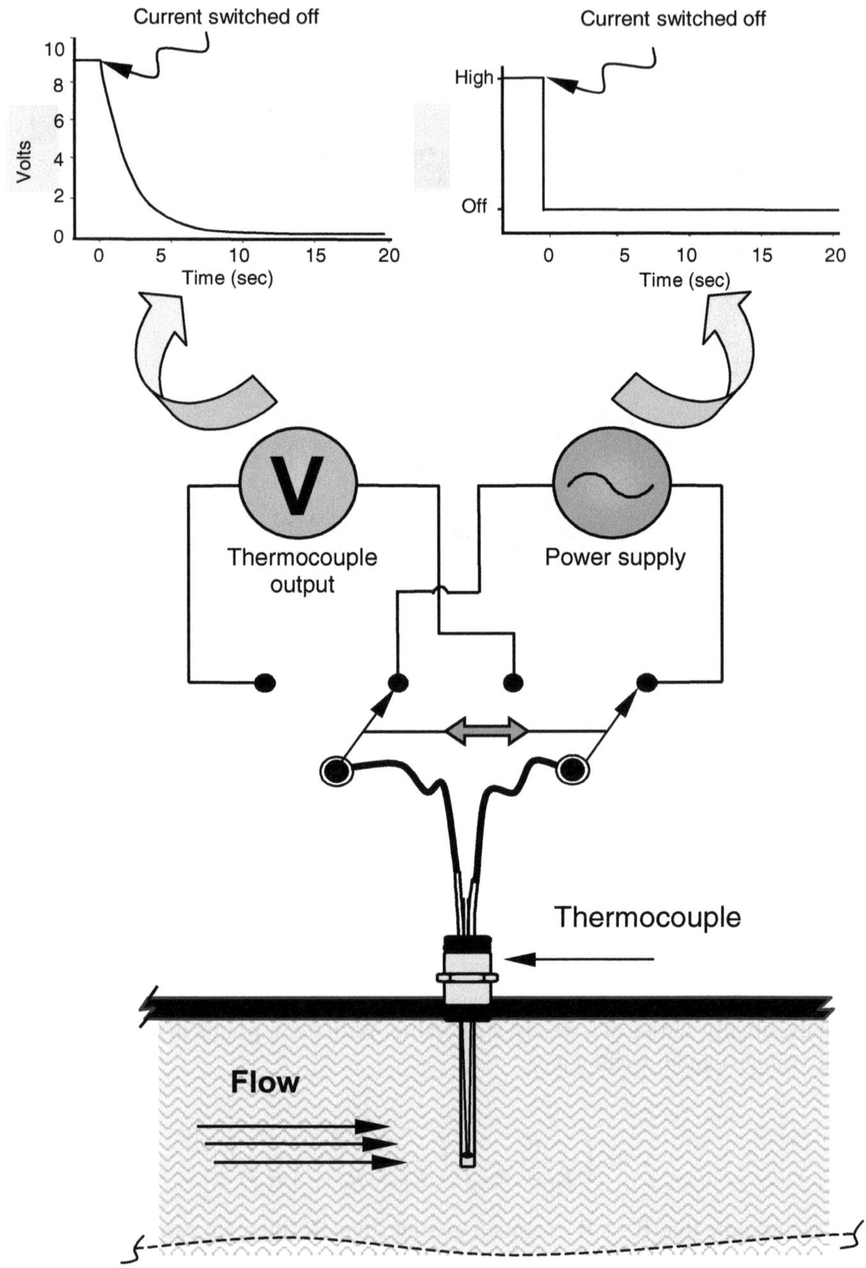

Fig. 6.5. Equipment setup for LCSR testing of thermocouples

is exposed. Therefore, the assumption that V changes linearly with δR is easily met. If necessary, R_1 can be made large enough to satisfy the linearity assumption, but a value of 100 to 200 ohms for R_1 is usually adequate.

In the past, the LCSR test was performed on one RTD at a time. In early 2000, however, new equipment was developed to test multiple RTDs simultaneously. Typically, four to six RTDs can be LCSR tested together. Fig. 6.3 shows a diagram of a multichannel LCSR test unit. Note that the system is shown to include not only the Wheatstone bridges but also the digital data acquisition unit that is used to collect and analyze the LCSR data.

Fig. 6.4 shows the screen of the LCSR test equipment on which is displayed the LCSR data for four RTDs. The LCSR data for RTDs is usually sampled at 0.001 to 0.04 second intervals, depending on the RTD's expected response time and the conditions under which the RTD is tested. Typically, 2 to 20 seconds of data is sampled for direct-immersion RTDs and 20 to 60 seconds for thermowell-mounted RTDs when these sensors are used in flowing water.

For thermocouples, the LCSR test procedure is quite different than for RTDs, although the principle of the test and data analysis is the same. More specifically, for LCSR testing of thermocouples, an AC signal is more often used than a DC signal to heat the sensor for the test. This is because AC current cancels the Peltier effect at the thermocouple junction. Furthermore, for LCSR testing of thermocouples, higher currents (e.g., 500 mA) are usually needed. This is because a thermocouple circuit's resistance is distributed along the length of the thermocouple. This is in contrast with RTDs, in which the circuit resistance is dominated by the resistance of the sensing element. Therefore, in LCSR testing of thermocouples, the whole thermocouple heats up when the LCSR test current is applied.

In LCSR testing of RTDs, the data is collected as current is running through the circuit and the RTD is heating up. For thermocouples, the LCSR data is collected after the current is cut off and while the thermocouple is cooling down. More specifically, in LCSR testing of thermocouples, the current is applied for a few seconds and then switched off while the thermocouple output is monitored as it cools to the ambient temperature (Fig. 6.5).

The LCSR transient that results from thermocouple cooling is dominated by the temperature transient at the thermocouple junction, although the thermocouple wires also heat up and cool down during the LCSR test.

To obtain the thermocouple response time, the LCSR transient must be analyzed using the same procedure as RTDs. Therefore, the LCSR cooling transient for thermocouples is often inverted when presenting the plot of the raw data. Fig. 6.6 shows typical LCSR transients from laboratory and in-plant testing of thermocouples.

6.3.2 LCSR Transformation

The LCSR test is based on causing a step change in temperature inside the sensor, while a sensor's dynamic response is obtained from measuring the reaction to a step change in temperature outside the sensor. Fortunately, there is a way to convert the LCSR transient from the internal heating of a sensor to yield the sensor's dynamic

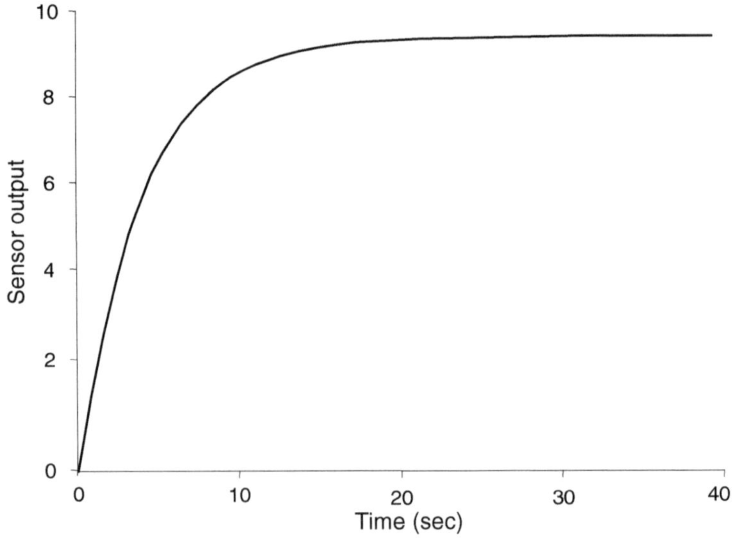

(a) Laboratory test in water at 1 meter/second

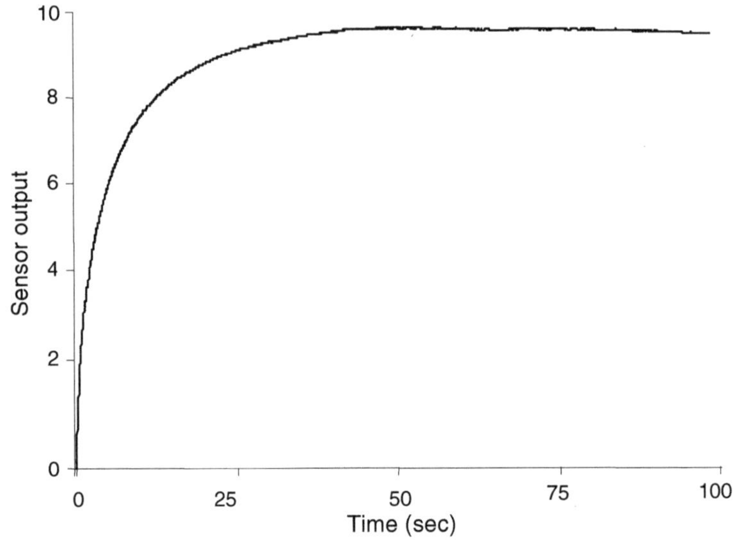

(b) In-plant test in a PWR

Fig. 6.6. LCSR transients from laboratory and in-plant testing of thermocouples

response to a step change outside the sensor. Fig. 6.7 compares a raw LCSR transient for an RTD, the converted LCSR transient, and the RTD's response to a step change in temperature outside the RTD, obtained using a plunge test. This data is derived from a laboratory test in which the RTD is placed in a rotating tank of water at room temperature, ambient pressure, and a flow rate of 1 meter per second.

The process of converting the LCSR data is referred to as the *LCSR transformation*. This transformation is the same for RTDs and thermocouples and was developed in the mid-1970s by Professor T. W. Kerlin of the University of Tennessee and validated by the author.[12] This work produced the following formula, which is referred to as the *LCSR equation*:

$$T(t) = A_0 + A_1 e^{-t/\tau_1} + A_2 e^{-t/\tau_2} + \ldots + A_n e^{-t/\tau_n} \tag{6.4}$$

where T(t) represents the LCSR transient, and $\tau_1, \tau_2, \ldots, \tau_n$ are referred to as *modal time constants*. These modal time constants are related to the sensor's response time (τ) by the following equation:[13]

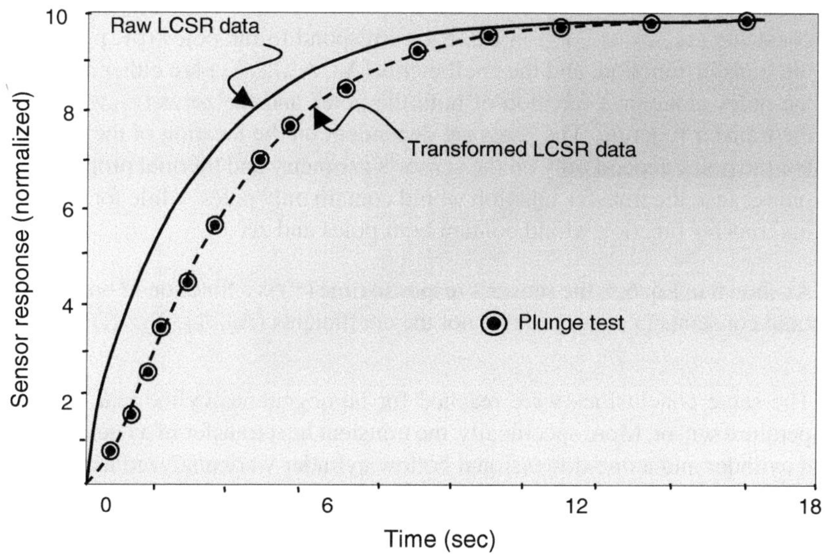

Fig. 6.7. Comparison of raw and transformed LCSR data with corresponding plunge-test transient from laboratory testing of an RTD

$$\tau = \tau_1 \left[1 - ln\left(1 - \frac{\tau_2}{\tau_1}\right) - ln\left(1 - \frac{\tau_3}{\tau_1}\right) - \ldots - ln\left(1 - \frac{\tau_n}{\tau_1}\right)\right] \tag{6.5}$$

where ln is the natural logarithm operator and n is the number of modal time constants.

The τ in Eq. 6.5 is equal to the sensor's response time to a step change in temperature of the fluid outside the sensor. That is, τ is equal to the time that it would take for the sensor's output to reach 63.2 percent of its final steady-state value following a step change in the temperature of the fluid outside the sensor.

Eq. 6.4 was arrived at using a nodal heat-transfer model. In particular, heat balance equations for a nonhomogeneous cylindrical model representing a temperature sensor's sensing section were written in a finite-difference form and solved. The following conclusions were produced:

- The dynamic response of a nonhomogeneous cylindrical temperature sensor such as an RTD or a thermocouple is given by an exponential series of the form shown in Eq. 6.4. The form of this equation is the same whether the sensor is perturbed from the inside, as in the LCSR test, or from the outside, as in the plunge test.

- The exponents in Eq. 6.4 (τ_1, τ_2, ..., τ_n) depend on the sensor geometry and thermal properties but not on whether the sensor is perturbed from the inside or the outside. However, the coefficients ($A_0, A_1, \ldots, A_n$) are dependent on the forcing function (i.e., on whether the sensor is heated from the inside or the outside).

- In terms of a transfer function representation of dynamic response, the modal time constants (τ_1, τ_2, ..., τ_n) in Eq. 6.4 correspond to the poles (p_1, p_2, ..., p_n) of the transfer function, and the coefficients (A_0, $A_1, \ldots, A_n$) are either a function of the poles alone or a function of both the poles and the zeros (z_1, z_2, ..., z_n) of the transfer function. The zeros are dependent on the location of the heat source, but the poles depend only on the sensor's geometry and thermal properties. For a plunge test, the transfer function would contain only poles, while for a LCSR test the transfer function would contain both poles and zeros.

- As shown in Eq. 6.5, the sensor's response time (τ) is a function of only the modal time constants (τ_1, τ_2, ..., τ_n), not the coefficients (A_0, A_1, A_2, ...) in Eq. 6.4.

The same conclusions were reached for homogeneous cylindrical models of a temperature sensor. More specifically, the transient heat transfer of a one-dimensional solid cylinder and a one-dimensional hollow cylinder were analyzed using a continuum analytical approach, as opposed to a nodal approach. The numerical results of the nodal approach and the continuum approach for numerous simulations turned out to be within 2 percent of each other. This indicates that the two solutions are essentially the same. In particular, the continuum approach showed that the eigenvalues for a sensor response to a step change in temperature at the sensing element inside the sensor (LCSR test) are the same as for the sensor response to a step change in temperature on the surface of the sensor (plunge test). This is expected because the boundary conditions for the two responses are the same. Furthermore, the continuum approach showed that the expansion coefficients in the analytical results for homogeneous cylinders are not the same for the LCSR and the corresponding plunge test. This is also expected because the forcing functions for the two tests are different.

The continuum analysis also showed that the Biot Modulus (N_{Bi}) plays an important role in describing the thermal process in a homogeneous model of a sensor. The Biot Modulus is the ratio of a sensor's internal heat-transfer resistance to its surface heat-transfer resistance and is given by:

$$N_{Bi} = \frac{Internal\ Heat\ Transfer\ Resistance}{Surface\ Heat\ Transfer\ Resistance} = \frac{h}{k/R} = \frac{hR}{k} \tag{6.6}$$

where h is the heat-transfer coefficient on the surface of the sensor's sensing tip, R is the outside radius of the sensing tip, and k is the thermal conductivity of the sensor internals at the sensor's sensing tip.

Table 6.2. Relationships between Biot Modulus and modal time constants of a hypothetical temperature sensor

Biot Modulus	τ_1/τ_2	**Normalized Response Time (sec)**
0.4	21.69	2.0040
1	10.49	1.000
2	7.18	0.6713
5	5.64	0.4740
10	5.34	0.5049
20	5.28	0.3719

Response time is normalized to 1 when the Biot Modulus is equal to 1.
Source: AMS Topical Report to NRC for Approval of LCSR Technology.[2]

The Biot Modulus helps us determine such things as how fluid properties (e.g., fluid flow, temperature, or pressure) affect the sensor's response time. Furthermore, the Biot Modulus affects the spacing between eigenvalues ($\lambda_1, \lambda_2, \ldots, \lambda_n$) in the solution to continuum heat-transfer models for temperature sensors. On the other hand, the modal time constants (τ_i) that are used to calculate a sensor's response time (τ) using Eq. 6.5 are inversely proportional to $(\lambda_i)^2$. From these two preceding statements, we infer that the Biot Modulus is related to the ratio of modal time constants. This is shown in the numerical results listed in Table 6.2 for a hypothetical temperature sensor. When the Biot Modulus is small (e.g., 0.1 to 0.5), the sensor's overall heat-transfer resistance is dominated by the film resistance, and τ_1/τ_2 is large. This means that higher-order time constants ($\tau_2, \tau_3, \ldots$) are not as important as τ_1 in determining the sensor's overall response time (τ). Furthermore, the RTD response time for a small Biot Modulus is dominated by the fluid flow rate, temperature, and pressure.

For a larger Biot Modulus (e.g., 10), the overall heat-transfer resistance is dominated by the internal heat-transfer resistance, and the ratio of τ_1/τ_2 is smaller. This means that: 1) the fluid flow rate and other surface conditions do not affect the sensor's overall response time; and 2) the higher-order time constants ($\tau_2, \tau_3, \ldots$) are important for determining the sensor's overall response time (τ).

The Biot Modulus was calculated for the Rosemount Model 104 RTD (a thermowell-mounted sensor with large internal heat-transfer resistance), and the Rosemount Model 176 RTD (a direct-immersion RTD whose sensing element is attached to the inside of the sheath, giving the sensor a very small internal heat-transfer resistance). The results are shown in Table 6.3 for laboratory and plant conditions.

Table 6.3. Biot Modulus calculated for two Rosemount RTDs

RTD Surface Condition	Biot Modulus (hR/k)	
	Rosemount Model 104	Rosemount Model 176
Laboratory water flowing at 1 m/sec, at 80°C, and ambient pressure	27	0.34
PWR service conditions (water flow rate of about 10 meter/second, at 300°C, and 150 bar)	300	3.8

Source: NUREG-0809

Table 6.4. Relation between the number of eigenvalues and accuracy of LCSR transformation

Number of Eigenvalues (N)	τ_{LCSR}/τ_{True}	
	hR/k = 1	hR/k = 100
1	0.84	0.69
2	0.93	0.83
3	0.95	0.88
4	0.97	0.91
5	0.97	0.93
20	0.99	0.93

Source: EPRI Report NP-459

The results in Table 6.3 show that in this example, as the surface heat transfer increases from laboratory conditions to plant conditions, the Biot Modulus increases for both RTDs by a factor of about ten. We should note, however, that both calculations of this type and sensor response-time measurements in various media have shown that a sensor's dynamic response in one heat-transfer regime says little about the sensor's response in a different heat-transfer regime.

The Biot Modulus also helps us determine the number of eigenvalues (λ_is) that are needed to obtain correct results from a LCSR test. Table 6.4 shows the relationship between the number of eigenvalues (N) and the ratio between a hypothetical sensor's true response time (τ_{true}) and the response time as obtained from the LCSR test (τ_{LCSR}). The information in Table 6.4 is presented in terms of the ratio of τ_{LCSR}/τ_{true} for two different values of Biot Modulus.[14] These results show that with two terms, the LCSR results come within 7 percent of true response time for a Biot Modulus of 1 while it comes to within 17 percent of true response time for a Biot Modulus of

100. That is, the higher modes are more important when the Biot Modulus is large. Furthermore, higher modes are less sensitive to Biot Modulus than are lower modes.

Numerous references dating back to the late 1970s detail the development of the LCSR theory. Recently, the author compiled and updated these details for a finite difference analysis in *Sensor Performance and Reliability*, published by ISA in 2005.

6.3.3 Analyzing LCSR Data

We can use the theoretical development discussed in the preceding section to calculate the response time of hypothetical sensors in various fluid conditions (i.e., for different values of Biot Modulus). We can also use them to calculate response time for different forcing functions (e.g., a step change in temperature occurring in the inside of the sensor, as in the LCSR test, or a step change on the outside of the sensor, as in the plunge test). However, a real sensor's actual response time cannot be obtained from theory. It is impossible to know the exact geometries or dimensional and physical properties of a sensor or its material. For that reason, theory is only useful for determining how to use the data. In particular, the LCSR data is used with Eqs. 6.4 and 6.5 to identify a sensor's response time under the actual installation and process conditions tested. The procedure is as follows:

1. Perform the LCSR test and digitize the LCSR transient.
2. Fit the LCSR transient to Eq. 6.4, and identify $\tau_1, \tau_2, \ldots, \tau_n$ (it is not necessary to identify the coefficients $A_0, A_1, A_2, \ldots, A_n$).
3. Substitute $\tau_1, \tau_2, \ldots, \tau_n$ in Eq. 6.5 and calculate τ.

This simple procedure should normally yield a temperature sensor's response time with good accuracy, provided that: 1) the LCSR transient is smooth enough to be easily fit to Eq. 6.4; 2) the modal time constants ($\tau_1, \tau_2, \ldots, \tau_n$) are accurately identified; and 3) the heat-transfer assumptions for the LCSR transformation are satisfied for the sensor being tested. Each of these conditions presents challenges that must be overcome in order to obtain reliable response-time results from the LCSR test. The following paragraphs describe how to overcome each of these challenges.

Ensemble Averaging

Fig. 6.8 shows three LCSR transients: a single one, an average of 10 LCSR transients, and an average of 40 LCSR transients. These are for an RTD as installed in a nuclear power plant during power operation. Fig. 6.8 makes it clear that the single LCSR transient is not smooth. The reason for this is inherent temperature fluctuations, which can be caused by such effects as stratification of temperature and flow, action by the process temperature controller, random heat transfer, and vibration. To obtain a smooth LCSR transient, follow these two steps: 1) use the maximum allowable test current (i.e., 50 mA) to improve the signal-to-noise ratio, and 2) repeat the LCSR test 10 to 50 times on the same RTD and average the results (by ensemble averaging, as illustrated in Fig. 6.9).

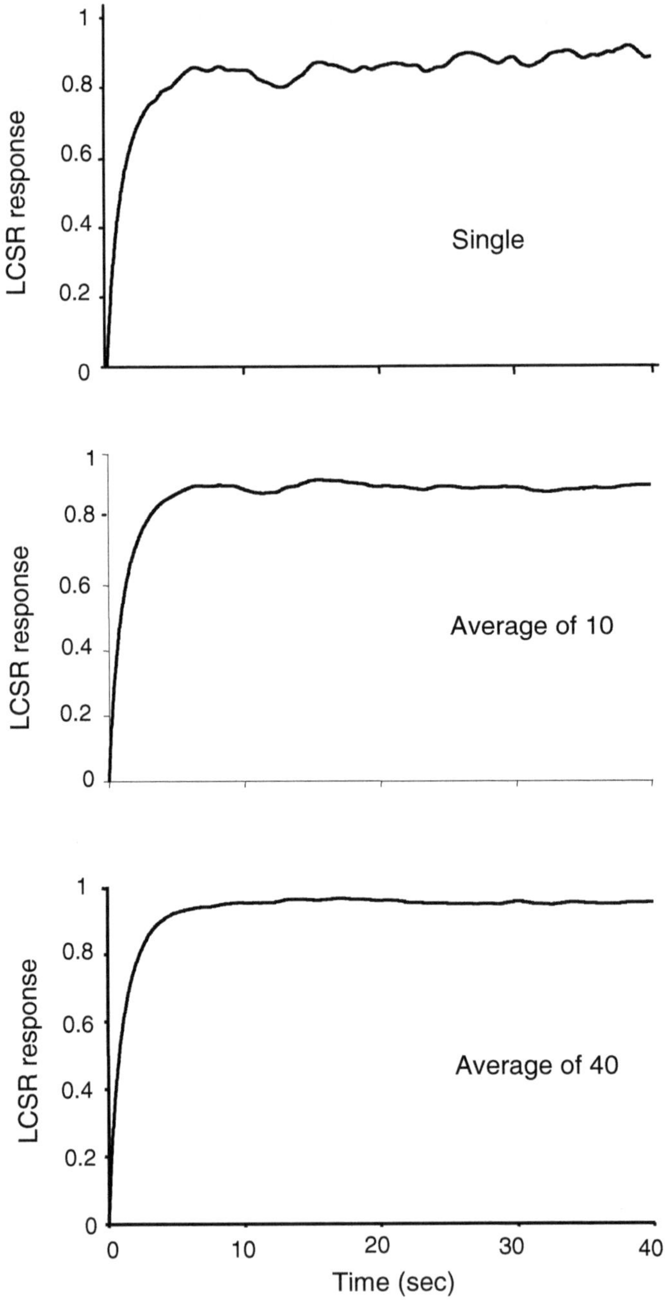

Fig. 6.8. Single and averaged LCSR transients

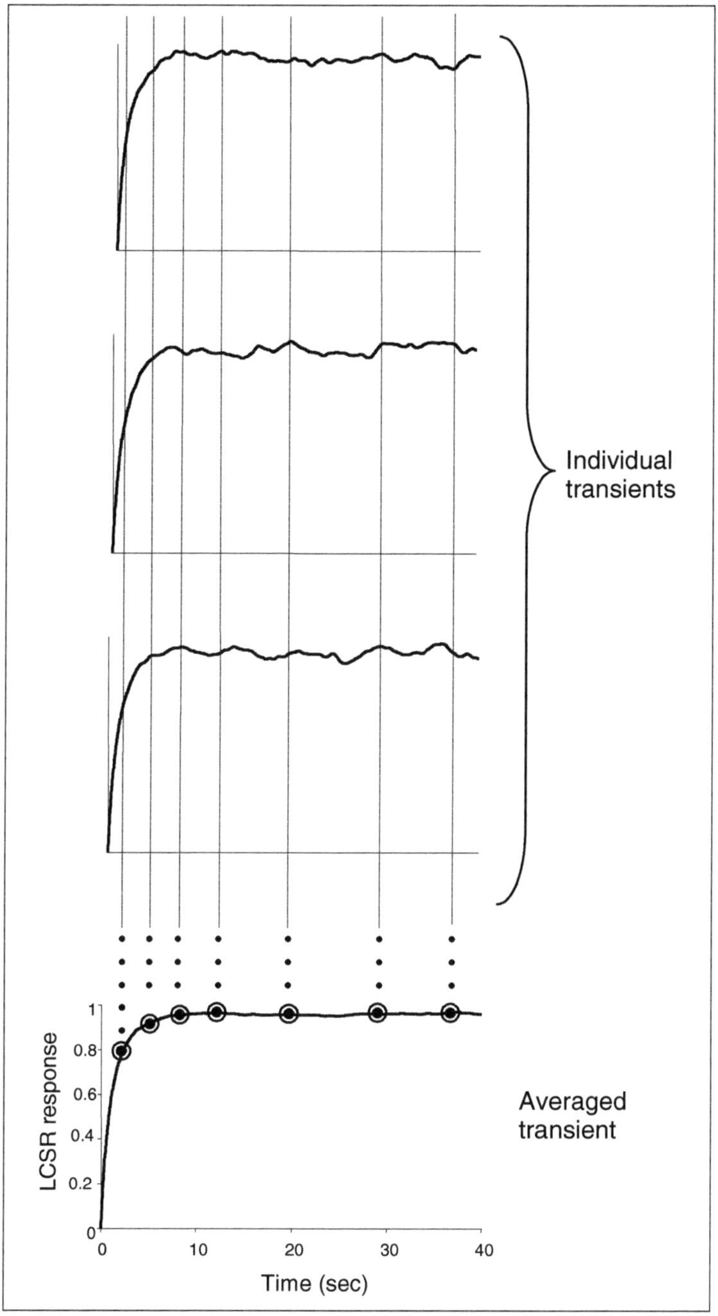

Fig. 6.9. Ensemble averaging of LCSR transients

Normally, filtering the data to smooth the LCSR fluctuations should not be used in lieu of averaging because filtering may affect the response-time results. However, a low-pass filter may be used to eliminate high-frequency noise caused by vibration, electrical interferences, or grounding problems that cannot be resolved by other means.

Effect of Higher-order Modes

To arrive at the modal time constants $\tau_1, \tau_2, \ldots \tau_n$, the smooth LCSR transient is fit to Eq. 6.4. Once identified, these modal time constants are used in Eq. 6.5 to identify the sensor response time (τ). Although this is a simple procedure in principle, identifying the modal time constants beyond τ_1 and τ_2 is very difficult. This problem was recognized in the late 1970s after the LCSR transformation was initially developed. In response, in a new R&D effort initiated at the University of Tennessee, W. P. Poore determined how to account for the effect of higher-order modes of the LCSR transient (i.e., $\tau_3, \tau_4, \ldots \tau_n$). Poore's method involved performing a comprehensive heat-transfer simulation based on a cylindrical sensor assembly.[13] He performed both a lumped-parameter nodal analysis (finite difference) of nonhomogeneous models and a continuum analysis of homogeneous models. For each case, he calculated the response time of hypothetical sensors for simulated plunge and LCSR tests. Poore used the results to determine the relationship between a sensor's overall response time (τ) based on a plunge test and the response time as calculated using only τ_1 and τ_2 and truncated Eq. 6.5. This showed that a temperature sensor's overall response time is related to the ratio of τ_2/τ_1 by the following equation:

$$\tau = f\ (\tau_2/\tau_1)\ [\tau_1 (1 - ln\ (1 - \tau_2/\tau_1)] \qquad (6.7)$$

where $f(\tau_2/\tau_1)$ is given by the correlation shown in Fig. 6.10. A fifth-order polynomial was then fit to the data, as shown by the solid curve in Fig. 6.10 to obtain the equation for $f\ (\tau_2/\tau_1)$. This equation is referred to as the *LCSR correction factor* (CF) and is written as:

$$CF = 1.0043 + 0.05578 \left(\frac{\tau_2}{\tau_1}\right) + 19.590 \left(\frac{\tau_2}{\tau_1}\right)^3 - 238.38 \left(\frac{\tau_2}{\tau_1}\right)^3 + 1352.2 \left(\frac{\tau_2}{\tau_1}\right)^4 - 2622.9 \left(\frac{\tau_2}{\tau_1}\right)^5 \qquad (6.8)$$

In summary, the CF was determined by mathematically computing the values of τ as well as $\tau_1[1 - \ln(1 - \tau_2/\tau_1)]$ for several different hypothetical (theoretical) sensors and then plotting the ratio of the results of the two computations versus τ_2/τ_1. The hypothetical sensors used for this development had a variety of sizes and geometries.

If only τ_1 can be identified from the LCSR data, then the sensor's response time is calculated from the following formula:

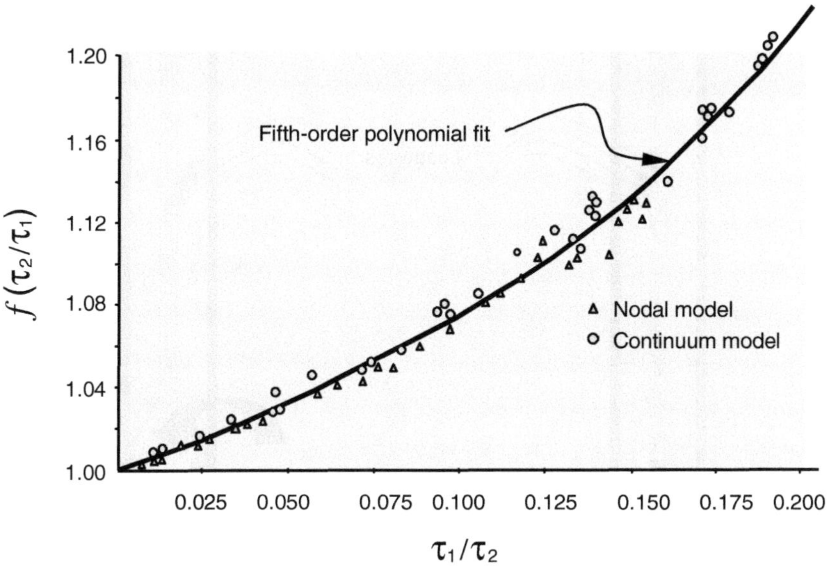

Fig. 6.10. LCSR correction factor

$$\tau = (\tau_1)(1.5) \tag{6.9}$$

The factor 1.5 was arrived at based on theoretical studies and has been validated in laboratory testing of typical sensors.

Meeting LCSR Assumptions

The LCSR method's validity depends on several assumptions about the construction details and geometry of the sensor's sensing tip. In particular, for the LCSR test to yield accurate results, the sensor's sensing element must be located at the center of a cylindrical sensor assembly. In addition, there should be no significant heat capacity between the sensing element and the centerline of the sensor (Fig. 6.11). Furthermore, the heat transfer between the sensing element and surrounding medium should be predominantly one-directional (radial), as illustrated in Fig. 6.12.

In Chap. 4, we showed x-rays of representative nuclear-grade RTDs. These x-rays were taken at the time when LCSR was developed in order to examine the validity of the assumptions just stated. It was later determined that the validity of the LCSR assumptions is best established by laboratory experiments. That is, in order for each sensor design to have its response time tested in-situ using the LCSR method, a representative sample of the sensor and its matching thermowell (if one is used) should be tested in a laboratory. In these laboratory tests, the standard plunge test method should be used to measure the sensor's response time; then an LCSR test of the sensor should be performed under the same conditions. This procedure should show that the LCSR test provides comparable results to that of the plunge test. More

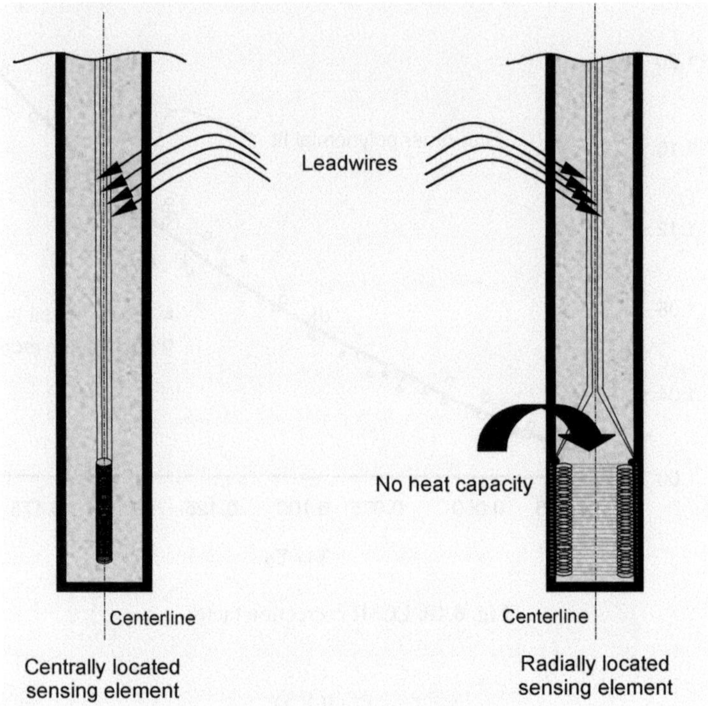

Fig. 6.11. Central geometry of sensing element

specifically, it has been established that if the plunge and LCSR results are within 10 percent of each other, then the sensor is said to have met the LCSR assumptions and is labeled as "LCSR testable."

6.3.4 LCSR Validation for RTDs

The LCSR method has been validated for several RTD models of the type used in nuclear power plants. When the LCSR method was initially developed in the late 1970s, it was validated for Rosemount RTDs (Table 6.5). At that time, U.S.-made PWRs predominantly used Rosemount RTDs to measure reactor coolant temperatures. Since then, Rosemount has gradually reduced its presence in the area of safety-related temperature measurements in nuclear power plants, and other manufacturers have entered the market. For example, Weed Instrument Company and RdF Corporation in the United States and Sensycon and others in Europe are now supplying nuclear-grade RTDs to the industry. Many of these RTDs have been validated for LCSR testing. Tables 6.6 and 6.7 show representative results. Each table shows the RTD's response time as measured first by the plunge test and then by the LCSR test under the same conditions. Both the plunge and LCSR tests for the validation data were performed in a rotating tank of room-temperature water flowing at 1 meter per second.

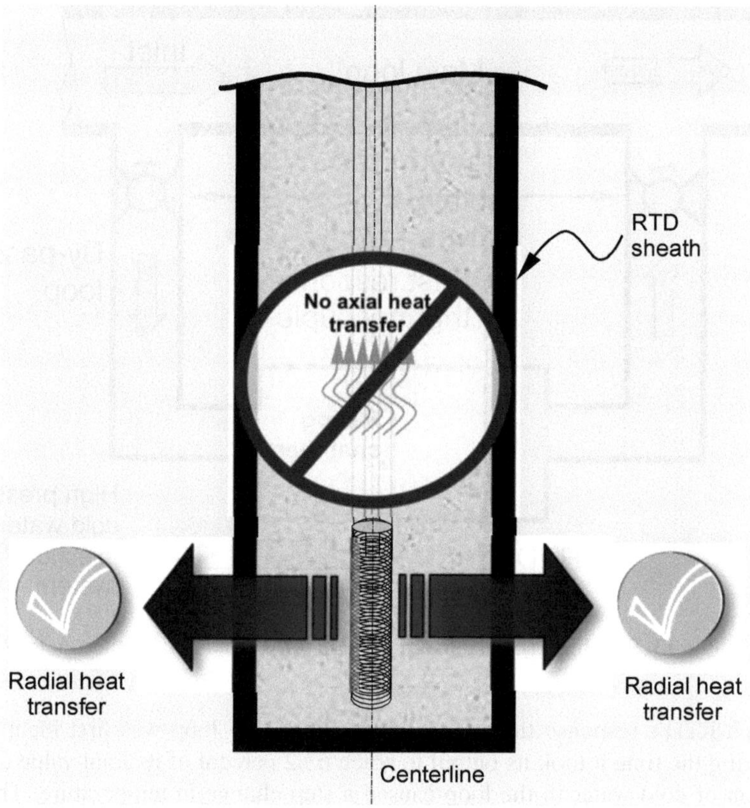

Fig. 6.12. Illustration of radial heat transfer from RTD sensing element

During the development of the LCSR test, in addition to laboratory work, validation tests were conducted at PWR operating conditions. This work was performed in the late 1970s by the author and his French colleagues at the Renardières laboratory of EdF and also at the Saclay laboratory of CEA, both near Paris, France.[13] The work used representative PWR RTDs from Rosemount. The main effort was conducted at the EdF facilities in Renardières in a loop that simulated the operating conditions of a PWR plant. In particular, the loop was operated at a temperature near 300°C, a pressure near 150 bars, and a flow of about 10 meters per second. At that time, EdF used this loop to test nuclear reactor components such as pumps and valves. For the LCSR validation work, EdF designed a test section in the loop as shown in Fig. 6.13. One RTD at a time was installed in the loop, and its response time was measured by direct exposure to a step change in temperature of the water in the loop. EdF designed a cold-water injection system to provide the step change in temperature under the loop's high-temperature, pressure, and flow conditions. A thermocouple was installed in the loop across from the RTD with its tip near the RTD's tip so as to provide the timing signal for measuring the RTD's response time.

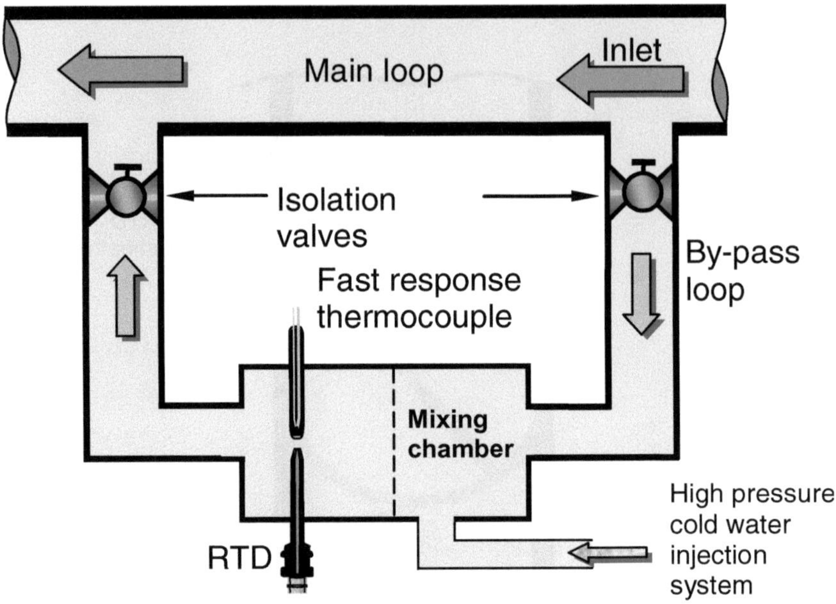

Fig. 6.13. Simplified schematic of EdF loop for validating LCSR technology

Each RTD's response time as installed in the EdF loop was first identified by measuring the time it took its output to reach 63.2 percent of its final value after the injection of cold water in the loop caused a step change in temperature. The RTD was then tested under the same conditions using the LCSR method, and the data was analyzed to provide the RTD response time. Representative results of these tests are presented in Table 6.8. Figs. 6.14 and 6.15 show photographs of the EdF loop.

6.3.5 LCSR Validation for Thermocouples

The LCSR test has also been validated for thermocouples in flowing water and flowing air under laboratory conditions.[15,16] Tables 6.9 and 6.10 show representative results of these tests, which used the same procedure as was used for the LCSR validation of RTDs.

Thermocouples are not routinely tested for response time in nuclear power plants because response time is not as important for nuclear power plant thermocouples as it is for RTDs. Rather, thermocouples are LCSR tested for maintenance, to manage degradation due to aging, to separate cable problems from sensor problems, for troubleshooting, and to verify proper thermocouple installation in thermowell. For example, in Russian-made PWR reactors (i.e., VVER reactors), the LCSR method has been used to determine whether thermocouples reached the end of their thermowell. In these plants, thermocouples are installed in long thermowells (e.g., 10 to 20 meters) that extend to various locations within the reactor core. Using the LCSR method,

Table 6.5. Results of laboratory validation of LCSR method for Rosemount RTDs

Response Time (sec)

	RTD Model	Sheath O.D. (mm)	Thermowell O.D. (mm)	**Plunge Test**	**LCSR Method**	Sensing Elements	Extension Wires	$R_0(\Omega)$
1.	176 KF	9.5	N/A	**0.38**	**0.40**	1	4	200
2.	104 ADA	3.2	6.4	**7.1**	**7.2**	1	2 + dummy	200
3.	104 ADA Bare	3.2	N/A	**3.1**	**3.1**	1	2 + dummy	200
4.	104 VC	3.2	6.4	**5.3**	**5.5**	1	2 + dummy	200
5.	104 VC Bare	3.2	N/A	**2.3**	**2.1**	1	2 + dummy	200
6.	177 GY	8.5	N/A	**6.1**	**6.3**	2	8 (4 per element)	100
7.	177 HW	7.4	10.4	**11.7**	**12.3**	2	8 (4 per element)	100
8.	104 AFC	6.1	9.6	**5.3**	**5.2**	1	2 + dummy	200
9.	104 AFC Bare	6.1	N/A	**3.0**	**3.1**	1	2 + dummy	200
10.	104 AFC and NEVER-SEEZ	6.1	9.6	**3.9**	**3.9**	1	2 + dummy	200

1. Source: EPRI Report NP-1486.
2. All tests are in room-temperature water flowing at 1 meter per second.
3. Bare means without thermowell.
4. Dimensional information is given in approximate, rounded numbers.
5. O.D.: outside diameter.
6. mm: millimeter.
7. NEVER-SEEZ is a thermal coupling compound that is sometimes used in the tip of RTD/thermowell assemblies to improve response time.

Table 6.6. Results of LCSR validation of Weed RTDs under laboratory conditions

	RTD Number	Sheath O.D. (mm)	Thermowell O.D. (mm)	Response Time (sec)		Sensing Elements	Extension Wires	$R_0(\Omega)$
				Plunge Test	**LCSR Method**			
1.	N9004	6.4	9.5	**2.9**	**2.8**	1 or 2	4 or 8	200
2.	N9004 Bare	6.4	N/A	**1.6**	**1.7**	1 or 2	4 or 8	200
3.	N9019	9.5	N/A	**2.9**	**2.8**	1 or 2	4 or 8	200
4.	SP612	6.4	9.5	**4.1**	**3.9**	1 or 2	4 or 8	200

1. All tests are in room-temperature water flowing at 1 meter per second.
2. Bare means without thermowell.
3. Dimensional information is given in approximate, rounded numbers.
4. O.D.: outside diameter.
5. mm: millimeter.
6. Weed RTDs are offered as both single-element and dual-element sensors with four or eight extension leads

Table 6.7. Results of laboratory validation of LCSR method for RdF RTDs

	Model Number	Sheath O.D. (mm)	Thermowell O.D. (mm)	Response Time (sec)		Sensing Elements	Extension Wires	$R_0(\Omega)$
				Plunge Test	**LCSR Method**			
1.	21204 Bare	9.5	N/A	**2.0**	**2.1**	1	4	200
2.	21232	6.4	9.5	**4.9**	**5.0**	1	4	200
3.	21232 Bare	6.4	N/A	**0.9**	**1.0**	1	4	200
4.	21458	6.4	9.5	**5.1**	**5.2**	1	4	200
5.	21458 Bare	6.4	N/A	**1.8**	**1.9**	1	4	200
6.	21459	6.4	9.5	**4.9**	**5.2**	2	8	200
7.	21549 Bare	6.4	N/A	**2.0**	**2.2**	2	8	200

1. All tests are in room-temperature water flowing at 1 meter per second.
2. Bare means without thermowell.
3. Dimensional information is given in approximate, rounded numbers.
4. O.D.: outside diameter.
5. mm: millimeter.
6. RdF RTDs are offered in single- or dual-element sensors with four or eight extension leads.
7. RdF 21204 RTDs are direct immersion.

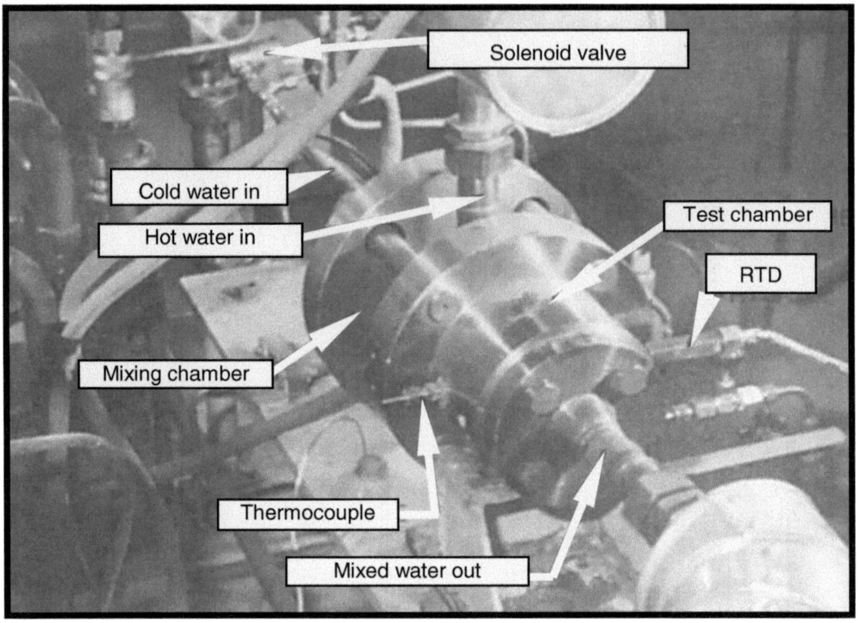

Fig. 6.14. RTD and thermocouple installation in the EdF loop

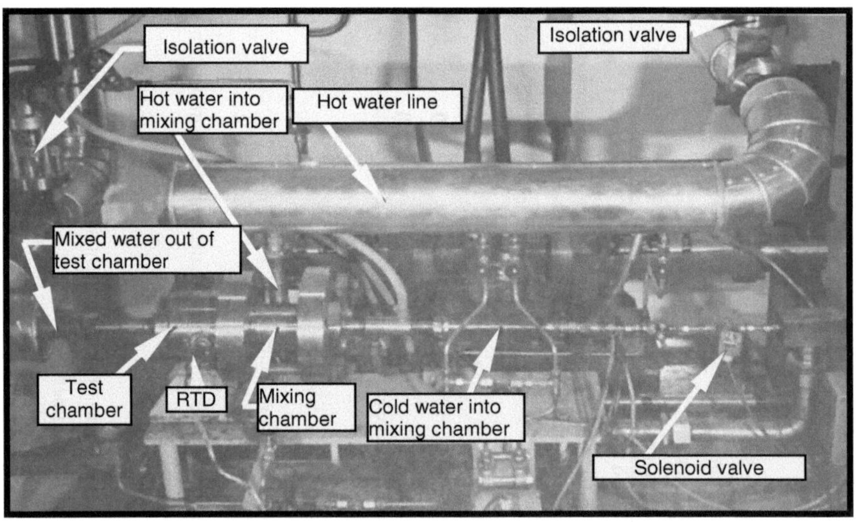

Fig. 6.15. Test section of EdF loop used in LCSR validation tests

Table 6.8. Representative results of LCSR validation of Rosemount RTDs under PWR operating conditions at EdF loop

RTD Model	Response Time (sec)	
	Direct Test	LCSR Test
176KF	0.14	0.13
177HW	8.8	8.4
104-AFC	6.2	5.9
104-AFC with Never-Seez	4.1	3.7

Source: EPRI Report NP-1486.

VVER plants have been able to verify that thermocouples have indeed reached the end of the thermowell.

Table 6.9. LCSR validation results for thermocouples tested in flowing water

Thermocouple Tag Number	Response Time (sec)	
	Plunge Test	LCSR Method
29	1.40	1.10
27	2.00	1.99
43	0.37	0.37
44	2.10	2.19
46	1.98	2.39
36	1.43	1.33
38	1.90	1.98
40	0.43	0.43
04	3.06	2.83
07	2.72	2.96
09	0.76	0.49
13	0.27	0.29

These results are from laboratory testing in room-temperature water flowing at 1 meter per second.

6.3.6 Optimizing LCSR Parameters

The accuracy of LCSR results depends on how well the sensor meets the LCSR assumptions and also on the parameters used for data acquisition and data analysis. For example, the LCSR data for testing RTD response times in nuclear power plants is usually sampled with a sampling time of 0.001 to 0.04, depending both on the RTD and the test conditions. Furthermore, the number of samples that are collected must be selected correctly if LCSR test is to provide accurate results. Depending on the

RTD and its test conditions, between 1,000 to 2,000 samples are usually collected. The product of sampling time and the number of samples is the duration of the LCSR data acquisition, which varies from typically about 5 seconds for direct-immersion sensors to about 50 seconds for thermowell-mounted sensors in an operating plant.

Table 6.10. LCSR validation results for thermocouples tested in flowing air

	Response Time (sec)	
Thermocouple Tag Number	**Plunge Test**	**LCSR Method**
40	3.20	3.63
38	9.90	9.48
52	1.28	1.54
13	3.66	7.03
09	10.03	14.68
07	17.13	18.27
51	1.12	1.10
43	3.88	3.90
29	10.55	8.61
27	17.10	19.45
20	0.16	0.10
18	0.14	0.12
23	0.50	0.56

The above results are from laboratory testing in an air loop at a flow rate of 5 meters per second.

In addition to sampling rate and sampling time, LCSR analysis parameters must be optimized for each RTD design. The analysis parameters are those variables that are iterated in fitting the LCSR data to the LCSR equation. They are iterated during computer analysis of the LCSR data until the difference between the raw data and the fit to the LCSR equation is minimal. This effort, together with selecting the best data-sampling parameters, is referred to as *LCSR parameter optimization*. More specifically, optimizing LCSR parameters entails performing a comprehensive set of laboratory tests on representative sensors and thermowells (if one is used) for any new sensor design. These tests typically involve performing numerous plunge and LCSR tests under the same conditions. The LCSR data obtained from these tests is stored and later analyzed, iterating the LCSR sampling and analysis parameters. This process is continued until the response-time results from the LCSR tests are less than 10 percent of the plunge tests. At that point, the LCSR sampling and analysis parameters are recorded and referred to as the optimum LCSR parameters for the sensor. The optimum parameters are then used as the default starting point from which to acquire and analyze LCSR data for all the sensors having the design for which the optimum parameters were identified.

6.3.7 Accuracy of LCSR Results

Generally, the accuracy of in-situ response-time measurements made with the LCSR test is ± 10 percent. This is provided that: 1) the LCSR method has been validated for the sensor under test; 2) the LCSR data is clean and smooth; and 3) optimum analysis parameters are used to process the data. Furthermore, the plant's conditions must be suitable for the LCSR test. This means that the plant temperature must be stable and undergo no significant fluctuations or drift during the LCSR tests.

Three types of errors can occur in LCSR testing: 1) errors arising because the sensor design does not meet the assumption of the LCSR transformation; 2) errors in arriving at correct values of modal time constants; and 3) data acquisition and data analysis errors. We discuss each of these errors in the following paragraphs.

Errors from LCSR Assumptions

The LCSR transformation assumes that there are no zeroes in the sensor's transfer function. This assumption is satisfied if the sensor meets two conditions involving the position of the sensing element at the sensor's tip: 1) the heat that is generated in the sensor during the LCSR test dissipates to the outside of the sensor radially (i.e., there is little or no axial heat transfer); and 2) there is no significant heat capacity between the sensing element and the sensor's centerline. These assumptions are necessary to ensure that the heat transfer between the sensing element and the fluid around the sensor are the same in both the plunge and LCSR testing.

To ensure that the LCSR assumptions are met, each sensor design must be validated for response-time testing by the LCSR method.

Errors in Calculating Modal Time Constants

Normally, only two modal time constants (τ_1 and τ_2) can be identified from the LCSR data. As such, these two time constants must be accurately determined to arrive at a correct value for a sensor's overall response time (τ).

Two types of errors can affect the proper determination of τ_1 and τ_2: 1) noise either due to electrical pickup (high-frequency noise) or process fluctuations (low-frequency noise); and 2) drift in the process. The noise problem can be overcome by using larger heating currents and/or by averaging multiple LCSR data sets. In some plants, adequate signal-to-noise ratios are obtained by using moderate heating currents (e.g., 30 to 50 mA). In others, the noise is too high to overcome with a moderate heating current. In these instances, in addition to using the highest possible current (i.e., 50 mA), the LCSR test is repeated up to 50 times, and the results are averaged to minimize the effect of process noise on the LCSR data.

Drift is easily removed by implementing simple software to determine the drift rate and subtract it from the data. If the drift on the LCSR data is in a positive direction (data drifts up), the LCSR results tend to be larger than the sensor's actual response time, and if the drift is in a negative direction (data drifts down), the LCSR results will be faster than the true response time of the sensor.

Data Acquisition and Data Analysis Errors

Data acquisition and data analysis errors arise from: 1) improper sampling and analysis parameters in collecting and analyzing LCSR data; and 2) finite resolution in data sampling and computer calculations. To avoid errors due to sampling parameters, a rule of thumb is to use a sampling rate that is at least 100 to 1,000 times faster than the expected response time of the sensor (e.g., 0.02 seconds for a sensor with a response time of 4 seconds), sample the LCSR data for at least five times the expected response time of the sensors (e.g., at least 20 seconds for a sensor with a response time of 4 seconds), and repeat the LCSR test as many times as necessary (e.g., 10 to 50 repeats depending on plant temperature fluctuations) to provide a smooth LCSR transient by ensemble averaging. To avoid errors due to improper analysis parameters, use the optimum analysis parameters that have been identified for the sensors, as described in Sect. 6.3.6.

The finite resolution in data sampling and computer calculations is unavoidable given the existing state of the art. This situation will improve with further advances in data acquisition and data analysis technologies.

6.3.8 Effect of LCSR Heating Current

Temperature sensors such as RTDs that are used in power plants routinely experience a current of a few milliamperes (e.g., 1 to 2 mA) when their resistance is being measured. RTD manufacturers sometimes specify the maximum currents that can be used before self-heating will cause significant temperature measurement errors. A typical maximum recommended value is 10 mA. This would give a measurement error of about 0.5°C to 1°C for a typical RTD in a PWR plant. However, routine currents for measuring resistance are usually about 1 mA, which gives negligible measurement errors.

When the LCSR method was being developed, several manufacturers of nuclear plant RTDs were asked to examine the question of maximum allowable currents. As a result, a consensus was reached that currents of up to 80 mA for LCSR testing cause no deleterious effects in RTDs.[2] Furthermore, ORNL produced evidence that reasonable amounts of Joule heating needed for LCSR testing does not harm RTDs. This question was also examined in an EPRI-funded research program at UT. In this program, RTDs were injected with thousands of step changes in currents of up to 100 mA, simulating thousands of LCSR tests. This showed no measurable change in RTD calibration, response time, or other characteristics.

6.3.9 Effect of Temperature Stratification

Temperature stratification (also referred to as *temperature streaming*) is an inherent phenomenonin PWR plants. It occurs because the reactor coolant is unevenly heated as it passes through the core. As a result, the streams of water that exit the core and enter the hot-leg pipes are at different temperatures. Typically, the temperature around a ring in a hot-leg pipe could vary by 3°C to 10°C, depending on the plant.

In addition, swirling effects are sometimes encountered, as illustrated in Fig. 6.16. The combination of temperature streaming and swirling can cause large temperature fluctuations and temperature nonuniformity. These problems do not normally subside as the reactor coolant travels through the system. In fact, the problem often continues on into the cold legs, although not nearly as prominently as in hot legs.

Fig. 6.17 shows temperature-monitoring data for six hot-leg RTDs in a PWR plant. These graphs show the reading of the redundant hot-leg RTDs plotted in terms of the deviation of each individual RTD from the average reading of the six RTDs. Clearly, there is about ± 2°C difference (a total of about 4°C) between the readings of the six RTDs. Fig. 6.18 shows the same type of data plotted for a hot-leg RTD as a function of reactor power. This data was collected both during plant heatup to full power and plant cooldown from 100 percent to zero power. The data represents the deviation of one of the hot-leg RTDs from the average of all the hot-leg RTDs in the same loop plotted versus power. As the reactor power increases from zero to 100 percent, the RTD error increases from zero to about 1.5°C and vice versa.

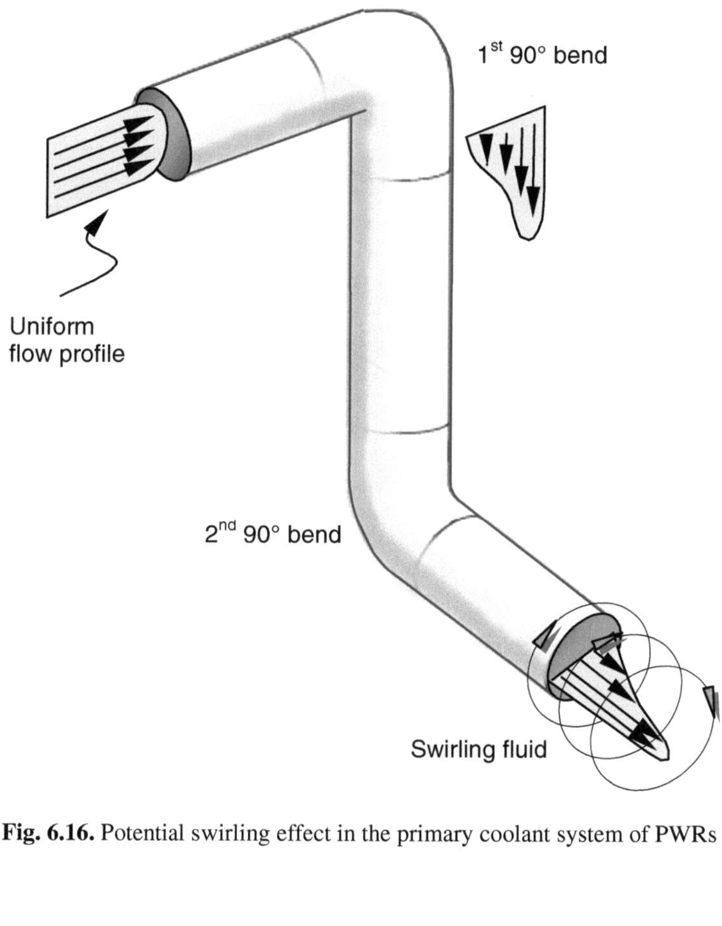

Fig. 6.16. Potential swirling effect in the primary coolant system of PWRs

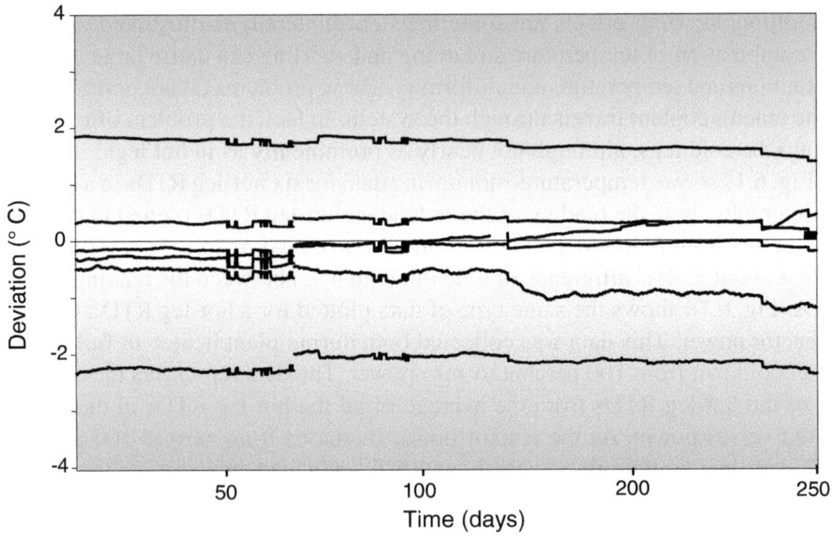

Fig. 6.17. Deviation of redundant hot-leg RTDs due to temperature stratification

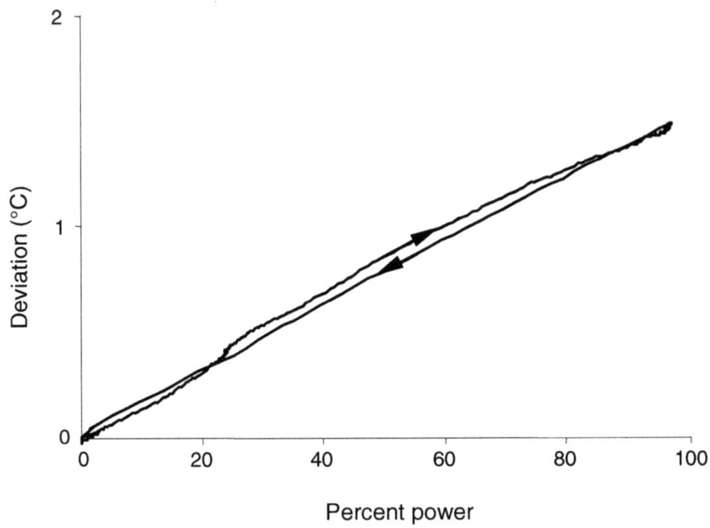

Fig. 6.18. Temperature stratification error as a function of reactor power

The temperature stratification and swirling problems were first recognized in the mid-1960s at the San Onofre nuclear power plant Unit 1 in California. This plant, which is now retired, was a two-loop Westinghouse PWR. To resolve the problem, the San Onofre Unit 1 plant was retrofitted with bypass loops and RTD bypass manifolds, as shown in Fig. 6.19. Subsequent to the San Onofre experience, almost all Westinghouse PWRs were retrofitted or built with bypass loops and RTD bypass manifolds. The bypass loops helped mix the reactor water, improving measurements of the bulk temperature.

Each bypass loop in a PWR plant typically uses sampling scoops (Fig. 6.20). These scoops are installed in the reactor's coolant pipes to sample the reactor coolant from different streams around the hot-leg pipe and to transport the samples through bypass pipes to the manifolds, where temperature is measured using RTDs. To compensate for the transport time delay, direct-immersion, fast-response RTDs are used in the bypass manifolds. The transport delay, which is typically about two seconds, is the time it takes for the water to travel from the primary coolant pipes through the bypass loops to the RTD bypass manifolds.

The bypass manifolds resolved the temperature measurement problems but added about 100 meters of extra piping and other hardware to the reactor coolant system, including more than 100 pipe hangers, more than 60 snubbers, 60 to 70 valves, and so on. Over time, the bypass loops and manifolds caused maintenance problems, increased radiation exposure to maintenance personnel, and raised the probability of forced outages. As a result, the nuclear industry decided to remove the bypass manifolds beginning in the mid-1980s. That is, after about 10 to 20 years of operating with bypass manifolds, Westinghouse PWRs began, in the mid-1980s, to remove the RTD bypass manifolds and install new thermowell-mounted RTDs directly in the hot-leg and cold-leg pipes (Fig. 6.21). In doing this, the thermowells for the new RTDs were installed inside the existing sampling scoops wherever possible (Fig. 6.20).

Installing thermowell-mounted RTDs directly in the reactor coolant pipes–as opposed to installing direct-immersion RTDs in bypass manifolds–has consequences for the overall response time of the corresponding temperature measurement channel. Table 6.11 shows data on the dynamic response of thermowell-mounted RTDs installed directly in the reactor coolant pipes versus direct-immersion RTDs installed in bypass manifolds. In this example, to meet a response-time requirement of 6.0 seconds, a faster RTD (3.0 seconds or less) is needed with a bypass manifold to compensate for the two-second delay that is encountered when bringing the reactor coolant water from the reactor coolant pipe to the RTD in the bypass manifold. This delay decreases to 0.25 seconds (mixing time in the scoops) when RTDs are installed directly in the reactor coolant pipe.

In addition to affecting the dynamics of temperature measurements, removing bypass manifolds affects the accuracy requirements for the primary coolant RTDs. More specifically, with bypass manifolds, the water is already well mixed when it reaches the manifolds, and therefore the bulk temperature is more accurately measured. When RTDs are installed directly in the reactor coolant pipes, the bulk temperature can typically be deduced by averaging the readings of redundant RTDs. The latter temperature measurement is obviously not as accurate as the former because the water is not as

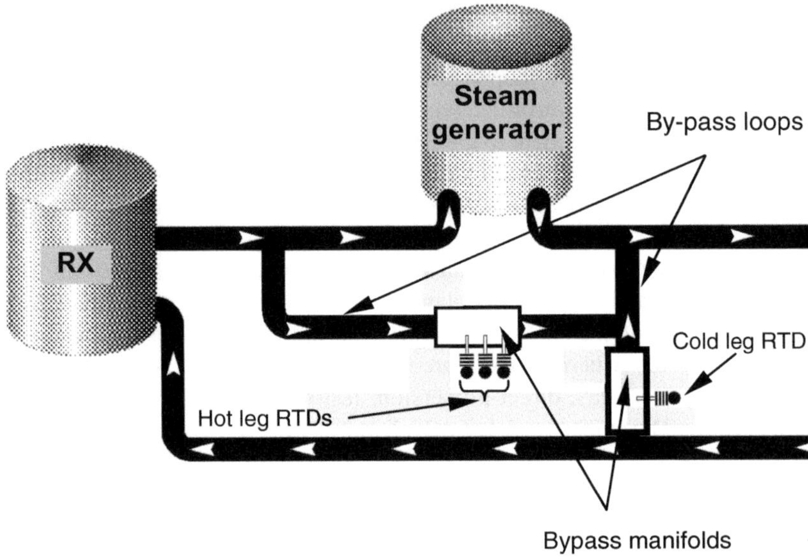

Fig. 6.19. Primary coolant system of a PWR plant with RTD bypass manifolds

well mixed in the sampling scoops as it would be in the bypass manifolds. As a result, the requirements on calibrating thermowell-mounted RTDs installed directly in the reactor coolant pipes are more stringent. In Chap. 5, we described the cross-calibration method, which was developed to verify the calibration of primary coolant RTDs in PWR plants. In particular, we described data processing and data analysis techniques and correction algorithms to ensure that the accuracy of installed primary coolant RTDs is assessed correctly.

Today, many Westinghouse PWRs have removed their bypass manifolds and installed redundant thermowell-mounted RTDs around the reactor coolant pipes. This has an effect on LCSR testing of the RTDs. More specifically, the temperature fluctuations caused by temperature stratification and swirling in the reactor coolant system affect the LCSR data. Fig. 6.22 shows normal LCSR data from a nuclear power plant RTD and data for a nuclear power plant RTD in which temperature stratification and

Table 6.11. Example of system response time with and without RTD bypass manifolds

	Response Time (sec)	
Component	**With Manifold**	**Without Manifold**
RTD	3.0	4.75
Electronics	1.0	1.0
Transport/Mixing	2.0	0.25
Total	6.0	6.0

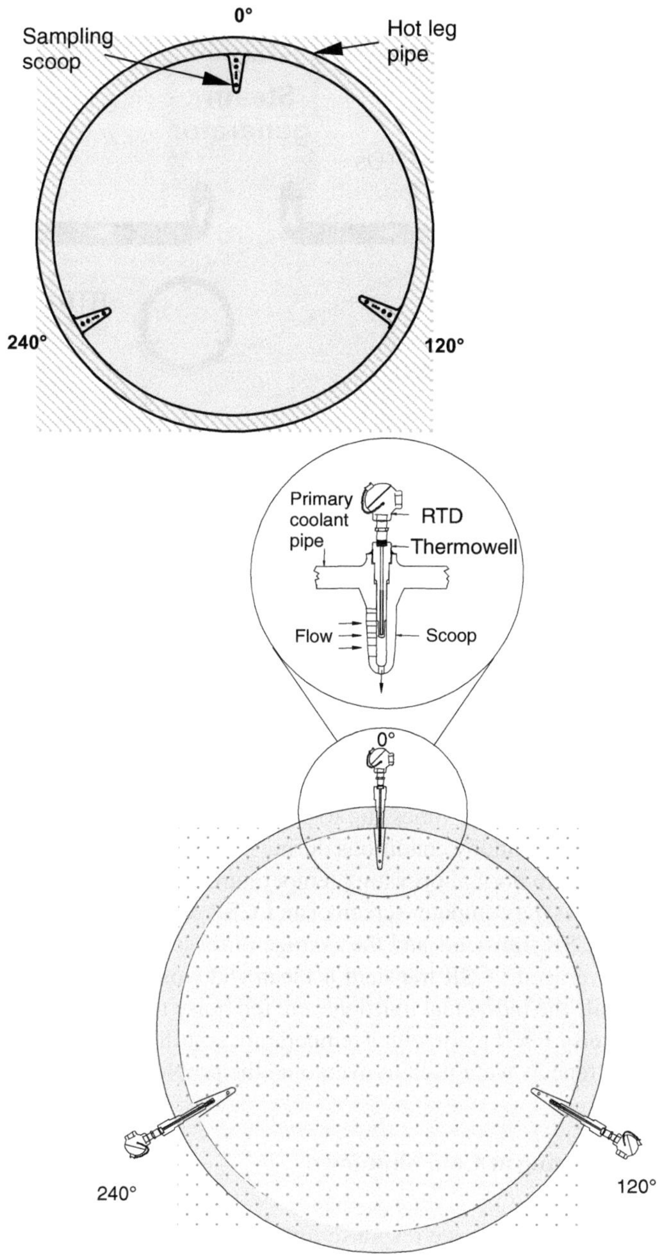

Fig. 6.20. Sampling scoops in the primary coolant pipes of Westinghouse PWRs

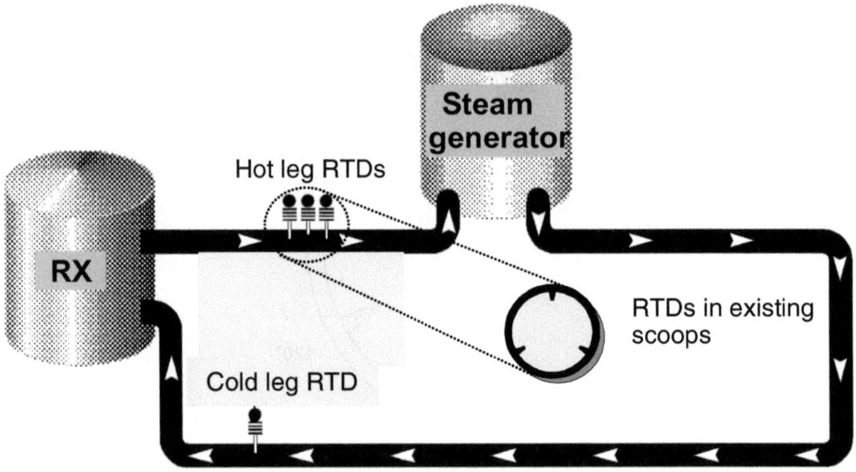

Fig. 6.21. Primary coolant system of a PWR plant after removal of RTD bypass manifolds

swirling causes fluctuations (noise) on the LCSR data. The effect depends on the plant. Some plants have large temperature fluctuations, and others have very little. Sometimes, the same RTD in the same plant experiences the problem during one operating cycle but not another (Fig. 6.23). At other times, two loops of the same plant have the process noise problem for different RTD orientations in the pipe, as shown in Fig. 6.24. Of course, the problem has to be resolved in order for the LCSR test to produce reliable response-time results. There are two solutions for this problem: 1) repeat the LCSR tests many times (up to 50) and average the results; or 2) perform the LCSR test at hot standby conditions during startup when the plant temperature is stable and as close to normal operating temperature as possible.

Fig. 6.25 shows the computer screen of an LCSR data acquisition system in which 40 individual LCSR transients and the average of these transients are displayed. It is clear that the averaged LCSR transient is clean and would yield a reliable response-time result, while the individual transients are too noisy to be analyzed. This data was obtained in a noisy PWR plant with a temperature stratification problem. Both higher current and 40 repeats were used to overcome the problem.

6.3.10 LCSR Testing at Cold Shutdown

To obtain a sensor's in-service response time, the test must be performed at or near normal operating temperature, pressure, and flow while the plant is at power or during hot standby conditions. When new sensors are installed in the plant, there is no way to know their in-service response times until after the plant is at or near the hot standby conditions or when the plant is operating at power. If a new sensor then fails to meet its response-time requirement, the plant may have to shut down to replace the RTD. To minimize this possibility, LCSR testing may be performed on new sensors at cold

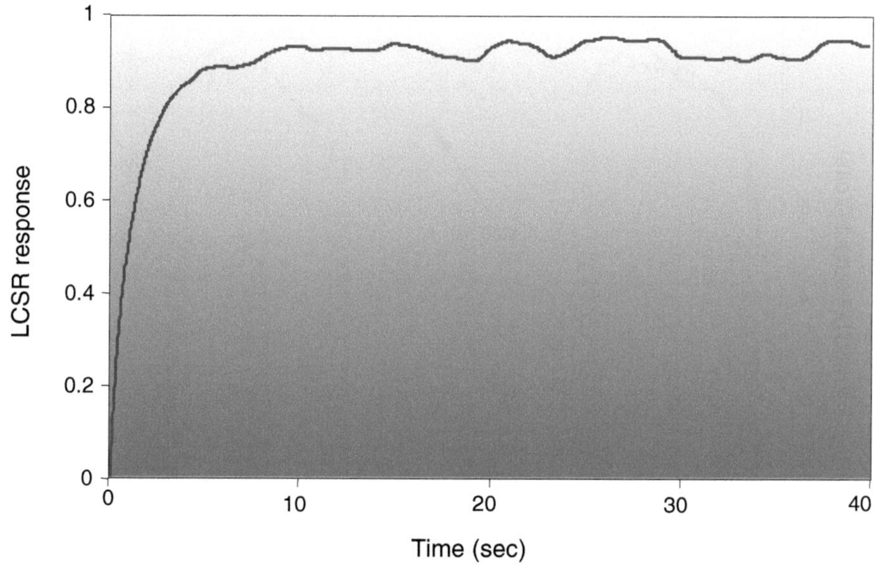

(a) Normal LCSR transient

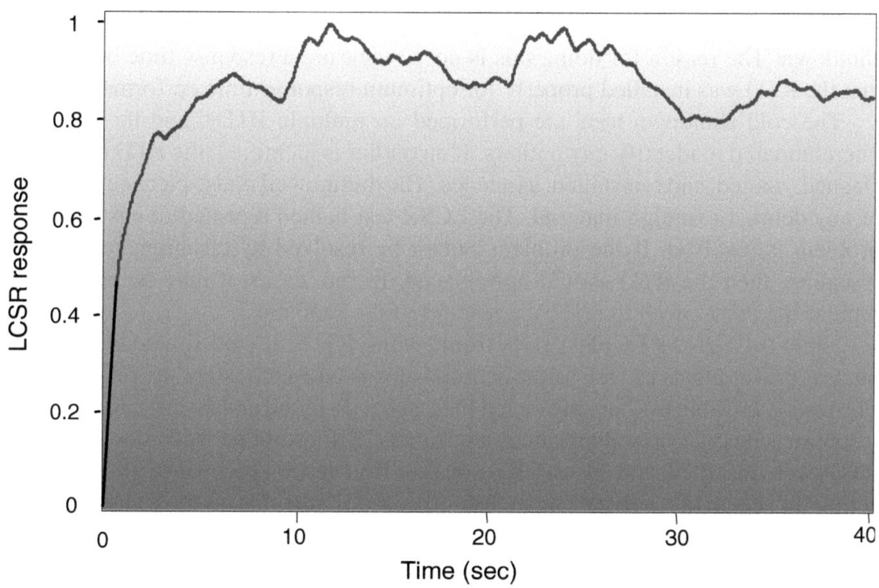

(b) Noisy LCSR transient

Fig. 6.22. Temperature stratification effect on LCSR data

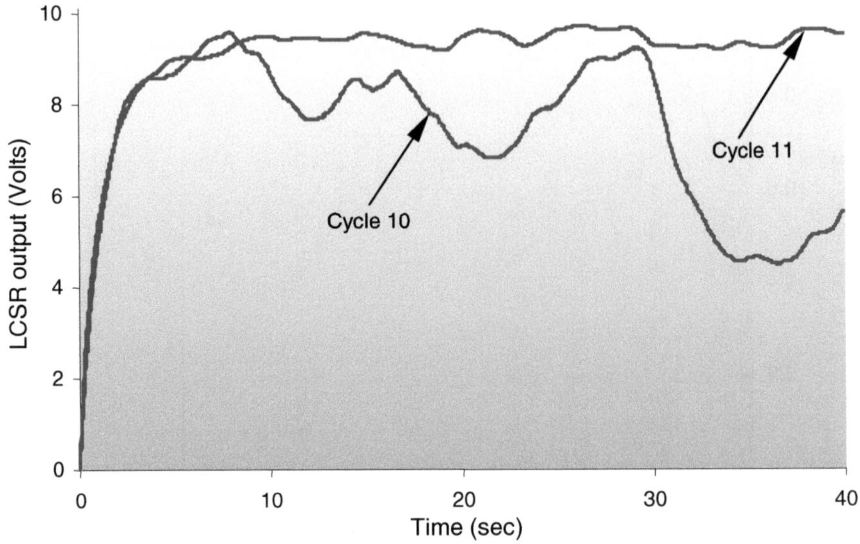

Fig. 6.23. LCSR transients for an RTD in two different operating cycles in a PWR plant

shutdown. The reason for doing this is not to measure a response time but to verify that the RTD was installed properly for optimum response-time performance.

The cold shutdown tests are performed on multiple RTDs, and the results are intercompared to identify any outliers. If an outlier is identified, the RTD is removed, cleaned, rotated, and reinstalled, as needed. The thermowell is also cleaned and purged of any debris or foreign material. The LCSR test is then repeated to ensure that the problem is resolved. If the problem cannot be resolved by cleaning, rotating, and reseating, then the RTD should be replaced. In rare cases, it may be necessary to replace the thermowell to resolve a response-time problem.

Table 6.12 shows LCSR results from testing RTDs at cold shutdown in various nuclear power plants as well as the action that was taken to restore the response time. The results in Table 6.12 are presented in terms of as-found and as-left values for RTD response times. The as-found values are from LCSR testing performed when the new RTDs were installed, and the as-left results are from testing performed after corrective action was taken to resolve the response-time issue. Note that, in most cases, cleaning alone resolved the problem and that in only a few cases did the RTD itself have to be replaced. In one instance, the problem could not be resolved by cleaning or replacing the RTD; the thermowell had to be replaced.

6.4 Self-Heating Test

The self-heating test is performed to detect gross changes in response time. It is useful only for RTDs; it does not apply to thermocouples.

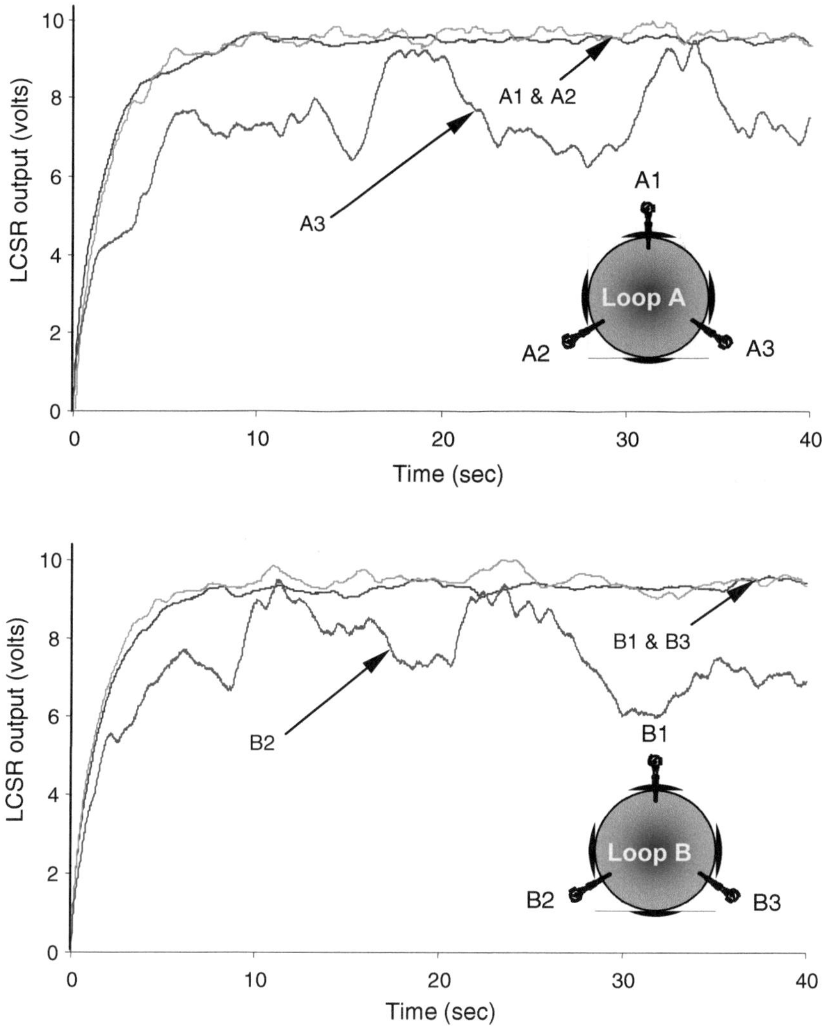

Fig. 6.24. Effect of temperature stratification on LCSR data depending on orientation of the RTD in the pipe

The self-heating test does not measure the response time of an RTD; it provides a means for detecting RTD response-time degradation. Typically, both LCSR and self-heating tests are performed on RTDs because the two tests complement each other to provide a complete picture of the RTD's dynamics. The self-heating tests are typically performed together with LCSR tests both when response time is measured at hot standby or at operating conditions and during testing at cold shutdown to verify RTD installation.

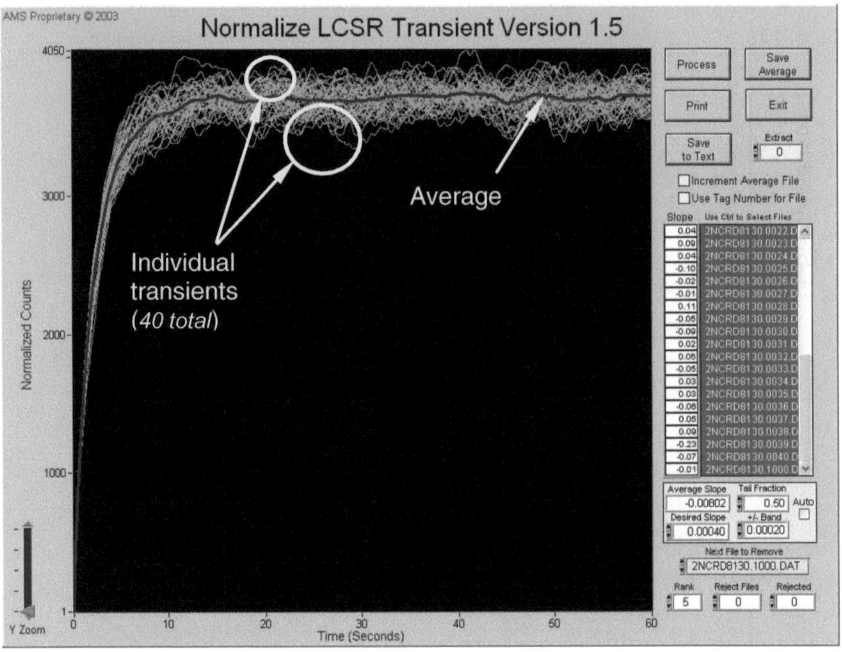

Fig. 6.25. LCSR data acquisition screen showing individual LCSR transients and the average of these transients

Table 6.12. RTD response-time problems resolved at cold shutdown

Response Time (sec)		
As Found	**As Left**	**Action Taken to Resolve the Problem**
11.6	4.7	Cleaned thermowell
22.5	7.5	Cleaned thermowell
14.7	5.9	Cleaned thermowell
37.4	13.0	Cleaned thermowell
9.0	5.0	Reseated RTD
18.0	14.0	Reseated RTD
19.2	9.5	Reseated RTD
14.5	5.4	Installed new RTD
24.0	7.8	Installed new RTD
24.0	17.0	Debris removed
27.9	5.8	Replaced thermowell

6.4.1 Test Description

Like the LCSR method, the self-heating test is based on heating the RTD with a small DC current (I). It is performed using the same Wheatstone bridge as in the LCSR test. For the self-heating test, the steady-state increase in the RTD resistance (ΔR) as a function of input electric power ($P = I^2R$) is measured. The result is referred to as the

Self-Heating Index (SHI) of the RTD and is expressed in the units of ohms per watt (Ω/ω).

The following derivation shows the correspondence between an RTD's SHI and its response time (τ). For simplicity, the description is given for a first-order system, though RTDs are not typically represented by first-order dynamics. The steady-state relation between temperature and I^2R heating generated in an RTD is given by:

$$Q = UA(T - \theta) \tag{6.10}$$

where:

Q = Joule heating generated in the RTD by applying I^2R heating
U = overall heat-transfer coefficient at the RTD's sensing tip (includes both internal and surface heat transfer)
A = heat-transfer area
T = RTD temperature
θ = temperature of fluid in which the RTD is installed

For constant fluid temperature, Eq. 6.10 may be written as follows:

$$\Delta Q = UA\,\Delta T \tag{6.11}$$

Therefore, the temperature rise per unit power generated in the RTD is:

$$\frac{\Delta T}{\Delta Q} = \frac{1}{UA} \tag{6.12}$$

The resistance of the RTD's platinum element is approximately proportional to its temperature (i.e., $\Delta R = \alpha \Delta T$, where α is the temperature coefficient of resistance). Thus:

$$\frac{\Delta R}{\Delta Q} = \frac{\text{Constant}}{UA} \tag{6.13}$$

On the other hand, an RTD's response time is approximately given by the following equation, assuming that the RTD is a first-order system (again, acknowledging that an RTD is not generally a first-order system):

$$\tau = \frac{MC}{UA} \tag{6.14}$$

where:
M = mass of the sensing tip of the sensor
C = specific heat capacity of the sensor material

If the heat capacity C remains constant, then:

$$\tau = \frac{\text{Constant}}{UA} \tag{6.15}$$

Comparing Eqs. 6.13 and 6.15 leads to the conclusion that τ is proportional to $\frac{\Delta R}{\Delta Q}$. That is:

$$\tau \propto \frac{\Delta R}{\Delta Q} \quad or \quad \tau \propto SHI \tag{6.16}$$

where $\frac{\Delta R}{\Delta Q}$ is equal to SHI, and α represents proportionality. This equation shows that a change in an RTD's response time can be identified from a change in its SHI.

6.4.2 Test Procedure

A self-heating test is typically performed using a Wheatstone bridge according to the following procedure:

1. Balance the bridge, and record the resistance of the RTD. The current is normally about 1 to 2 mA at this stage.

2. Switch the current "high." The high current at this stage could be about 5 mA.

3. Wait for the RTD resistance to settle (i.e., reach steady state).

4. Balance the bridge, and measure the new value of the RTD resistance.

5. Record the data in a chart that shows the RTD resistance in ohms, the current through the RTD in mA, and the power in the RTD sensing element in milliwatts (see Table 6.13).

6. Increase the current by about 5 to 10 mA, and repeat from Step 3 until at least four data points are recorded for a wide range of currents (e.g., 5 mA to 50 mA).

7. Plot the data on rectangular coordinates in terms of RTD resistance (R) versus power (P). This plot is referred to as the *self-heating curve* (Fig. 6.26).

8. Fit a straight line through the self-heating data, and calculate the slope of the line. The slope is the SHI.

This eight-step process has been automated. A computer performs the calculations, generates the self-heating curve, identifies the SHI, displays the results, and stores the data for trending. Fig. 6.27 shows a computer screen displaying the result of a self-heating test.

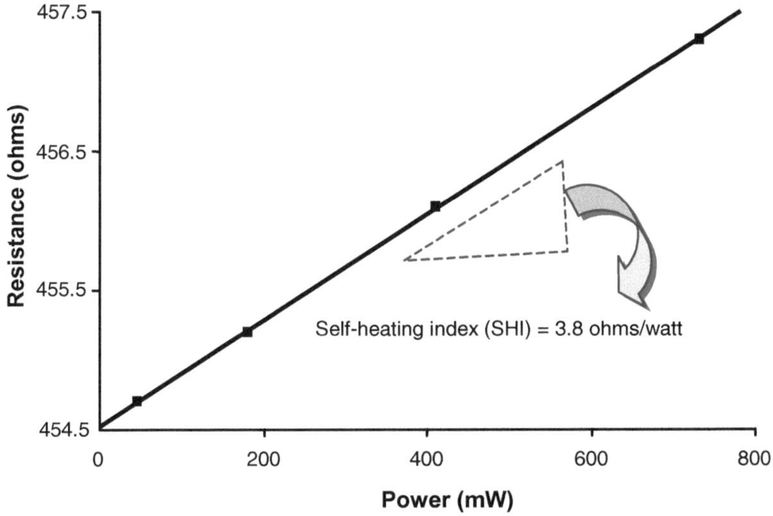

Fig. 6.26. Typical self-heating curve of an RTD from testing in a PWR plant

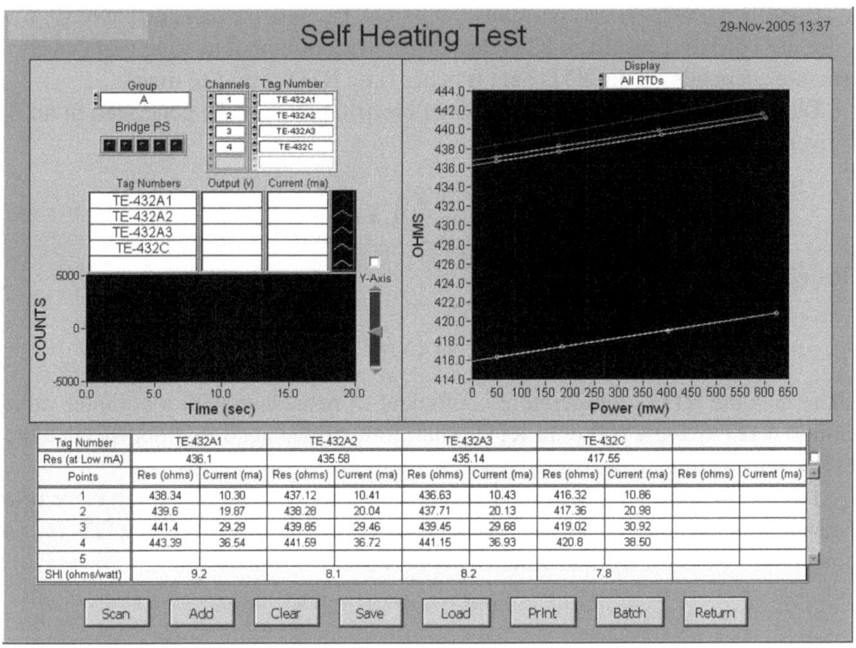

Fig. 6.27. Computer screen with results of a self-heating test

Table 6.13. Self-heating data

Data Point	R Resistance (Ω)	I Current (mA)	P Power = (mW)
1	442.72	10.72	50.88
2	443.88	20.61	188.55
3	445.84	29.99	400.99
4	447.72	37.09	615.91

Data is from in-plant testing of an RTD at operating conditions.
mA: milliamperes
mW: milliwatts

6.4.3 Self-Heating Error in RTDs

All temperature measurements with RTDs have inherent error because of self-heating. This is one of the main reasons why RTDs are not normally used to measure temperature in poor heat-transfer media such as air. For temperature measurement in air and other poor heat-transfer agents, thermocouples are usually used.

The self-heating error is normally very small because a small current (e.g., 1 mA) is usually used to measure an RTD resistance. The amount of heating that this current generates depends on the RTD heat transfer (i.e., RTD response time).

The temperature rise (ΔT) per unit of electric power (ΔP) generated in an RTD is given by:

$$\frac{\Delta T}{\Delta P} = \left(\frac{\Delta R}{\Delta P}\right) \bullet \left(\frac{\Delta T}{\Delta R}\right) \tag{6.17}$$

That is:

$$\frac{\Delta T}{\Delta P} = (SHI)(\alpha) \tag{6.18}$$

where α is the RTD's temperature coefficient of resistance with a nominal value of about 0.4 Ω /°C for a 100-ohm RTD. Therefore, all we need in order to calculate an RTD's self-heating error is its SHI.

Table 6.14 shows the self-heating index as well as the self-heating error of representative nuclear-grade RTDs. These results are from tests of the RTDs in room-temperature water flowing at 1 meter per second.

6.5 Noise Analysis Technique

The output of all process sensors in nuclear power plants normally contains fluctuations due to random flux, random heat transfer, turbulence, vibration, and other mechanical and thermal hydraulic phenomenon. These fluctuations (noise) can be extracted from the sensor output and analyzed to yield the sensor's response time. The

Table 6.14. Self-heating error of representative nuclear-grade RTDs

RTD Model	$R_0(\Omega)$	SHI (Ω/watt)	Self-Heating Error (^{o}C/watt)
	Rosemount		
176KF	200	6.1	7.6
104AFC	200	5.6	7.0
177HW	100	7.3	18.3
177GY	100	8.7	21.8
	Weed		
9004	200	8.5	10.6
9019	200	8.0	10.0
	Other		
Sensycon 1703	100	22.0	55.0
RdF 21458	200	4.6	5.8
RdF 21459	200	4.6	5.8
RdF 21232	200	3.0	3.8
Conax 7N13-1000-02	200	22.0	27.5
Conax7RB4-10000-01	200	11.7	14.6

noise analysis technique is normally used for in-situ response-time testing of pressure, level, and flow transmitters. Furthermore, the technique is used to detect blockages, voids, and leaks in pressure sensing lines (see Chap. 9), to measure vibration of reactor internals using neutron sensors, and to detect core flow anomalies by cross-correlating noise signals from neutron detectors and core exit thermocouples.[17]

To test the response time of temperature sensors, the LCSR method is the most commonly used technique because it can yield a sensor's response time so accurately. If accuracy is not critical (e.g., if a large margin exists between the sensor's nominal response time and the plant's required response time), then noise analysis may be used. Alternatively, noise analysis can be used to identify if a sensor's response time has changed. If the sensor's response time has changed significantly, then the exact response time can be measured using the LCSR method. The advantage of the noise analysis technique is that it does not require that the sensor be removed from service, and many sensors can be tested simultaneously using a single multichannel noise data acquisition system.

6.5.1 Laboratory Validation

The noise analysis method for testing RTD response times was first validated by the author and his French colleagues at the EdF laboratory in Renardières and CEA

Table 6.15. Results of validation of noise analysis performed at EdF's Renardières laboratory

	Response Time (sec)	
RTD Model	**Injection Test**	**Noise Analysis**
176 KF	0.14	0.18
177 HW	8.8	7.7
104 AFC	6.2	5.1
104 AFC w/Never-Seez	4.1	3.7

laboratory in Saclay in France.[13] These facilities were described in Sect. 6.3.4. The results of the validation tests are shown in Table 6.15.

Fig. 6.28 shows the PSD of a Rosemount 177 HW RTD used to validate the noise analysis technique at Renardières. The fit to the PSD, also shown in the figure, was used to calculate the RTD's response time. The response time was also calculated using time domain analysis of noise data by autoregressive modeling (AR). The PSD and AR results were then averaged to arrive at the response-time values shown in Table 6.15 for noise analysis.

The noise analysis technique has been further validated using data from a laboratory test loop at low pressure and low flow conditions. The results are shown in Table 6.16 for four Rosemount RTDs that were the most commonly used sensors in the nuclear power industry at the time of the validation work.

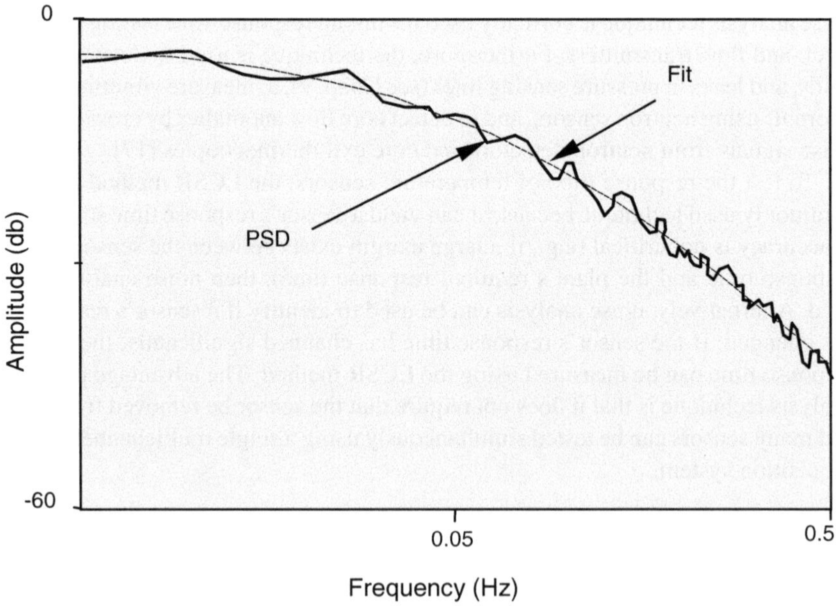

Fig. 6.28. PSD of Rosemount 177 HW RTD from data acquired at the EdF loop

Table 6.16. Laboratory validation of noise analysis technique for RTDs

	Response Time (sec)	
Item	**Direct Test**	**Noise Analysis**
		4.5
2	2.2	2.0
3	3.9	4.0
4	3.0	3.2

6.5.2 In-Plant Validation

In a few nuclear power plants, RTDs have been tested using both the LCSR and noise analysis techniques, performed under the same conditions. The results of these measurements are shown in Table 6.17. A representative PSD for these tests is shown in Fig. 6.29. Fig. 6.29 also shows a PSD of a thermocouple tested in a PWR plant.

6.6 NRC Regulations

The NRC Regulatory Guide 1.118, NUREG-0800, and NUREG-0809 all relate directly or indirectly to sensor response-time testing. For example, Regulatory Guide 1.118 provides criteria, requirements, and recommendations for the periodic testing of plant protection systems in nuclear power plants. It states that "safety system response time measurements shall be made periodically to verify the overall response time (assumed in the safety analysis of the plant) of all portions of the system from and including the sensor to operation of the actuator."

Regulatory Guide 1.118 refers to two IEEE standards (namely, IEEE Standard 279 and IEEE Standard 338). More specifically, Regulatory Guide 1.118 states that the criteria, requirements, and recommendations in IEEE Standards 279 and 338 are generally acceptable, notwithstanding several exceptions and/or clarifications.

Regulatory Guide 1.118 was originally issued in the mid-1970s. It gave a boost to the development of new methods for in-situ testing of response time in RTDs and pressure, level, and flow transmitters in nuclear powerplants. In particular, the development of the LCSR test was in large part stimulated by Regulatory Guide 1.118. When the LCSR test was completed, a topical report was submitted to the NRC on the LCSR technique. This resulted in a SER designated as NUREG-0809. Through this SER, the NRC approved the LCSR method for measuring the in-service response times of RTDs in nuclear power plants.[11]

In addition to the LCSR method, the SER includes a review of the self-heating and noise analysis methods. In particular, the SER introduces the self-heating and noise analysis methods as acceptable means for qualitatively monitoring the degradation of a sensor's response time.

The noise analysis technique is not used as much to test RTD response time because the LCSR method has fulfilled the needs of the nuclear industry. As for testing the

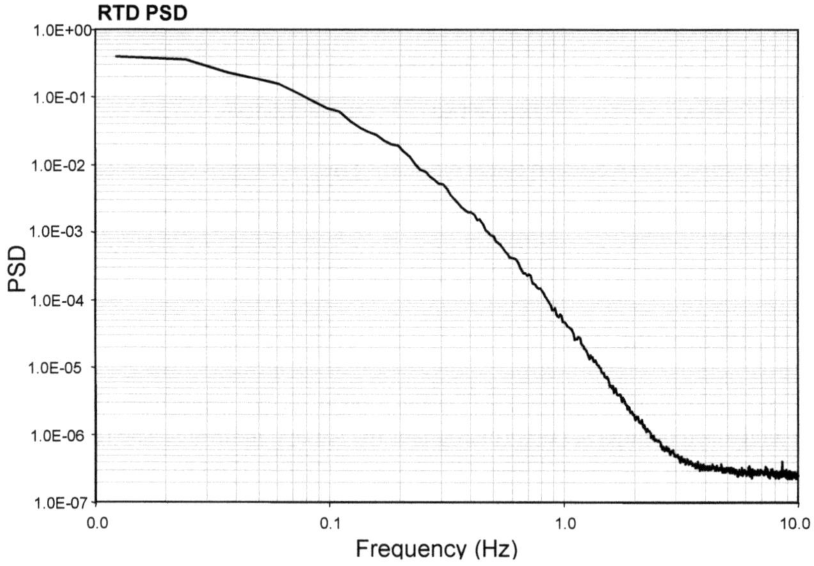

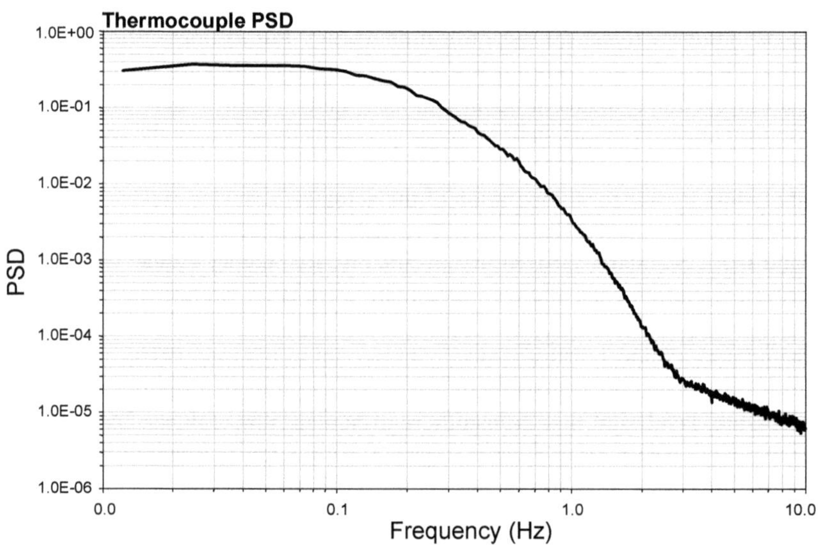

Fig. 6.29. PSDs of an RTD and a thermocouple from testing in a PWR plant at normal operating conditions

Table 6.17. Results of in-plant testing of RTDs using LCSR and noise analysis techniques

	Response Time (sec)	
Item	**LCSR Method**	**Noise Analysis**
1	3.7	3.4
2	3.5	3.3
3	3.6	2.9
4	4.0	4.5
5	2.9	3.3
6	4.5	4.4
7	3.4	3.0
8	4.6	4.4
9	4.6	4.4
10	6.5	6.8
11	2.4	2.8
12	2.7	2.6
13	2.4	2.1
14	2.4	1.7
15	0.9	0.9
16	3.5	3.2
17	3.7	4.0
18	2.7	2.7

response time of thermocouples, in spite of the success of the LCSR method, the nuclear industry prefers the noise analysis technique and uses it more often. This is because the LCSR test for thermocouples, although more accurate, requires high heating currents (> 500 ma), which could cause problems in thermocouples such as degradation of the thermocouple seal and/or insulation material.

In Appendix 13 of Chapter 7 of the NUREG-0800, the NRC has stated its position on RTD calibration and response-time testing.[10] Two statements indicative of the NRC's stance on sensor response-time testing are:

1. "Performance of an RTD is characterized by its accuracy and response time. To ensure adequate performance of the RTD, its accuracy and response time should be verified;" and
2. "Response time of RTDs may be verified using the LCSR method. The LCSR method should use an analytical technique such as the LCSR transformation."

Appendix B describes these points in more detail.

6.7 Factors Affecting Response Time

The response time of nuclear plant temperature sensors is predominantly affected by environmental conditions such as fluid flow rate and temperature, installation into a thermowell (when one is used), and degradation due to aging. These effects on the response time of temperature sensors are described in this section.

6.7.1 Ambient Temperature Effect

Changes in ambient temperature can affect sensor response time because: 1) the heat transfer between the sensor and surrounding fluid is dependent on temperature; and 2) dimensional changes occur as temperature changes.

The heat-transfer coefficient at the sensor surface changes with temperature because water thermal conductivity, specific heat capacity, and viscosity are dependent on temperature. For example, the film heat-transfer coefficient for a temperature sensor typically decreases by about a factor of two as water temperature increases from 20°C to 300°C.

The effect of temperature on sensor dimensions is a more dominant factor. More specifically, a temperature sensor is composed of several layers of materials. Ideally, these materials are homogeneous and in perfect contact with one another. In actuality, cracks and gaps likely exist within regions and at boundaries. As temperature increases, the gaps and cracks may open or close depending on the temperature coefficient of expansion of the sensor material. Since gas-filled gaps and cracks have a large effect on the heat-transfer resistance, this could be a large (but unpredictable) factor whose net effect might be an increase or decrease in response time with temperature. As such, the effect of temperature on sensor response time can only be identified by in-situ testing of response time at process operating conditions.

6.7.2 Effect of Fluid Flow Rate

The film heat-transfer coefficient for temperature sensors is a function of fluid flow rate. The magnitude of this relation in determining the response time of a temperature sensor depends on the ratio of internal heat-transfer resistance to surface heat-transfer resistance (i.e., the Biot Modulus). For example, a sensor whose internal heat-transfer resistance is 90 percent of the total at low flow rates can experience only a maximum improvement of about 10 percent in response time even at very high flow rates. In contrast, a sensor with a large surface heat-transfer resistance can experience a decrease in response time by a factor of two or more as flow rate is increased. Therefore, unlike temperature, the effect of flow on response time is predictable. That is, increasing the flow always decreases the response time; however, the amount of change in response time depends on the Biot Modulus. If the Biot Modulus is large, the flow effect is minor. If the Biot Modulus is small, the flow effect is significant.

6.7.3 Ambient Pressure Effect

If the sensor sheath were compressible, then increased pressure would compact the sensor material, improve the heat transfer, and reduce the response time. However, the sheath of temperature sensors is not normally compressible, and the effect of pressure is thus insignificant.

Pressure also affects the thermophysical properties of water (density, specific heat capacity, thermal conductivity, and viscosity), but the effect is small. As such, the total effect of pressure on a temperature sensor's response time is negligible.

6.7.4 Aging Effects

Since the response time is controlled by heat diffusion, response time could degrade either because of changes in the overall heat-transfer resistance and/or effective heat capacity of the sensor material. Since response time generally degrades, it is useful to postulate possible causes:

1. Changes in the properties of filler or bonding material. Filler material and/or cement is used to hold the sensing element in place in temperature sensors. Tests in air have shown that the cement changes from a homogeneous, plasticlike material to a flaky, hard material when heated in air to about 300°C. This changes (increases or decreases) the sensor's response time.[13]

2. Material on sensor surface. If any material (such as corrosion products or crud) adheres to the sensor's surface, it would increase the heat-transfer resistance and therefore the sensor's response time.

Table 6.18. Examples of RTD response-time degradation in nuclear power plants

	Response Time (sec)	
End of Cycle One	**End of Cycle Two**	**Change**
	Thermowell-Mounted RTDs	
2.7	3.7	37%
4.0	5.9	48%
2.4	3.3	38%
	Direct-Immersion RTDs	
1.9	2.5	32%
2.8	3.9	39%
2.0	2.5	25%

Results in table are from LCSR testing performed on RTDs as installed in nuclear power plants under normal operating conditions.

Table 6.19. Typical results of periodic measurement of RTD response times in a nuclear power plant

RTD Tag Number	Response Time (sec) 1997	1999	1999*	2000
411A1	5.3	5.9		6.3
411A2	3.3	3.3		4.2
411A3	5.0	5.0		4.9
411B	4.9	5.6		5.7
410B	5.4	6.3	3.8	3.7
421A1	4.7	5.2		5.6
421A2	4.8	4.8		4.6
421A3	5.4	5.5		5.6
421B	3.9	4.3		4.9
420B	4.8	5.4		5.8
431A1	4.4	4.0		4.5
431A2	5.5	5.8		6.2
431A3	4.5	4.0		4.1
431B	4.7	5.4		5.4
430B	5.7	6.3	3.8	3.8
441A1	5.7	5.7		5.7
441A2	4.5	4.2		5.2
441A3	4.7	5.2		5.1
441B	5.8	6.3	5.7	6.1
440B	5.4	6.1	4.2	3.8

Results are from LCSR testing of the RTDs as installed in an operating PWR.
*RTDs exceeding the 6.0- second requirement of the plant were replaced in 1999.

3. Changes in contact pressure or contact area. In thermowell-mounted sensors, the contact pressure between the sensor sheath and the inside wall of the thermowell can affect the response time. A higher contact pressure gives a faster response. In spring-loaded sensors, a gradual relaxation of the spring can cause a gradual decrease in contact pressure and thereby an increase in response time. Also, some sensors use bushings that have points or grooves to establish contact between the sensor and the inside wall of the thermowell. If vibration causes relative motion between the sensor and the thermowell, then the resulting wear would diminish the contact and slow response time.

Table 6.18 shows selected results of in-situ response-time measurements of direct-immersion and thermowell-mounted RTDs in nuclear power plants. They show significant degradation over just a single operating cycle of 18 to 24 months. These results

Table 6.20. Example of results showing RTD response-time degradation over a single cycle in a PWR plant

RTD Tag Number	Response Time (sec) Initial Test	One Cycle Later
112HA	3.4	6.1
112HB	4.7	5.6
112HC	4.3	5.4
112HD	3.2	4.4
112CA	3.0	3.7
112CB	6.3	7.3
112CC	3.4	3.6
112CD	5.6	5.4
122HA	3.3	3.7
122HB	4.4	5.1
122HC	3.5	3.6
122HD	3.7	5.2
122CA	3.5	4.8
122CB	3.9	4.7
122CC	2.8	3.1
122CD	4.3	4.6

Results are from LCSR testing of the RTDs as installed in an operating PWR.

are not typical, but they show that sensor response time can degrade significantly even over a short period of time. Tables 6.19 and 6.20 show additional RTD response time results from tests in two different PWR plants performed over a period of time. These results were selected to demonstrate the level of RTD response-time degradation that have been seen in some plants.

6.8 Summary

The response time of a temperature sensor depends on installation and process conditions. In particular, the process temperature and flow have a significant influence on response time. The effect of flow is generally predictable, but the effect of temperature is not. That is, increasing the flow decreases the response time, but increasing the temperature can cause an increase or a decrease in response time, depending on how temperature affects the material properties and heat transfer inside the sensor. For thermowell-mounted sensors, the response time depends largely on the fit between the measuring tip of the sensor and its thermowell. This effect dominates the response time of thermowell-mounted sensors. That is, any sensor/thermowell mismatch can make a large difference in the sensor response time. As such, the response time of a thermowell-mounted sensor must be measured with the sensor installed in the thermowell in which the sensor is used.

The LCSR method was developed in the mid-1970s to measure the "in-service" response time of nuclear plant temperature sensors. The method takes into account the effects both of installation and process conditions on response time. It is used routinely in nuclear power plants and is currently the only method that has received formal approval from the U.S. NRC.

As for thermocouples, the LCSR method is also well developed but not formally reviewed or approved by a regulatory body. This is mainly because nuclear plant thermocouples are not currently subject to any stringent response-time performance requirements. Thermocouples are, however, tested for response time in some plants. The results are tracked along with other performance measures, such as cable tests and cross-calibration, to establish the health, reliability, and residual life of thermocouples.

In addition to the LCSR method, the self-heating test and noise analysis technique are available for testing a temperature sensor's response time. The self-heating test does not provide a response time but yields an index called the SHI that is proportional to response time. Therefore, the test may be used to identify degradation of sensors' response time. The sensitivity of a sensor's response time to the SHI is not always very high, and the self-heating test is only useful for RTDs. Thermocouples cannot be tested with this method.

The noise analysis technique provides the response time of a sensor, but it is not normally as accurate as the LCSR test. The advantage of the noise analysis technique over the LCSR method, however, is that it does not require that the sensor be taken out of service for response-time testing. Furthermore, the noise analysis test can be performed remotely and passively on multiple sensors.

7

Nuclear Plant Pressure Transmitters

7.1 Transmitter Types

Nuclear plant pressure transmitters are electromechanical systems that are designed to measure pressure and differential pressure (including level and flow). In this book, the term *pressure transmitter* is used to mean pressure, level, or flow transmitter.

A pressure transmitter may be viewed as a combination of two systems: a mechanical system and an electronic system.[18] The pressure transmitter's mechanical system contains an elastic sensing element (diaphragm, bellows, Bourdon tube, etc.) that flexes in response to pressure. The movement of this sensing element is detected using a displacement sensor and converted into an electrical signal that is proportional to pressure.

Typically, two types of pressure transmitters are used in most nuclear power plants' safety-related pressure measurements. These are referred to as *motion-balance* and *force-balance*, depending on how the movement of the sensing element is converted into an electrical signal. In motion-balance transmitters, the displacement of the sensing element is measured with a displacement sensor (e.g., a strain gauge or a capacitive detector) and converted into an electrical signal (e.g., 4-20 mA DC current) that is proportional to pressure. In force-balance transmitters, the applied pressure forces a sensing rod in the transmitter to deflect. This deflection is opposed by an electromechanical feedback system in the transmitter. The feedback system consists of a force motor that works to keep the sensing rod at an equilibrium position. The amount of electrical current supplied to the force motor is proportional to the applied pressure.

The transmitter's electronics consist of active and passive components and circuitry that perform signal conditioning, temperature compensation, and linearity adjustments on the output signal. Typically, the transmitter electronics for low- and high-pressure applications are the same, while the sensing element is different. For example, one manufacturer uses three different elastic elements to accommodate several pressure ranges, from 0 to a maximum of about 200 bars (about 3000 psi), using the same transmitter housing design.

7.2 Transmitter Population and Application

A nuclear power plant generally uses about 200 to 800 pressure and differential pressure transmitters to measure the process pressure, level, and flow in its primary and secondary systems. The specific number of transmitters used for important measurements in a plant usually depends on the type and design of the plant. For example, the number of transmitters used in PWRs depends on how many reactor coolant loops there are. Fig. 7.1 shows a primary coolant loop of a PWR plant and some of the important transmitters and other sensors that are typically found there. These transmitters include differential pressure transmitters (designated as *dp*), which are called *dp cells* or *dp transmitters*. A dp transmitter is used to measure fluid flow and level. A nondifferential transmitter is used to measure absolute and gauge pressure. Fig. 7.2 shows the principle behind absolute, gauge, and differential pressure measurements. For measuring absolute pressure, one side of the sensing element is opened to the process pressure and the other side is evacuated. For gauge pressure measurements, one side is opened to the process pressure, and the other side is left at the ambient pressure. In both absolute and gauge pressure transmitters, the side that is opened to the process pressure is referred to as the *high side*. In differential pressure measurements, however, both sides of the sensing element are connected to the process pressure, with one side arbitrarily marked high and the other side marked low. Any differential pressure transmitter can be configured to measure gauge pressure by connecting one side to the process line and opening the other side to the atmosphere.

The upper range of normal static pressures is typically about 200 bars (approximately 3,000 psi) in PWRs and about 100 bars (approximately 1,500 psi) in BWRs. Fig. 7.3 shows a simplified schematic of a BWR plant including some of the important pressure transmitters used in this type of plant.

The movement of the sensing element in nuclear plant pressure transmitters is normally converted into a DC current and transmitted in a two-wire circuit. This circuit consists of the transmitter in the field and its power supply, which is usually located remotely from the transmitter in an instrument cabinet in the control room area. The same two wires that are used to supply power to the transmitter electronics serve to provide the current loop on which load resistors are placed in series as shown in Fig. 7.4. The voltage drop across the resistors is used to measure or monitor pressure or differential pressure. Using a current loop allows the pressure information to be transmitted over a long distance without loss of signal and with reduced electrical noise and interferences.

7.3 Nuclear Qualification

Pressure, level, and flow transmitters provide most of the important signals for controlling and ensuring the safety of nuclear power plants. Depending on their location and service, some of these transmitters must be able to withstand and operate properly in any potential environment, including before, during, and after an accident.

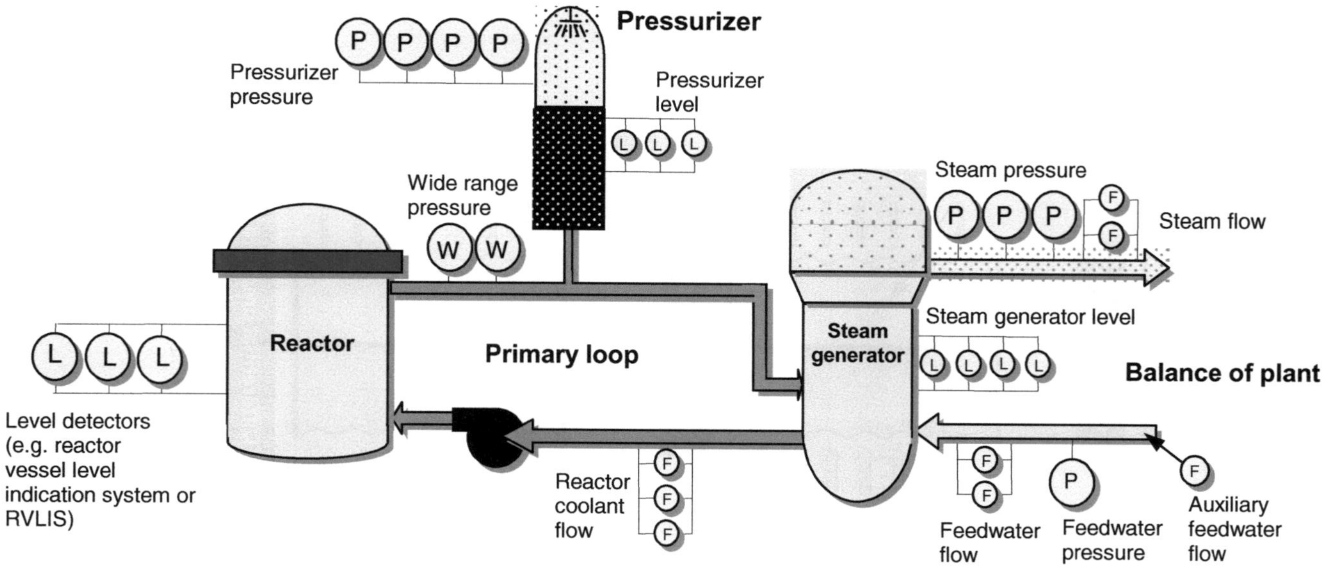

Fig. 7.1. Example of important pressure transmitters in a loop of a PWR plant

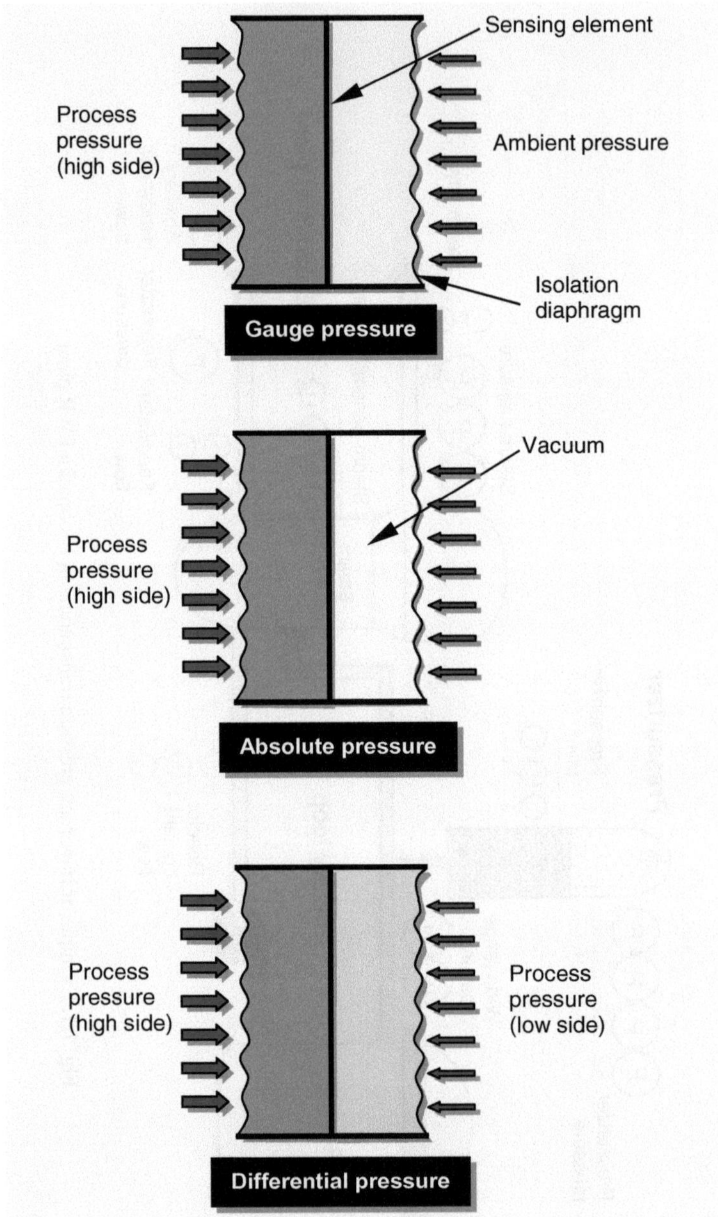

Fig. 7.2. Principle of gauge, absolute, and differential pressure measurement

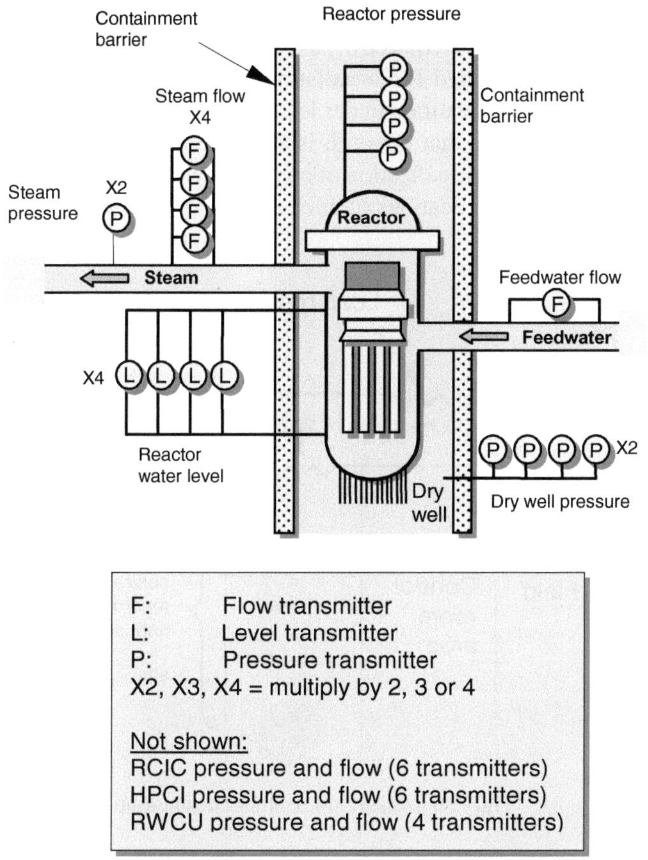

Fig. 7.3. Example of some of the important pressure transmitters in a BWR plant

For this reason, manufacturers type-test representative transmitters and generically qualify them under simulated accident conditions in a laboratory. The extent of the laboratory qualification is based on where and for what purpose the transmitter is to be used in a nuclear power plant.

The most extensive qualification is usually performed for transmitters that are to be used as "Class 1E" equipment for the reactor containment or other harsh environments of the plant. The term 1*E* is the safety classification given to electrical equipment and systems whose failure or damage could potentially release significant amounts of radiation into the environment. This class of equipment is qualified so as to assure its continued operation during and after a so-called design basis event (DBE).

A DBE consists of a hypothetical set of conditions that encompasses the worst-case events postulated for certain equipment in nuclear power plants. This includes the seismic conditions during an earthquake, the nuclear radiation environment during a LOCA, and the steam, temperature, and pressure environments of both a high-energy line break (HELB) and a LOCA. Any equipment that is qualified under these conditions can obviously be used for less severe services at any location in a plant. However, not all equipment is required to pass such a rigorous set of tests. For example, an instrument that is intended for installation in a mild environment, such as the auxiliary building, can be qualified under less stringent environments.

It should be pointed out that *Class IE* is a U.S. term; different organizations and countries define the safety classification of nuclear power plant equipment differently. Fig. 7.5 compares the classifications of nuclear power plant equipment.

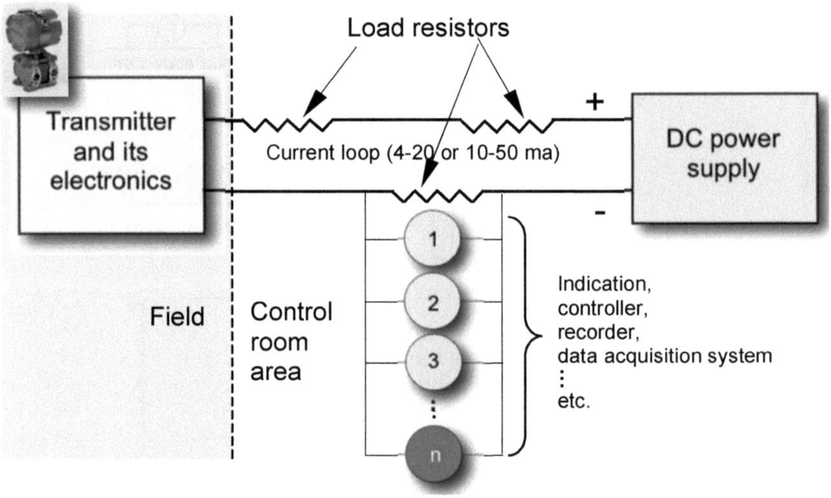

Fig. 7.4. Pressure transmitter current loop

According to the U.S. law designated in the Code of Federal Regulations (CFR) as 10 CFR 50.59, three categories of electric equipment that are important to safety must be identified and qualified according to their application and specified performance. These three categories are: (1) safety-related (Class 1E) equipment; (2) non-safety-related (non-Class 1E) equipment whose failure could adversely affect other safety-related equipment; and (3) certain postaccident monitoring equipment, as identified in the Regulatory Guide 1.97 of the U.S. NRC.

7.3.1 Qualification Procedure

The general procedure for qualifying equipment for nuclear power plants is to bring the equipment to the end of its intended design or qualified life through artificial aging. For pressure transmitters, the aging usually includes a conservative combination of lifetime plant radiation exposure, thermal aging, vibration, and pressure cycling. Aging may be accelerated using documented methods that are acceptable to the nuclear industry, such as the Ahhrenius theory of thermal aging. The synergistic effects of radiation dose rate, cycling, and elevated temperature are often accounted for when aging a specimen for qualification testing.

<table>
<tr><th>ORGANIZATION AND/OR COUNTRY</th><th colspan="7">EQUIPMENT CLASSIFICATION</th></tr>
<tr><td rowspan="2">International Atomic Energy Agency</td><td colspan="4">Systems important to safety</td><td colspan="3" rowspan="2">Systems not important to safety</td></tr>
<tr><td colspan="2">Safety system</td><td colspan="2">Safety-related system</td></tr>
<tr><td>International Electrotechnical Commission</td><td colspan="3">Category A</td><td>Category B</td><td colspan="2">Category C</td><td>Unclassified</td></tr>
<tr><td>France</td><td colspan="2">1E</td><td colspan="2">2E</td><td colspan="3">IFC/NC</td></tr>
<tr><td>European Utilities Requirements</td><td>F1A (Automatic)</td><td></td><td colspan="2">F1B (Automatic and Manual)</td><td colspan="2">F2</td><td>Not Classified</td></tr>
<tr><td>United Kingdom</td><td colspan="3">Category 1</td><td colspan="2">Category 2</td><td colspan="2">Not classified</td></tr>
<tr><td>United States of America</td><td colspan="2">1E</td><td colspan="5">Nonnuclear safety</td></tr>
</table>

Fig. 7.5. Safety classification of nuclear power plant equipment (Source: IAEA-TECDOC-1402, 2004)

Aging is usually followed by seismic testing. Recommended practices for seismic testing are provided by IEEE in its IEEE Standard 344. If the equipment is to be used in the reactor containment area or in a location where it can become submerged in water or exposed to high humidity or radiation during an accident, then the equipment must also undergo envi-ronmental qualification. Environmental qualification usually follows the seismic tests and is performed in accordance with the requirements of IEEE Standard 323. A pressure transmitter that is seismically and/or environmentally qualified for nuclear service is often referred to as a *nuclear-grade transmitter.*

7.3.2 Qualified Life

The qualified life of equipment is established based on the results of laboratory qualification tests. Fig. 7.6 shows the qualified life of a nuclear plant pressure transmitter based on the temperatures to which the transmitter may be exposed during normal plant operation. Two curves are given in Fig. 7.6 for this transmitter: one curve for the transmitter module (body and other hardware), another for its electronics. Clearly, the transmitter electronics have a shorter qualified life than the transmitter module under the same operating temperature.

The correlation between the qualified life of a sensor and the normal temperature to which it is exposed determines the limits of the ambient temperatures that must be maintained to allow the sensor to be used for the full length of its qualified life.

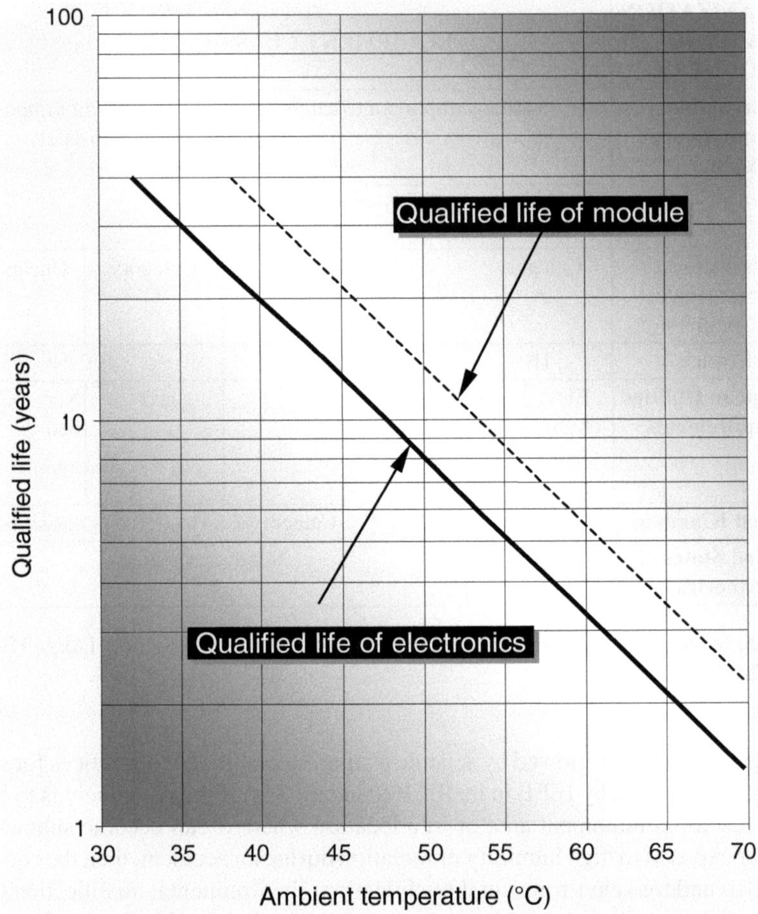

Fig. 7.6. Example of qualified life versus operating temperature for a nuclear-grade pressure transmitter

In the example in Fig. 7.6, if the transmitter's electronics are replaced at the end of their qualified life, then the transmitter's qualified life can be extended to that of the module's qualified life.

7.4 Transmitter Manufacturers

Only a few manufacturers (fewer than ten) provide most of the pressure transmitters used in the safety systems of nuclear power plants. For example, Barton, Foxboro, and Rosemount have provided the majority of pressure transmitters for U.S. plants since the 1960s. The representative transmitter models from these manufacturers and their environmental qualification status for nuclear services are given in Table 7.1. Note that some transmitters do not have environmental qualifications even though they are used for safety-related applications. These are safety-related transmitters that are located in the areas of the plant that are not subject to the consequences of a LOCA.

Table 7.1 also shows the manufacturers' specifications for their transmitters' response times. These are the nominal response-time estimates; the actual response times of individual transmitters may be significantly different. Also note that different manufacturers define *response time* differently. For example, Barton typically defines its transmitters' response times as the time it takes for the output of the transmitter to reach its final value after a step change in input pressure, from 10 percent to 90 percent. Rosemount generally uses time constant (i.e., the time required for the sensor output to reach 63.2 percent of its final value after a step change in pressure), and the response-time specification for Foxboro transmitters is often based on frequency response data.

Since early 2000, the Foxboro transmitters for nuclear power plants have been supplied by the Weed Instrument Company. In fact, the names of the manufacturers

Table 7.1. Representative nuclear plant pressure transmitters

Transmitter Manufacturer	Model Number	Range Code	Environmental Qualification	Nominal Response Time (sec)
Barton	752		No	N/A
	763		Yes	< 0.18
	764		Yes	< 0.18
Foxboro (Weed)	E11		No	< 0.30
	E13		No	< 0.30
	NE11		Yes	< 0.30
	NE13		Yes	< 0.30
Rosemount	1152	3	Yes	0.31
		4	Yes	0.13
		5	Yes	0.09
		6	Yes	0.06
	1153	3	Yes	2.0
		4	Yes	0.5
		5–9	Yes	0.2
	1154	4	Yes	0.5
		Others	Yes	0.2

1. Foxboro transmitters for nuclear power plant applications are now supplied by Weed Instrument Company.
2. The response-time values given in this table are from the manufacturers' specifications. Actual response times may be significantly different.
3. The definition of response time by different manufacturers may be different.

and suppliers of nuclear plant pressure transmitters have been changing over the years because of company mergers and acquisitions and for other reasons. Historically, in addition to the three manufacturers just mentioned, nuclear-grade pressure transmitters have been supplied under several names including the following: Westinghouse Veritrak, Tobar, Camille Bauer, Fischer & Porter (F & P), Gould, Hartmann & Braun (H & B), Schlumberger, Gulton-Statham, Bailey Sereg S.A., KDG Mobrey, and others. The Rosemount and KDG Mobrey transmitters are now supplied by the Emerson Process Management Company, the Veritrak and Tobar by Weed Instrument Company, and Gulton-Statham by AMETEK. There are also new names for suppliers of the other transmitters mentioned.

Of all these transmitter manufacturers, Foxboro and F & P have typically produced force-balance type transmitters for nuclear power plants while the remaining manufactures have typically supplied motion-balance transmitters for the nuclear power industry. The following sections provide a description of the four transmitter manufacturers whose products have been used in the U.S. nuclear industry since the mid-1960s.

7.4.1 Barton Transmitters

Four models of Barton transmitters are used for safety-related applications in nuclear power plants. These are Models 752, 753, 763, and 764. Figs. 7.7 and 7.8 show photographs of these sensors. (No photographs are shown for 753 and 763 because these two models are similar to Models 752 and 764, respectively, in physical configuration.) The principle of operation of the four Barton transmitters is essentially identical, except for the sensing element, which differs in each model.

The relevant specifications of the four Barton transmitters are summarized in Table 7.2. We will also describe them in the following four paragraphs, beginning with Barton Model 752. Because all four share similar principles of operation, only the operation of Barton Model 752 will be described in detail.

Barton Transmitter Model 752

Barton Model 752 is a differential pressure transmitter whose sensing element is made of dual bellows that are connected by a shaft. The bellows and the shaft are installed in a housing that is filled with silicone oil to protect against mechanical damage and to provide some damping of the pressure signal being measured (Fig. 7.9a). Additional damping can be provided electronically by adding a capacitor to the transmitter's circuitry. Barton transmitters are not normally supplied from the factory with a damping capacitor, but the capacitor can be added by the user, depending on the desired dynamic response and the level and frequency of noise that is to be removed.

The use of any damping device in a nuclear plant sensor must be done carefully to ensure that the dynamic response-time requirements are not exceeded. Appendix D shows an NRC information notice as an example of how the use of a damping device in pressure transmitters may have an undesirable consequence in a nuclear power plant. This NRC information notice warns the users of damping devices in nuclear power plants to pay close attention to the effect of damping devices on the response time of process sensors.

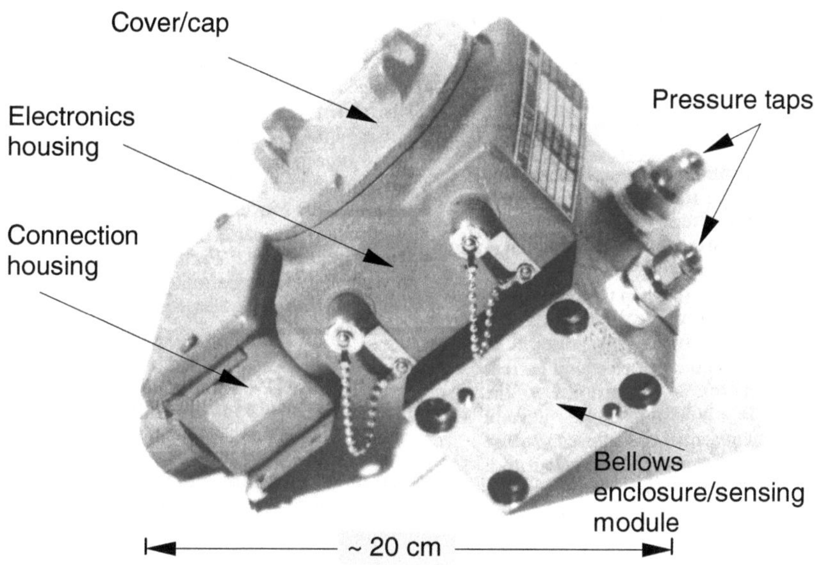

Fig. 7.7. Barton Model 752 transmitter (the electronics housing of a Barton Model 753 is similar in appearance)

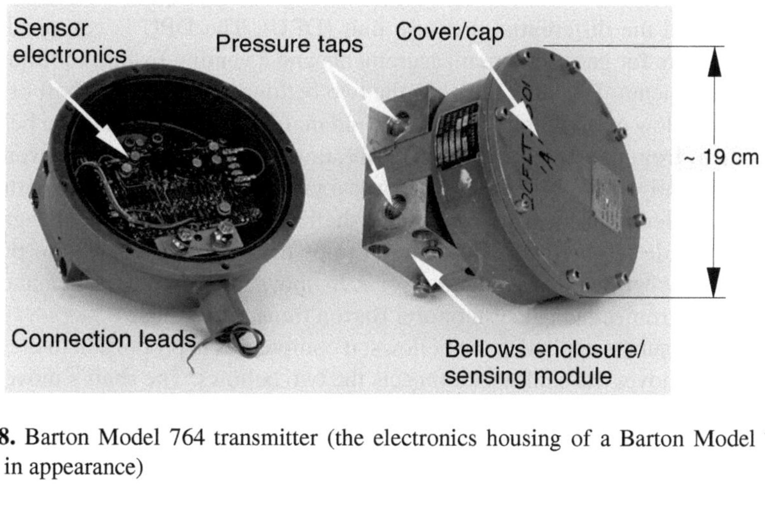

Fig. 7.8. Barton Model 764 transmitter (the electronics housing of a Barton Model 763 is similar in appearance)

Table 7.2. Manufacturer's specifications for Barton transmitters

Transmitter Characteristics	Transmitter Model Number			
	752	**753**	**763**	**764**
1. Type	Differential	Gauge	Gauge	Differential
2. Sensing Element	Bellows	Bourdon Tube	Bourdon Tube	Bellows
3. Accuracy (% of span)$^{(1)}$	± 0.25	±0.25	± 0.5	± 0.5
4. Range	0 – 50" wc 0 - 300 psid	0 - 25 psi 0 - 5000 psi	0 - 100 psi 0 - 3000 psi	0 – 100" wc 0 - 300 psid
5. Response Time	N/A	N/A	< 0.18 sec	< 0.18 sec
6. Drift (% of span/year)$^{(2)}$	N/A	N/A	± 1.0	± 1.0
7. Qualification Level $^{(3)}$	Seismic only	Seismic only	Qualified	Qualified

1. The accuracy statement for each transmitter includes linearity, hysteresis, and repeatability.
2. The response time for each transmitter is equal to the time that corresponds to 63.2 percent of the final output for a step change in input with an amplitude of 10% to 90%.
3. Nuclear qualification includes environmental and seismic qualification per the IEEE 323 and 344 standards for postaccident operation in nuclear power plants.

The two bellows, the shaft, and their housing in these Barton transmitters are together called the differential pressure unit (DPU). The DPU is equipped with a draining feature for gas pressure measurements and a venting feature for liquid pressure measurements. The shaft between the two bellows serves as the shaft of a valve that stops the flow of oil within the bellows and maintains enough oil in the bellows to protect them from damage or a shift in calibration when the bellows are overranged. More specifically, if the bellows experiences a pressure that is greater than the

DPU's differential pressure range, the shaft will move until it seals against its valve seat inside the bellows. This motion traps the oil in the bellows to protect it from rupture when it is overranged. Fig. 7.9b shows a simplified schematic of the complete electromechanical system of a Barton transmitter.

When pressure is applied to the bellows, it compresses in proportion to the applied pressure and moves the shaft that connects the two bellows. The shaft's movement is detected by the movement of a beam that fits in a square hole at the middle of the shaft (Fig. 7.10). Two strain gauges are bonded to the opposite sides of the beam to sense the movement. The beam and strain gauge assembly is housed in the DPU in the oil-filled

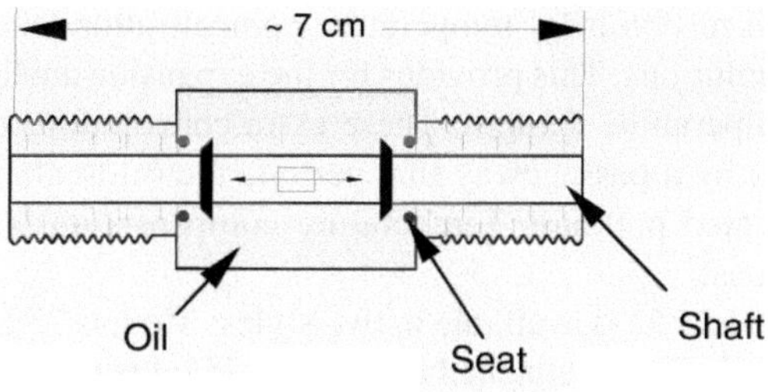

a) Oil-filled bellows

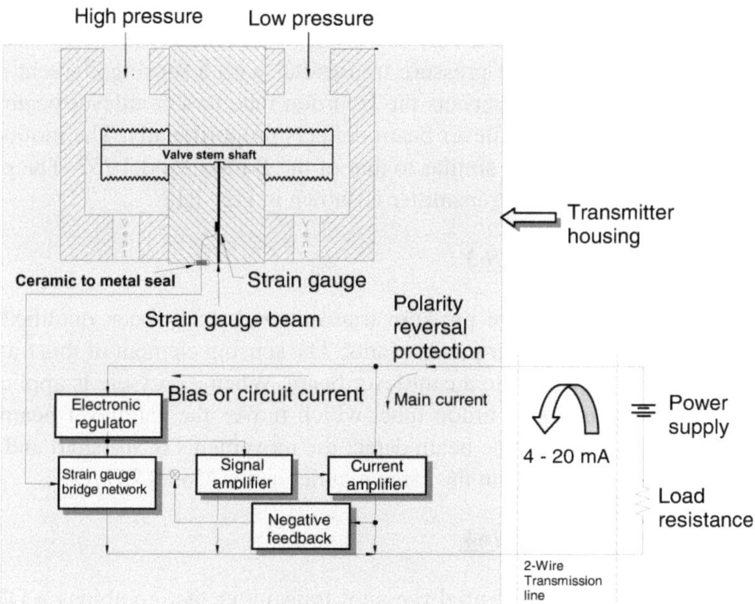

b) Complete system diagram

Fig. 7.9. Simplified diagram of a Barton double-bellows differential pressure transmitter

space. When the beam is moved, it applies tension to one of the two strain gauges and compresses the other. The resistance of the gauge that is under tension increases, and the resistance of the gauge under compression decreases. The two gauges are electrically connected in such a way as to form the two active arms of a Wheatstone bridge circuit. The output voltage of the bridge is converted by a current amplifier into a 4 to 20 or a 10 to 50 milliamperes (mA) current signal. In addition to the two strain gauges, the bridge circuit consists of zero and span adjustment potentiometers, bridge completion resistors, and temperature compensation components.

For temperature compensation, the high-pressure side of the DPU has a reservoir for the oil to flow to when it expands under temperature. This mechanism is illustrated in Fig. 7.11a; Fig. 7.11b contains a detailed drawing of the sensing module of a Barton transmitter. The oil reservoir for temperature compensation takes the form of a few extra bellows convolutions. This provides for the expansion and contraction of the oil as the ambient temperature changes. These extra convolutions are connected to the measuring bellows by a passageway that permits the oil to change volume without affecting the indicated pressure. Temperature compensation is also possible using electronic techniques.

The Barton Model 752 is available in two styles: Models 752-1 and 752-2. Model 752-2 has a readout indicator attached to it, while 752-1 is the standard model (called the "blind" model), with no local readout device.

Barton Transmitter Model 753

Barton Model 753 is a gauge pressure transmitter with a sensing element made of a Bourdon tube. A linkage connects the Bourdon tube to a cantilever beam. As the Bourdon tube flexes, the cantilever beam deflects proportionally. The motion of the beam is detected in a manner similar to that of the Barton Model 752. The principle of operation of a Barton 753 transmitter is shown in Fig. 7.12.

Barton Transmitter Model 763

Barton Model 763 is a gauge pressure transmitter that has been qualified for in-containment service in nuclear power plants. The sensing element of this transmitter is a Bourdon tube connected to a cantilever beam. When a pressure is applied to the transmitter, it deflects the Bourdon tube, which moves the cantilever beam. Strain gauges on opposite sides of the beam detect the movements of the shaft and convert them into an electrical signal in the same manner as for Model 753.

Barton Transmitter Model 764

Barton Model 764 is a differential pressure transmitter that combines a DPU with an electronic circuit. It works very much like the Barton Model 752, except that this transmitter is qualified for nuclear safety-related services.

7.4.2 Foxboro/Weed Transmitters

Four models of Foxboro transmitters are used in nuclear power plants: E11, E13, NE11, and NE13. The "N" in the model number indicates that the transmitter is qualified for nuclear service. The four models are all force-balance transmitters and are very similar in physical configuration and principle of operation.

Fig. 7.13 is a photograph of three different Foxboro force-balance transmitters. It includes the Model 611DM, which is an old design and was replaced by the Model NE11DM.

Table 7.3 summarizes typical specifications of Foxboro transmitters. The E13 and NE13 models can only be used for differential pressure measurements, while the E11 and NE11 models are suitable for absolute, gauge, and differential pressure measurements.

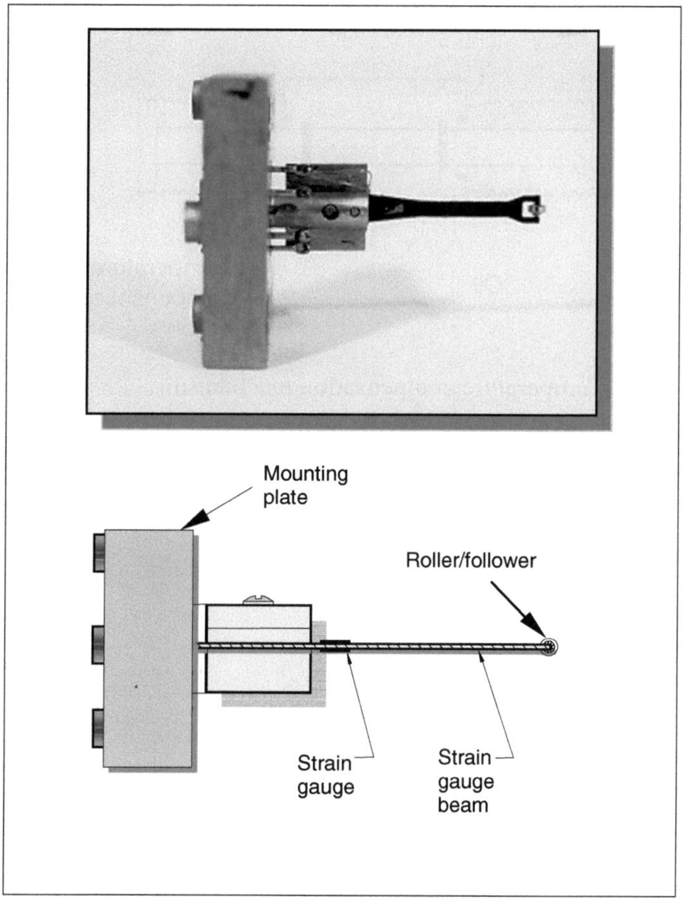

Fig. 7.10. Photograph and drawing of the displacement sensor in Barton transmitters

Three types of sensing elements are used in Foxboro force-balance pressure transmitters: Bourdon tube, bellows, and diaphragm. Figs. 7.14 through 7.16 show the schematics of the three Foxboro transmitters with these three different sensing elements. Note that except for the sensing element, the three transmitters are identical in their electromechanical structure and principle of operation. Each transmitter contains a force bar that deflects under pressure and a force motor that works to null (balance) the deflection of the force bar. The electric current that the motor uses to keep the force bar at equilibrium is proportional to the applied pressure. More specifically, the force that is produced by the applied pressure is balanced by an opposing force from the force motor.

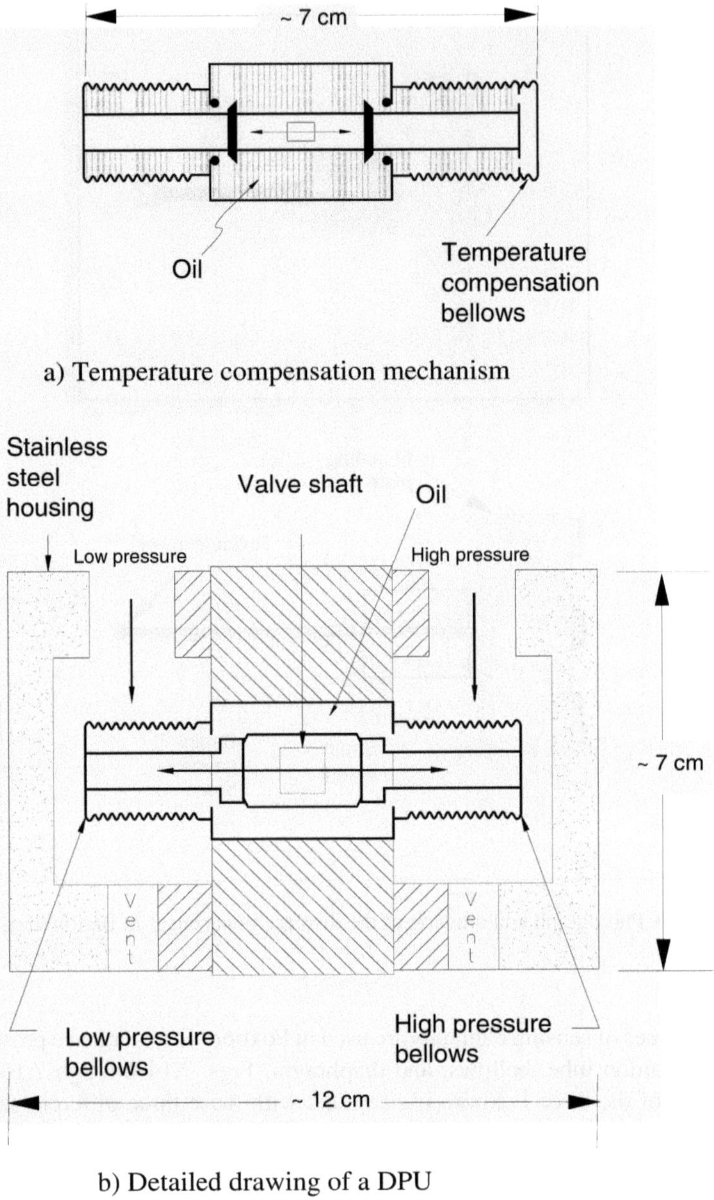

Fig. 7.11. Sensing module of Barton Transmitter Model 752

Table 7.3. Typical specifications of Foxboro force-balance pressure transmitters

Model	Variety	Type	Sensing Element	Range	Accuracy (±% of span)	Drift/Yr (% of span)
E11	E11AL	Absolute-Low Range	Diaphragm	0-90 mbar (0-70 mm Hg)	1.0	N/A
	E11AM	Absolute-Med. Range	Diaphragm	0-2 bar (0-1520 mm Hg)	0.5 to 1.25	N/A
	E11AH	Absolute-High Range	Bellows	0-50 bar (0-750 psia)	0.5	N/A
	E11DM	Differential	Bellows	-12 to 205 bar (-180 to 3000 psi)	0.5	N/A
	E11GM	Gauge	Bellows	1-205 bar (-15 to 3000 psi)	0.5	N/A
	E11GH	Gauge	Bourdon	1 to 830 bar (-15 to 12000 psi)	0.5 to 1.25	N/A
E13	E13DL	Differential	Diaphragm	-62.3 to +62.3 mbar (-25 to +25" wc)	0.5	N/A
	E13DM	Differential	Diaphragm	±0.5to ±2.2 bar (±205 to ±850" wc)	0.5 to 0.75	N/A
	E13DH	Differential	Diaphragm	±0.5to 2.2 bar (-205 to 850" wc)	0.5 to 0.75	N/A
NE11	NE11AL	Absolute-Low Range	Diaphragm	0-90 mbar (0-70 mm Hg)	1.0	1.2
	NE11AM	Absolute-Med. Range	Diaphragm	0-2 bar (0-1520 mm Hg)	0.5 to 1.25	0.5
	NE11AH	Absolute-High Range	Bellows	0-50 bar (0-750 psia)	0.5	0.33
	NE11DM	Differential	Bellows	-12-205 bar (-180 to 3000 psi)	0.5	0.25
	NE11GM	Gauge	Bellows	-1-205 bar (-15 to 3000 psi)	0.5	0.25 to 0.40
	NE11GH	Gauge	Bourdon	-1-414 bar (-15 to 6000 psi)	0.5 to 1.25	0.33
NE13	NE13DL	Differential	Diaphragm	-62.3 to +62.3 mbar (-25 to +25" wc)	0.5	0.33
	NE13DM	Differential	Diaphragm	-0.5 to 22 bar (-205 to 850" wc)	0.5	0.25
	NE13DH	Differential	Diaphragm	-0.5 to 22 bar (-205 to 850" wc)	0.5 to 0.75	0.25

1. N/A = Not available
2. Foxboro transmitters for nuclear power plants are now supplied by Weed Instrument Company

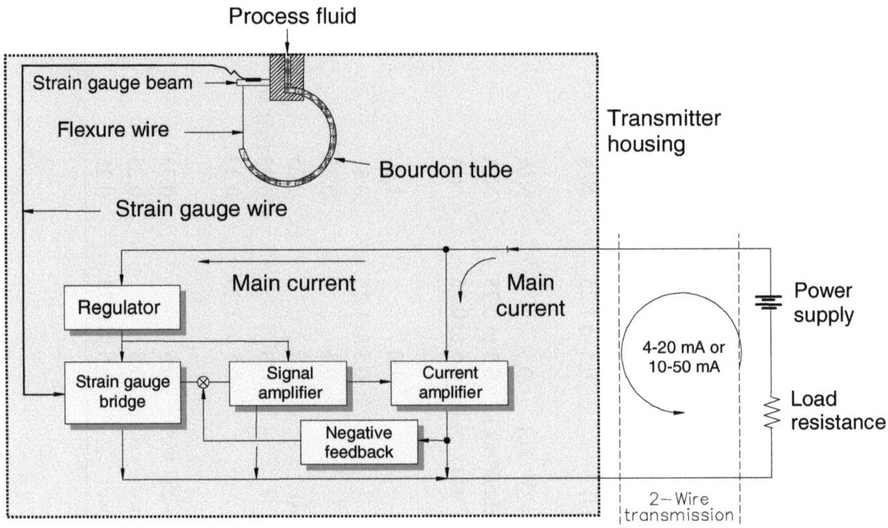

Fig. 7.12. Diagram of Barton Model 753 transmitter

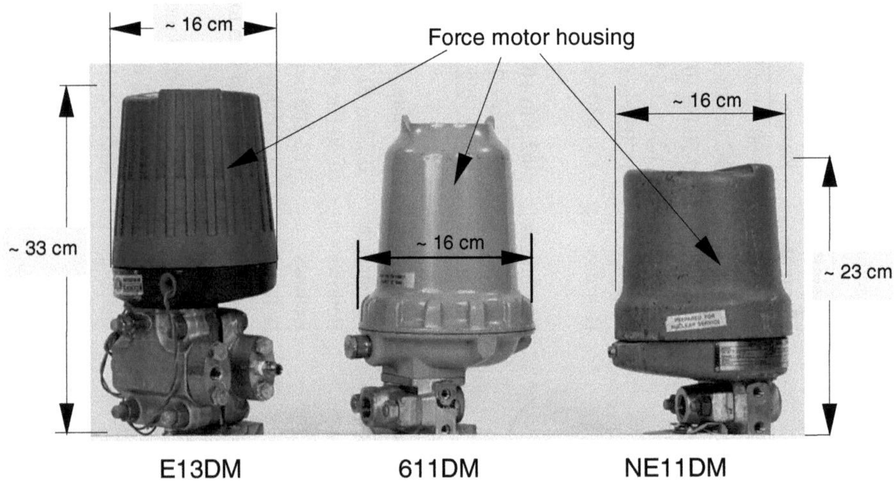

Fig. 7.13. Body styles of three models of Foxboro (Weed) transmitters

7.4.3 Rosemount Transmitters

Four models of Rosemount pressure and differential pressure transmitters are used to measure pressure, level, and flow in the primary and secondary systems of nuclear power plants. These are Models 1151, 1152, 1153, and 1154. Models 1152, 1153, and 1154 are qualified for nuclear safety-related service, while Model 1151 is a general-purpose transmitter that is used in nuclear power plants for non-safety-related

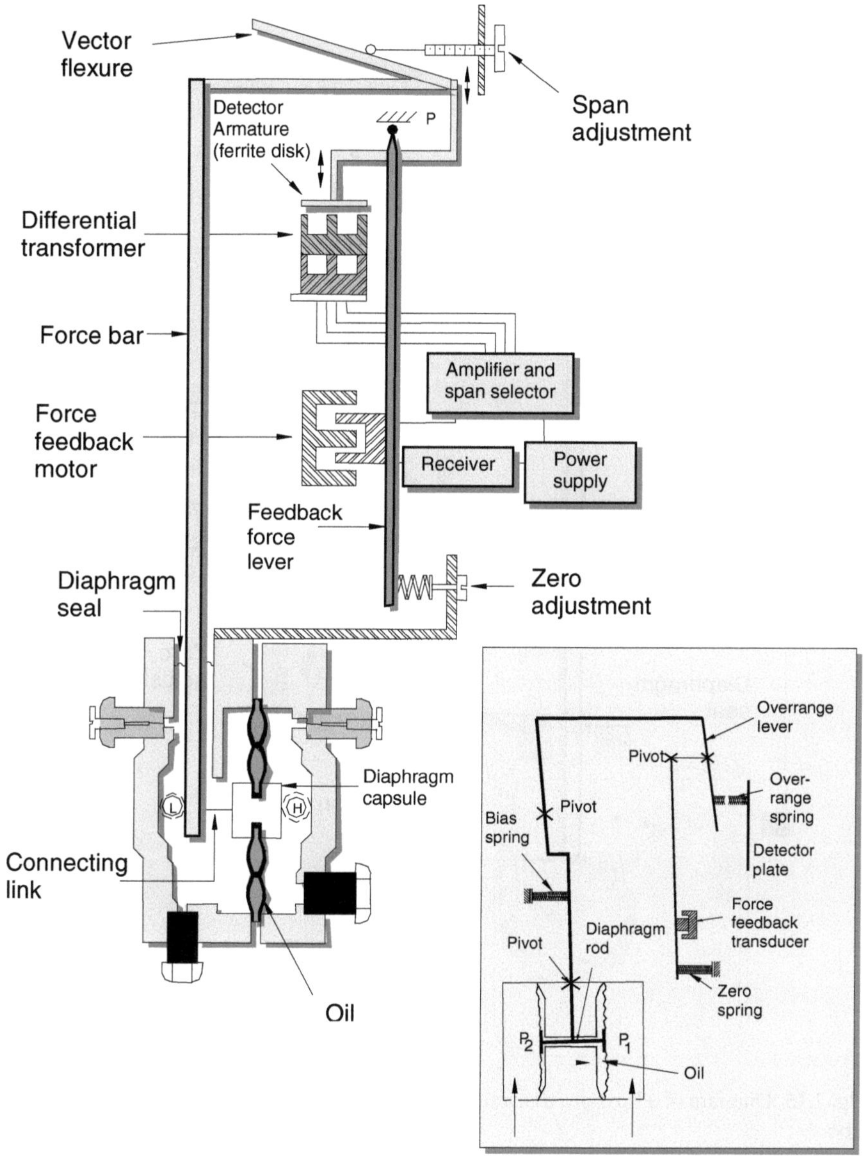

Fig. 7.14. Diagram of a Foxboro transmitter and its sensing element that is made of a diaphragm capsule

applications. The qualification status of the four Rosemount models is presented in Table 7.4.

The four Rosemount models are similar in physical configuration and principle of operation. Fig. 7.17 shows a photograph of a commercial and a nuclear-grade

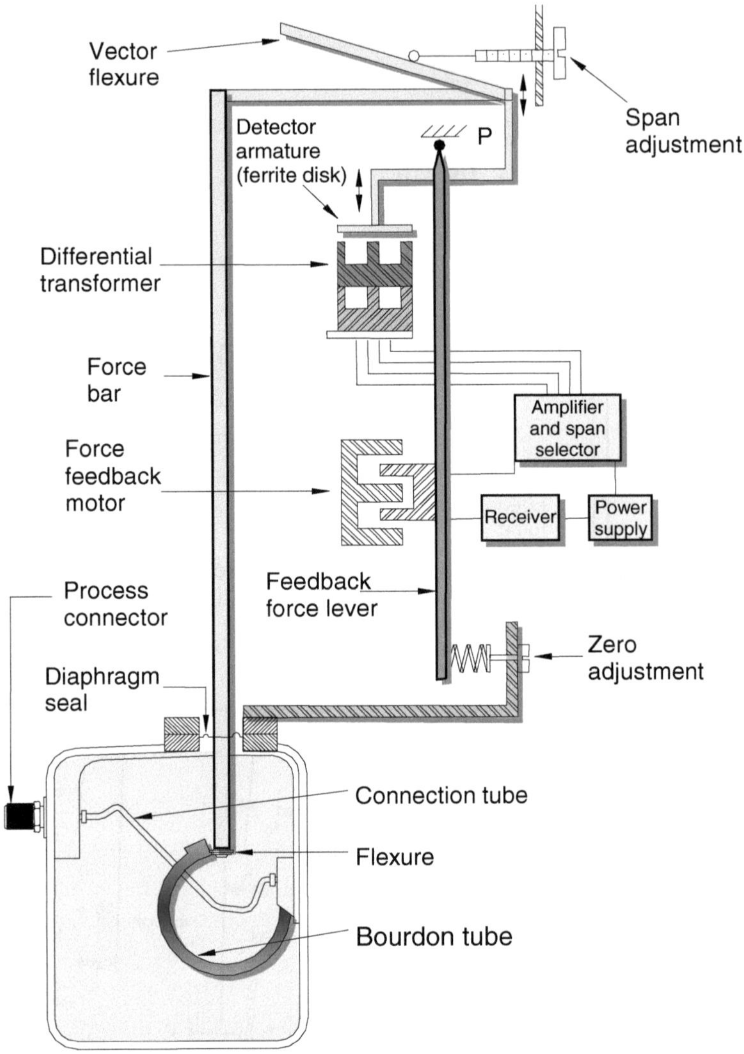

Fig. 7.15. Diagram of a Foxboro transmitter and its sensing element that is made of a Bourdon tube

transmitter, and Fig. 7.18 illustrates the principle of operation of these transmitters' sensing module. The sensing module is an oil-filled capacitance-type sensor called the *Delta-Cell* (δ-cell). The δ-cell is isolated from the process fluid by an isolation diaphragm, and silicone oil is used in the δ-cell to transmit the process pressure from the isolation diaphragm through a few capillary tubes to the sensing diaphragm at the center of the δ-cell. Fig. 7.18a shows one-half of a Rosemount sensing cell. The components that make up the Rosemount transmitter's sensing cell consist of two halves, center diaphragm, isolation diaphragms, and the oil fill fluid (see Fig. 7.18b).

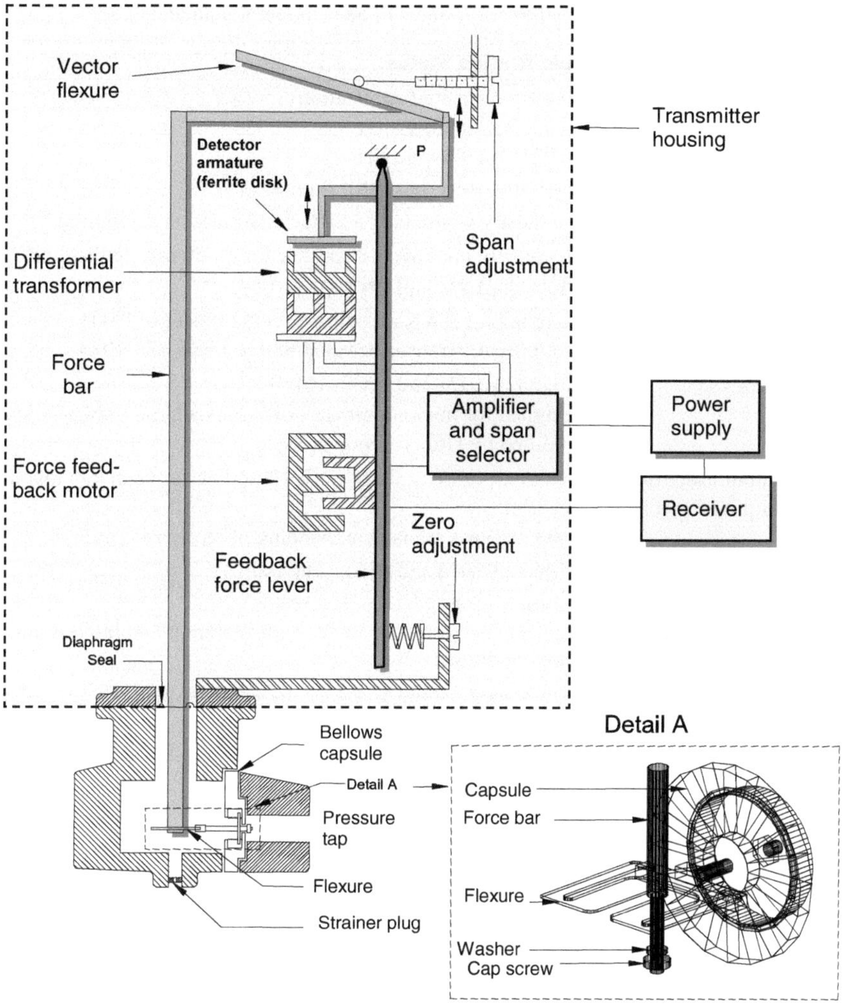

Fig. 7.16. Diagram of a Foxboro transmitter and its sensing element that is made of a bellows capsule

Each cell half consists of a metal cup filled with glass. A cavity is machined into the glass, and a metal film is deposited onto the glass to form a capacitor plate. A ceramic insert that has holes running through it exists between the cavity and the back of the cell cup. This insert provides a passageway through which the oil hydraulically transfers the pressure force from the process fluid to the center diaphragm. Each leadwire to the capacitor plate is actually a small-diameter tube through which each half of the cell is filled with oil after assembly. After the cell is filled with oil, the fill tube is pinched closed and soldered, thus serving as a sealed electrical lead to the capacitor plate.

Table 7.4. Qualification status of Rosemount transmitters

Transmitter Model	Qualification Status
1151	Non-safety-related applications No nuclear qualification 10CFR21 not applicable
1152	IEEE-323-1971 and IEEE-344-1975 Use mostly where only seismic qualification is required
1152-T1805	Seismically qualified transmitter for 10-50 mA applications
1153 Series B	IEEE-323-1974 and IEEE-344-1975 Designed for BWR and outside containment of PWR First pressure transmitter qualified to IEEE-323-1974
1153 Series D	IEEE-323-1974 and IEEE-344-1975 Designed for in-containment of PWR (stainless steel housing) Qualified by Utility Group at Wyle
“R” Output Electronics “R” Output Electronics	Improves performance of Series B and Series D under radiation conditions N0037, allows adjustable damping of electronics output
1154	IEEE-323-1974 and IEEE-344-1975 High-performing Class 1E transmitter Improves performance under high-radiation and high- temperature conditions
1154 Series H	Highest-performing Class 1E transmitter qualified per IEEE-323-1974 and IEEE-344-1975

The ability of the sensing cell to accurately sense different pressure ranges is a function of four parameters: the curvature of the cavity machined into the cell halves, the diameter of the capacitor plate deposited on the surface of the cavity, the stiffness (thickness) of the center diaphragm, and the stiffness of the isolating diaphragms. The last of these is significant only at the low pressure levels.

Displacement of the center diaphragm at its center is limited to a max- imum of approximately 0.101 millimeter (0.004 inches) by “bottoming” the sensing diaphragm against the back of the cell half. This feature provides overpressure protection for the cell. The sensing diaphragm differs in Rosemount’s transmitters, depending on the transmitter’s pressure range (range code). Generally, the higher the pressure range, the thicker the sensing diaphragm and the faster the transmitter’s response time.

The position of the sensing diaphragm is detected by capacitor plates on the two sides of the diaphragm. The capacitance between the diaphragm and either capacitor plate is about 150 picofarad (pF). The differential capacitance between the sensing diaphragm and the capacitor plate is converted electronically into a two-wire, 4-20 mA (or 10-50 mA) DC signal. The transmitter electronics include a diode bridge and a temperature compensation thermistor. The diode bridge rectifies the AC driving current into a DC output current.

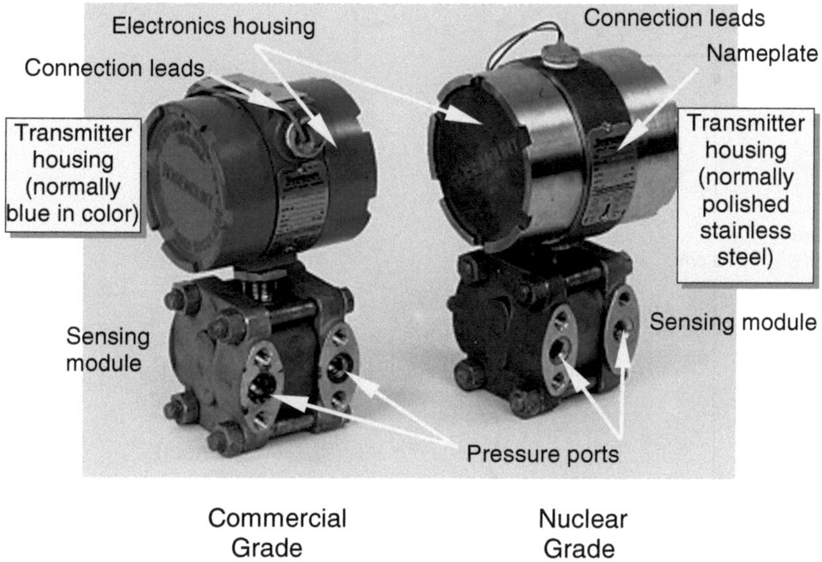

Fig. 7.17. Rosemount commercial and nuclear-grade transmitters

The pertinent characteristics of the four Rosemount transmitters, summarized in Table 7.5, are as follows:

1. Each model is available for absolute, differential, and gauge pressure measurements except Model 1154, which is available only for differential and gauge applications.
2. Model 1152 is qualified for nuclear safety-related applications but not for postaccident service. For postaccident service, Model 1153 or 1154 is used.
3. The electronic circuits of Models 1151 and 1152 feature a damping potentiometer to attenuate extraneous noise, if needed. The damping adjustment provides time-constant values between 0.2 to 2 seconds. The default setting is 0.2 seconds, which is set at the factory before the transmitter is shipped. According to Rosemount, since the transmitter calibration is not influenced by the time-constant setting, the user may set the damping adjustments at the site while the transmitter is installed in the process. The damping feature is available as an option on the 1153 and 1154 transmitters.
4. The Model 1153 and 1154 transmitters can be used as Class 1E equipment in nuclear power plants. In addition to having been qualified per IEEE 323 and 344 standards, Model 1153 has been qualified in environments typical of BWRs under accident conditions, and the same has been done with Model 1154 for PWRs.

Rosemount transmitters are available for a wide range of pressures, from a few millibars (a fraction of an inch of water) up to 200 or more bars (about 3,000 or more

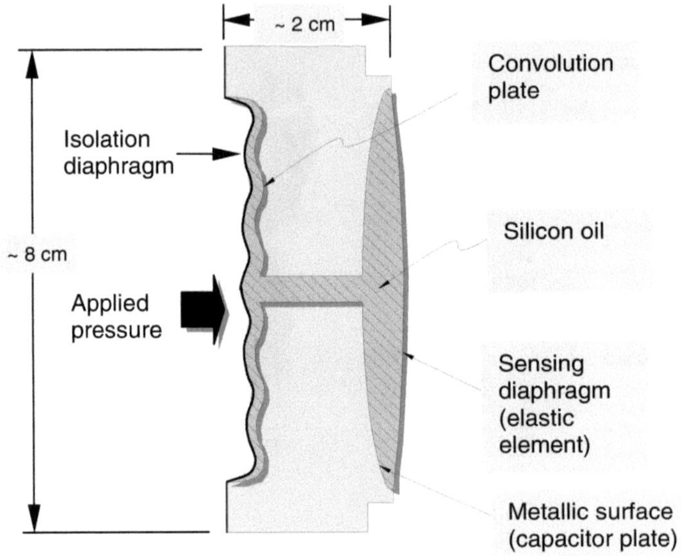

a) Cutaway of one-half of a Rosemount sensing cell (the other half is identical to this half)

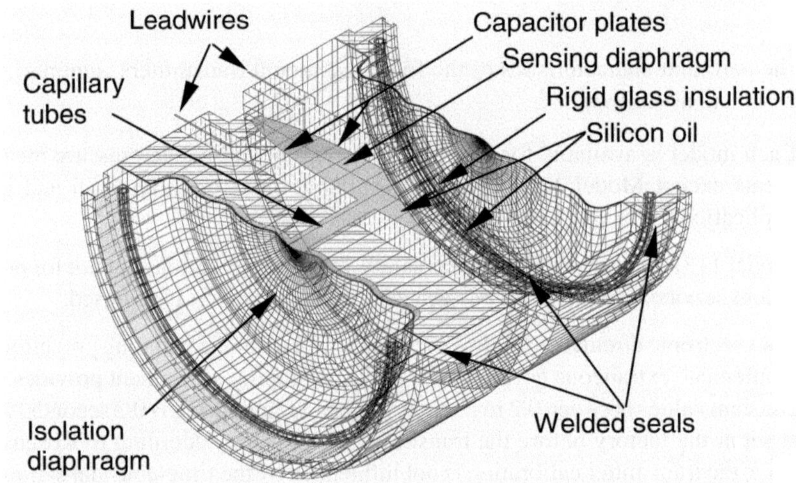

b) Complete sensing module of a Rosemount transmitter

Fig. 7.18. Diagram of sensing module of a Rosemount pressure transmitter

Table 7.5. Characteristics of Rosemount pressure transmitters

Transmitter Characteristics	Transmitter Model Number			
	1151	1152	1153	1154
Absolute	9	9	9	No
Differential	9	9	9	9
Gauge	9	9	9	9
Nuclear Qualified	No	9	9	9
Postaccident Qualified	No	No	9	9
Damping Adjustment	9	9	Option	Option
Accuracy (±% of span)	0.25 to 0.5	0.25	0.25	0.25
Drift (±% of upper range/ 6 months)	0.25 to 0.5	0.25	0.25	0.25

in psi). Each model comes in several range codes, depending on the transmitter's pressure range. An example, for Model 1153, is given in Table 7.6. In addition to the pressure range, the range code determines the transmitter's nominal response time. For Models 1151 and 1152, the response times are 0.2 to 2 seconds, depending on the range and damping. For Models 1153 and 1154, the response times are 2 seconds for range code 3, 0.5 seconds for range code 4, and 0.2 seconds for all other range codes.

For calibration, most Rosemount transmitters have a zero and a span adjustment that can be accessed from outside the transmitter. In addition, a linearity adjustment is available, located in the transmitter's electronics. The linearity adjustment is set at the factory and is not usually adjusted in the field.

It should be pointed out that all descriptions given in this chapter for the Rosemount and other transmitters are based on their conventional designs and historical performance specifications, which are representative of transmitters that are installed in the field and may not represent improvements or changes that might have been made in recent years. Furthermore, Rosemount and other manufacturers now produce smart transmitters and digital sensors for a variety of industrial applications including nuclear power plants. Both the conventional transmitters and smart transmitters from Rosemount and most other manufacturers have provided reliable service to the nuclear power industry, and their failure rates have been reasonably low in most cases.

7.4.4 Tobar Transmitters

Four models of Tobar transmitters have been used for safety-related measurements in nuclear power plants: Models 32DP1 and 32DP2, which are differential pressure

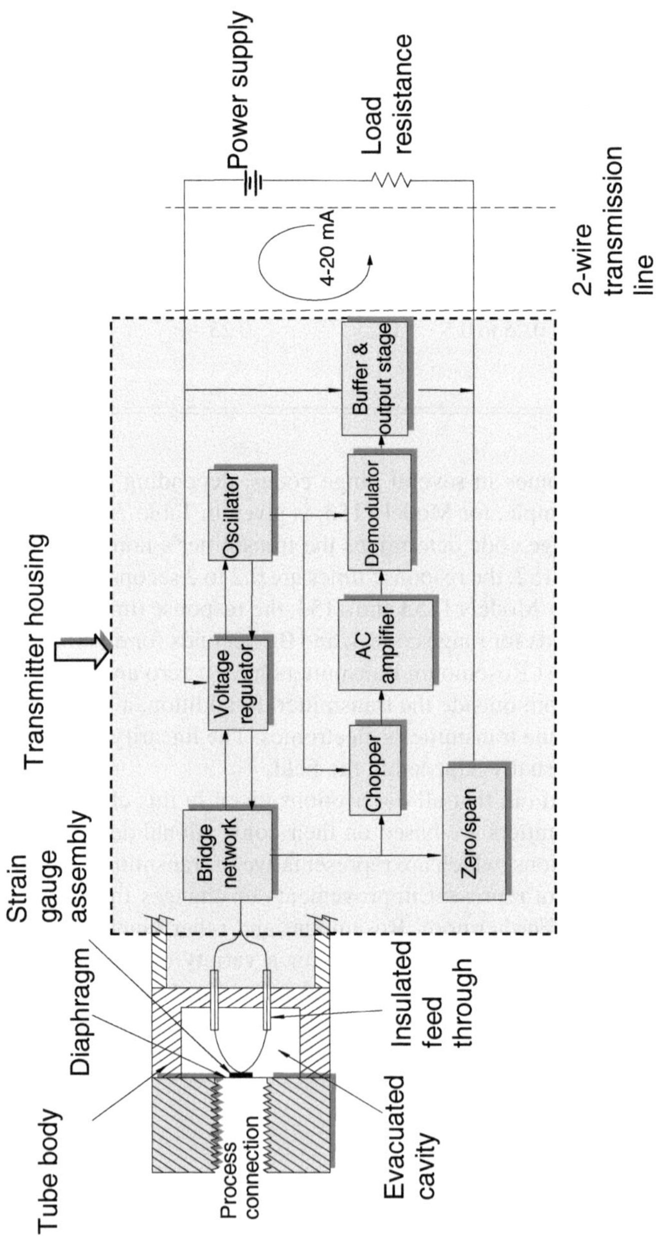

Fig. 7.19. Diagram of Tobar absolute pressure transmitter

Table 7.6. Typical specifications of a Rosemount 1153 transmitter Series B

Transmitter Model Numbers and Descriptions		
Model Number	**Description**	
1153AB	Absolute	
1153DB	Differential	
1153GB	Gauge	
1153HB	Differential, High Line Pressure	
Transmitter Range Codes and Corresponding Response Times		
Range Code	**Pressure Range**	**Response Time**
3	0 – 12 to 0 – 75 mBar	2 sec
4	0 – 62 to 0 – 374 mBar	0.5 sec
5	0 – 311 to 0 – 1868 mBar	0.2 sec
6	0 – 1.2 to 0 – 7 bar	0.2 sec
7	0 – 3.5 to 0 – 21 bar	0.2 sec
8	0 – 12 to 0 – 69 bar	0.2 sec
9	0 – 34 – 0 – 207 bari	0.2 sec
Transmitter Output Characteristics		
Model Number Suffix	**Output**	
P	4 – 20 mA, Standard	
R	4 – 20 mA, Improved Radiation Performance	

Table 7.7. Cross reference of Tobar and Veritrak model numbers

Item	Tobar Model Number	Equivalent Veritrak Model Number	Qualification Level
1	32DP1	76DP2	High-level Radiation
2	32DP2	76DP1	Low-level Radiation
3	32PA1	76PA2	High-level Radiation
4	32PA2	76PA1	Low-level Radiation

1."76" in Veritrak's model numbers corresponds to "32" in Tobar's model numbers; "2" in Veritrak's model numbers corresponds to "1" in Tobar's model numbers; and "1" in Veritrak's model numbers corresponds to "2" in Tobar's model numbers.
2. The Veritrak/Tobar transmitters are now supplied by Weed Instrument Company.

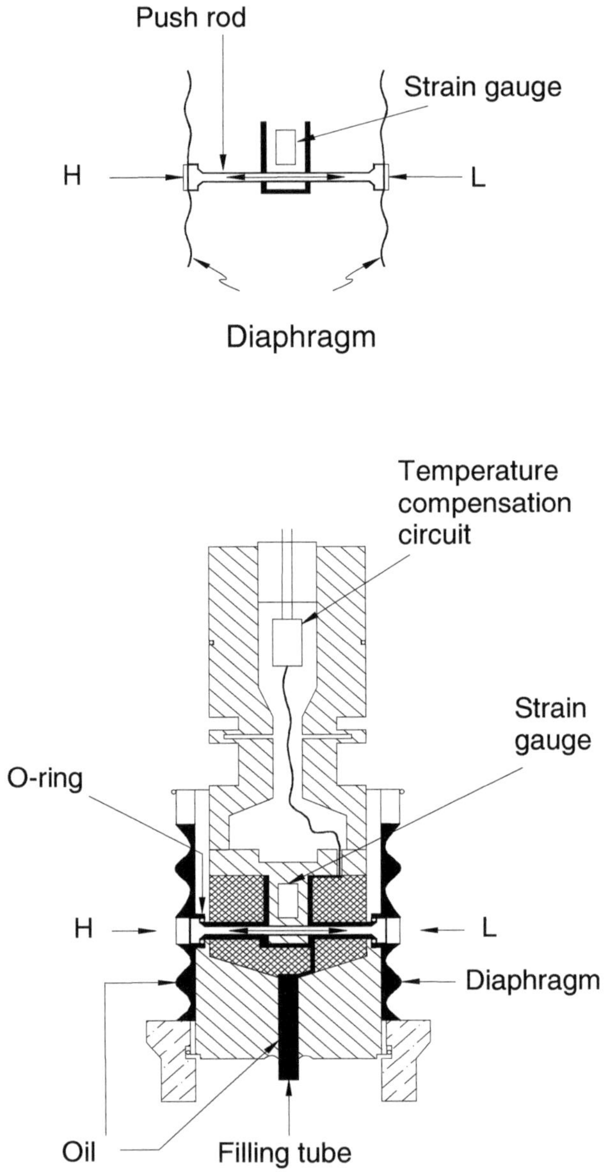

Fig. 7.20. Structure of sensing module of Tobar transmitters

transmitters, and Models 32PA1 and 32PA2, which are absolute pressure transmitters. Before Tobar began offering these models, these same transmitters were supplied by the Westinghouse Veritrak Company under a different set of model numbers, as shown in Table 7.7. The difference between the two Tobar differential pressure transmitters (32DP1 and 32DP2) lies in the level of nuclear qualification. The models that end

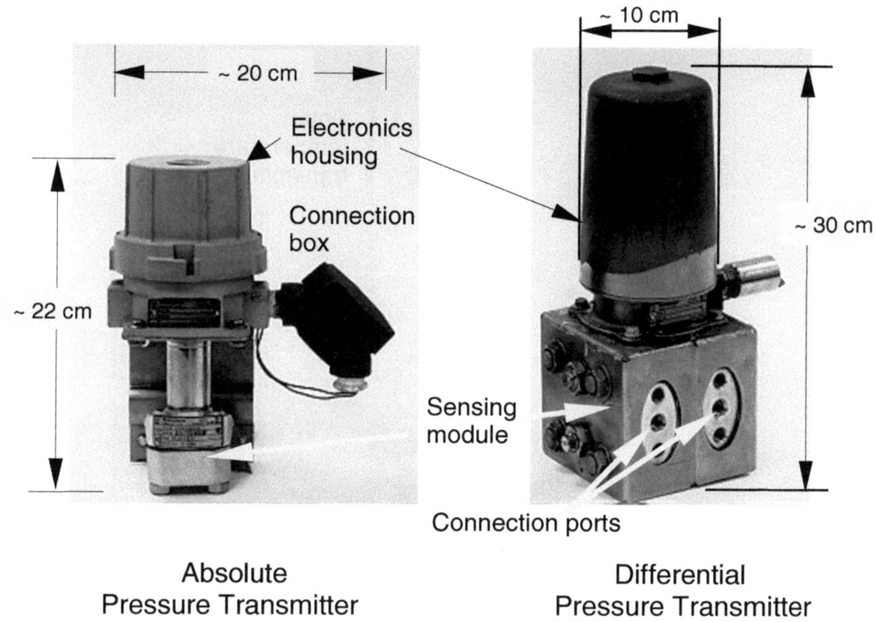

Fig. 7.21. Body styles of Tobar (Weed) transmitters

with a "1" are qualified for high-level radiation, and the models that end with a "2" are qualified for low-level radiation. The same is true for the two models of absolute pressure transmitters. In addition to the transmitters listed in Table 7.7, Tobar has absolute and differential transmitters that have the same type model numbers (but ending with a "5"), which are not qualified for nuclear safety-related service.

The sensing elements in the Tobar differential and absolute pressure transmitters are made of diaphragm capsules. In the absolute pressure transmitters (Fig. 7.19), the capsule assembly consists of a diaphragm on which a bridge network of strain gauges is deposited. The diaphragm is welded to the support and header assemblies, as shown in Fig. 7.20. The header assembly contains hermetic feedthrough leads to carry the electrical signal from the flexure to the amplifier.

The Veritrak/Tobar transmitters for nuclear power plants are currently supplied by the Weed Instrument Company under their model number DTN2010 for differential, absolute, and gauge pressure measurements. The DTN2010 has been qualified for nuclear service per IEEE Standards 323 and 344. The DTN2010 has new electronics that represent improvements over the original Veritrak/Tobar transmitters. The normal response time of the DTN2010 is electronically adjustable between 0.5 to 2.5 seconds.

In differential pressure transmitters, the process pressure lines are connected to the high- and low-pressure diaphragms. The space between the diaphragms is filled with a damping fluid (Fig. 7.20). The high- and low-pressure diaphragms are connected by pushrods to the capsule assembly sensor (strain-sensitive resistor array). The pushrods flex the capsule sensor up to about 0.101 mm (0.004 inches) to produce an electrical

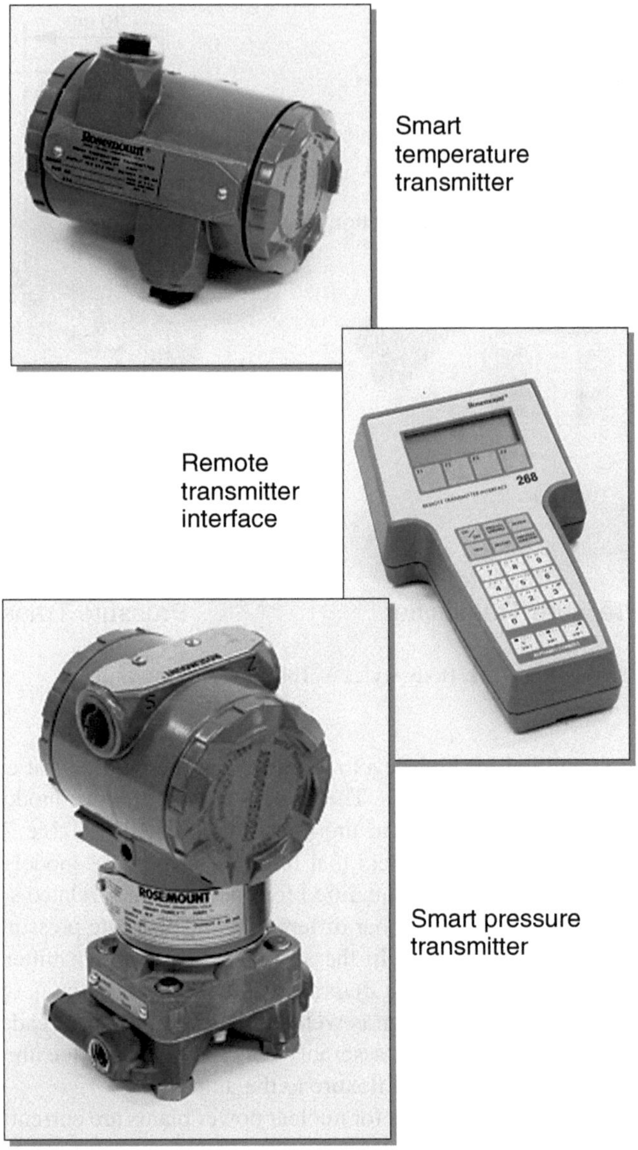

Fig. 7.22. Rosemount smart sensor modules

output. If an overpressure is applied, an O-ring within the diaphragm assembly seats against the capsule body. The fill fluid trapped between the diaphragm and capsule body will prevent further movement of the diaphragm. The capsule assembly consists of a flexure on which a strain gauge bridge network is deposited that detects the movement, as in the absolute pressure transmitter.

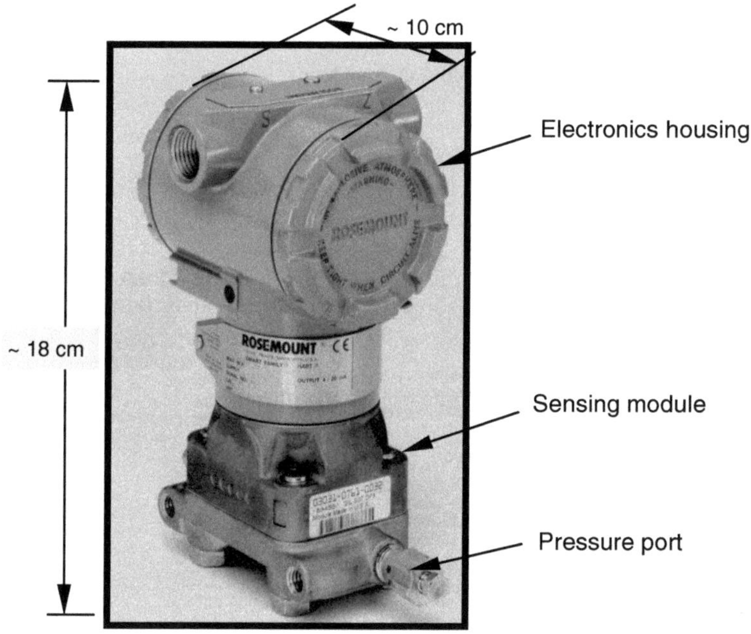

Fig. 7.23. Rosemount Model 3051N smart pressure transmitter for nuclear service

A photograph of two body styles of Tobar transmitters is shown in Fig. 7.21. It should be noted that nuclear power plants no longer use the Tobar transmitters under the name and model numbers mentioned in this book; Weed Instrument Company literature should be consulted for up-to-date information about these transmitters.

7.5 Smart Pressure Transmitters

Smart pressure transmitters found their way into the nuclear industry in the late 1990s. They are now used in a variety of services, including safety applications for which some smart sensors have been qualified by the manufacturer or the nuclear industry. Fig. 7.22 shows a photograph of physical configuration of smart temperature and pressure sensors by Rosemount.

Rosemount's smart pressure transmitters and other smart sensors are popular in nuclear power plants because of their ease of calibration (e.g., the sensor cap need not be removed), memory, ease of configuration, cost advantage compared to their conventional counterparts, and advanced features. As aging and obsolete transmitters are gradually replaced, the nuclear industry is depending more and more on smart transmitters. Fig. 7.23 shows a photograph of a smart pressure transmitter supplied by Rosemount (Model 3051N) for nuclear services. This transmitter is seismically qualified both for use in Class 1E safety-related applications per IEEE Standard 344 and for mild environments per IEEE 323.

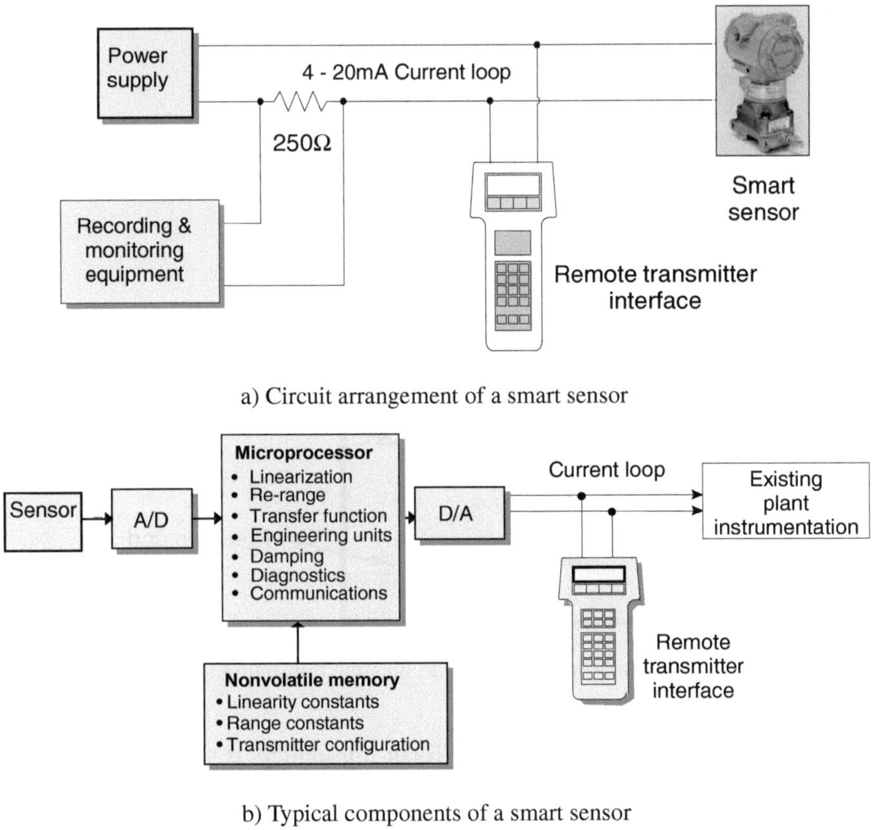

Fig. 7.24. Circuit arrangement and electronic components of a smart Rosemount sensor

Fig. 7.24 shows two block diagrams illustrating the circuit arrangement and the typical components of a smart sensor.

7.6 Fiber-Optic Pressure Transmitters

In addition to smart sensors, the nuclear power industry has strong interest in fiber-optic pressure transmitters. The industry has already taken advantage of fiber-optic cables and devices, especially in nonradiation environments within both the conventional and advanced reactors. As for radiation environments and in-containment use, fiber-optic sensors are still maturing, and no major transmitter manufacturer has yet shown an interest in developing or qualifying fiber-optic pressure transmitters for these applications. This is mainly because of technical concerns about the effect of radiation on fiber-optic components as well as business concerns, such as the small volume of the current nuclear market, the difficulty of introducing new products and new technology in nuclear power plants, and quality assurance issues.[19] However,

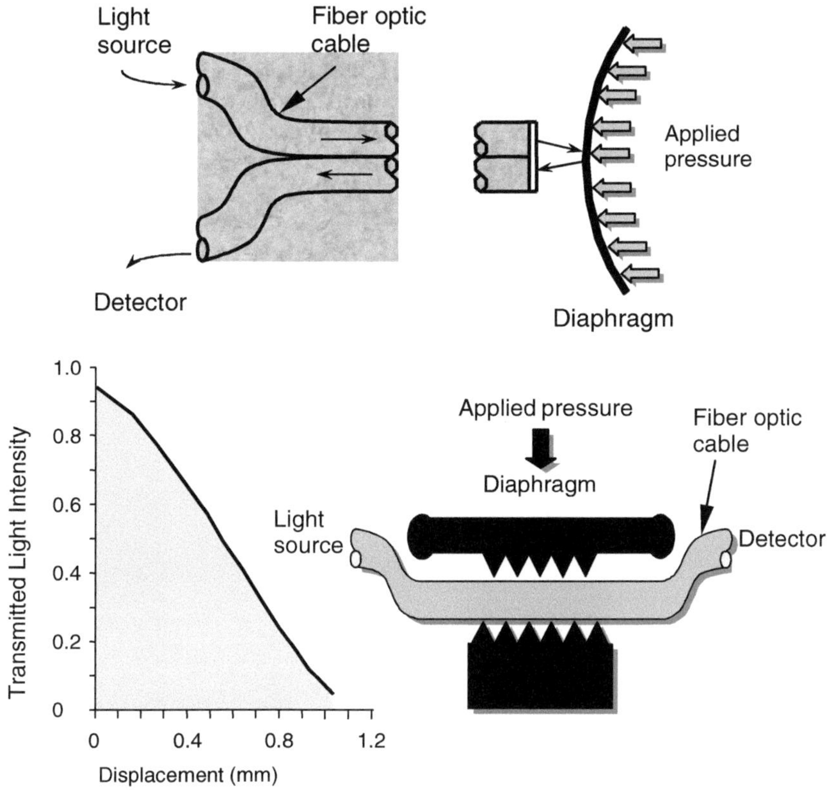

Fig. 7.25. Operation principle of simple fiber-optic pressure sensors

fiber-optic sensors are used successfully in other industries for their immunity to electromagnetic and radio frequency interference (EMI/RFI), ground loop immunity, small size, high sensitivity, and multiplexing capability. For example, the automotive industry uses fiber-optic sensors for their small size, the aerospace industry for their light weight and noise immunity, and the petrochemical industry for their immunity to explosion. The operation principle of simple fiber-optic pressure sensors is illustrated in Fig. 7.25. In addition to pressure measurement, fiber-optic sensors are available for measuring temperature, strain, vibration, and other parameters.

7.7 Wireless Pressure Transmitters

Wireless pressure transmitters are slowly finding their way into nuclear power plants, but not yet for any safety critical measurements in radiation areas or for in-containment use. A number of obstacles and challenges must be resolved before wireless sensors can be used for routine measurement of important process parameters in nuclear power plants. With time, however, wireless sensors are expected to play a major role in nuclear power plants as they provide substantial savings in wiring costs, and as importantly, facilitate data collection for remote diagnostics and on-line monitoring to verify the performance of the plant equipment and processes. It is expected that the next generation of advanced nuclear power plants, which will probably be deployed by the year 2020, and the so called Generation Four (Gen. IV) reactors to be deployed by the year 2030, will incorporate wireless sensing technologies in both their primary and secondary systems.

8

Characteristics of Pressure Sensing Lines

In nuclear power plants, pressure transmitters are usually located away from the process in order to reduce the effect of ambient temperature on the transmitter's operability and qualified life. High ambient temperatures (70EC and above) can affect the transmitter's mechanical components and shorten the life of its solid-state electronics. Other reasons for locating a transmitter away from the process are to reduce the adverse effects of radiation and vibration and to make it easier for personnel to access the transmitter for replacement or maintenance purposes.

To transport a pneumatic or hydraulic signal from the process to a transmitter, sensing lines are used to connect the pressure transmitter to the process piping, reactor vessel, or primary flow elements. Depending on the application, there will be one or two sensing lines for each transmitter.

Sensing lines are also referred to as *impulse lines* or *instrument lines*. Both liquid-filled and gasfilled sensing lines can be found in nuclear power plants. Liquid-sensing lines are typically filled with either the process liquid or oil, depending on the sensing line's design and application. Gas-sensing lines are filled with steam, air, nitrogen, or other gases, and there is sometimes a point in these lines where they transition to another medium such as oil or water. To achieve this transition, a diaphragm, bellows, or condensate pot is installed in the sensing line.

8.1 Design and Installation

The pressure sensing lines in the primary systems of nuclear power plants are usually made of stainless steel tubing that has a wall thickness typically of about 2.5 mm and diameter of about 10 to 15 mm. In the plants' secondary systems, however, most sensing lines are made of carbon steel and, in some rare cases, copper to prevent corrosion.

The sensing line length typically ranges from less than about 10 meters to over 200 meters, depending on the transmitter's location and the nature of its service in the plant. Tubing is preferred over piping in most applications because it may be installed in one piece, reducing the possibility of leaks. Since the sensing line's length affects the overall response time of a pressure sensing system, nuclear power plants try to make the sensing lines as short as possible. For that reason, the average length of sensing lines for safetyrelated pressure transmitters is usually about 35 meters or less.

Sensing line installations are designed to allow for thermal expansion and vibration without deformation, to ensure gravity-induced drainage, and to provide for selfventing. For liquidfilled sensing lines, selfventing is accomplished by sloping the sensing line downward to allow any gas or air in the line to vent to the process. The slope of sensing lines is typically about 30 mm per 30 cm. When this slope is not practical, the sensing line is sloped as much as possible, but usually not less than 3 mm per 30 cm. If the sensing line cannot be sloped as required, a high-point vent will be needed for liquid sensing lines and a low-point drain for gas-sensing lines.

The criteria for designing and installing sensing lines in nuclear power plants are presented in a number of industrial documents and standards such as the ISA Standard S67.02 entitled, "Nuclear SafetyRelated Instrument Sensing Line Piping and Tubing Standards for Use in Nuclear Power Plants." This standard shows several designs for the sensing lines used in a range of applications inside and outside reactor containment. Some examples of these designs are summarized in the following sections.

8.2 Sensing Lines for Transmitters Inside Containment

Fig. 8.1 shows the main components of a typical sensing line for a transmitter located inside the containment of a nuclear power plant. These components are as follows:

- **Root Valve** The root valve, the first valve in a pressure sensing line, is located at a point after the sensing line taps off the main process.
- **Condensate Pot** This is often installed in sensing lines for differential pressure transmitters that are used for level measurements. The condensate pot's purpose is to ensure that the reference leg leading to the transmitter is always filled with water as opposed to steam or air.
- **Restriction Device:** This device is installed as close to the process as possible to reduce the loss of process fluid if there is a break in the line downstream of the restriction. One problem with restriction devices is that they can increase the response time of the pressure sensing system. For that reason, in applications where a fast response is essential, it's best to avoid a restriction device.
- **Isolation Valve:** An isolation valve is always installed in pressure sensing lines in a location where maintenance personnel can readily access it; sometimes even during the plant's operation. This valve is usually installed in addition to the root valve. Depending on the pressure sensing system's location and nature of service, some root valves can perform the function of the isolation valve.
- **Equalizing Valve:** For differential pressure transmitters, an additional valve called an *equalizing valve* is used at the instrument manifold between the two sensing lines. Its purpose is to equalize the pressure to the transmitter as needed during calibration and maintenance activities.

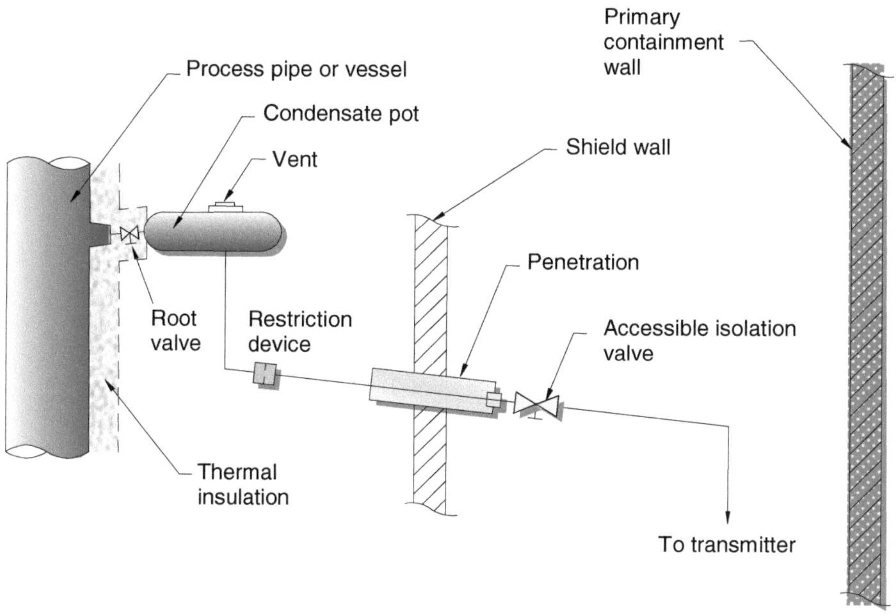

Fig. 8.1. Typical pressure sensing line for steam and water service inside a nuclear reactor containment

Fig. 8.2 shows another sensing-line design for water or steam service inside containment. In this sensing line, a condensate pot is not used. The diaphragm shown in the figure is an isolation diaphragm that is used to keep the process fluid from entering the transmitter. The line to the right of the isolation diaphragm in the figure is filled with a suitable fluid (as indicated by the x's on the line in Fig. 8.2). The isolation diaphragm is useful when the process fluid is corrosive, radioactive, or otherwise harmful to the transmitter or maintenance personnel.

8.3 Sensing Lines for Transmitters Outside Containment

Sensing lines that extend outside containment are typically designed like sensing lines for incontainment services, except that they are equipped with an additional device called a "selfactuating excess flow checkvalve." This device is used to automatically shut off the sensing line if the line ruptures downstream of the checkvalve (see Fig. 8.3).

An example of a sensing line that penetrates containment is one that leads to a containment pressure transmitter. These transmitters are often located outside containment, although they measure the ambient pressure inside containment. The containment pressure sensing lines may be installed in any one of the three ways shown in Fig. 8.4. Note that when the sensing line contains air, it is sloped upward, as shown

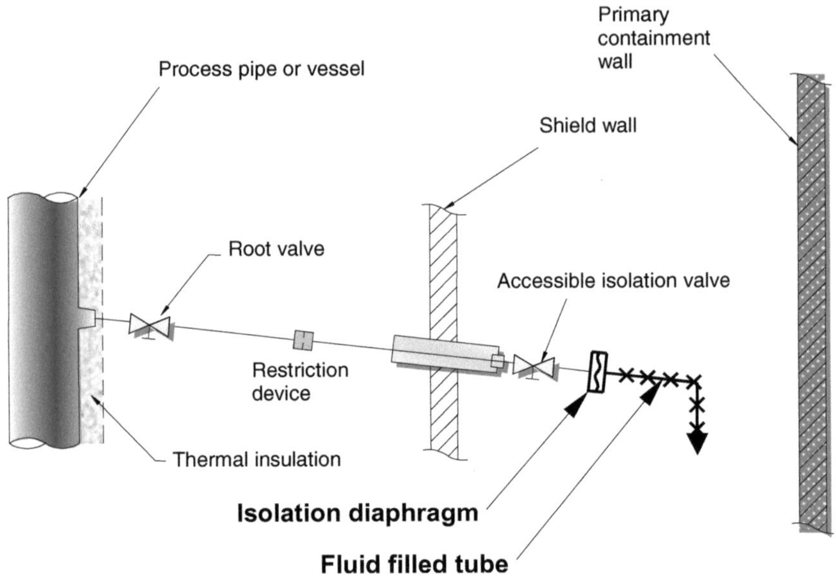

Fig. 8.2. Typical pressure sensing line with a provision to isolate the transmitter from the process fluid

in the topmost of the three drawings in Fig. 8.4. In most containment pressure transmitters, the sensing line is filled with oil running from the isolation diaphragm to the transmitter.

8.4 Sensing-Line Problems

Sensing lines can encounter a number of problems that can affect the pressure sensing system's accuracy and response time. We will discuss these problems in this section.

8.4.1 Blockages, Voids, and Leaks

Some examples of the sensing-line problems that have occurred in nuclear power plants are:

- Blockages due to sludge, boron, or deposits
- Air or gas entrapped in low-pressure sensing lines
- Frozen sensing lines (e.g., due to problems with insulation material or heat trace on the lines)
- Improper line-up or seating of isolation and equalizing valves
- Leakage in sensing lines (e.g., due to valve problems)

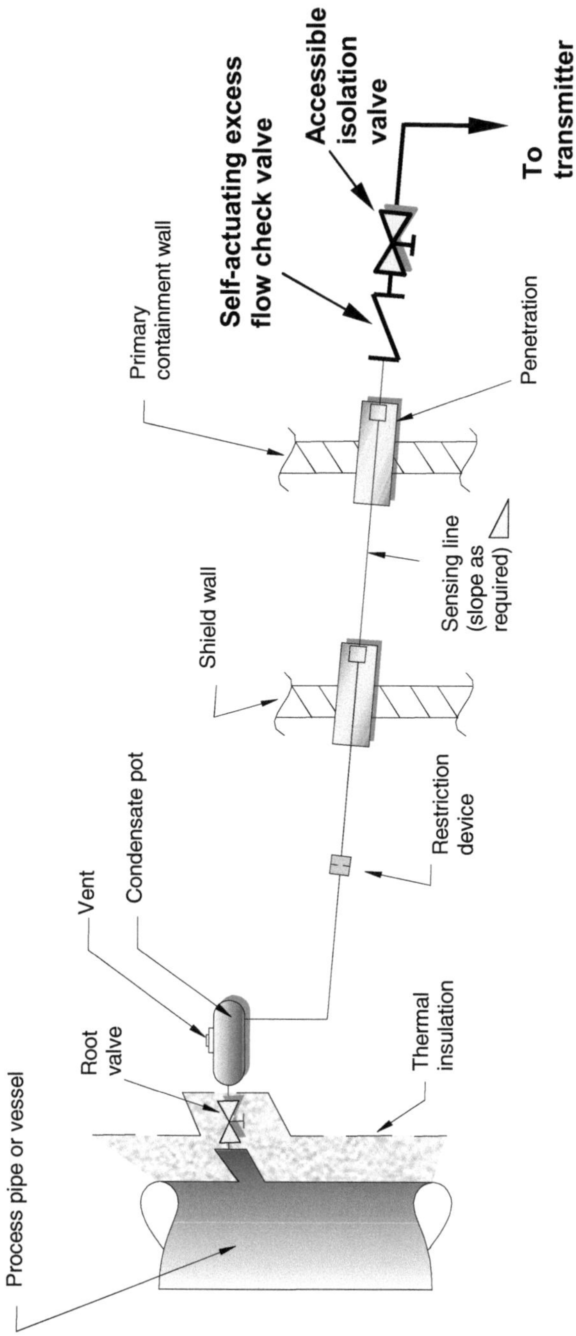

Fig. 8.3. Sensing line for water and steam service outside containment

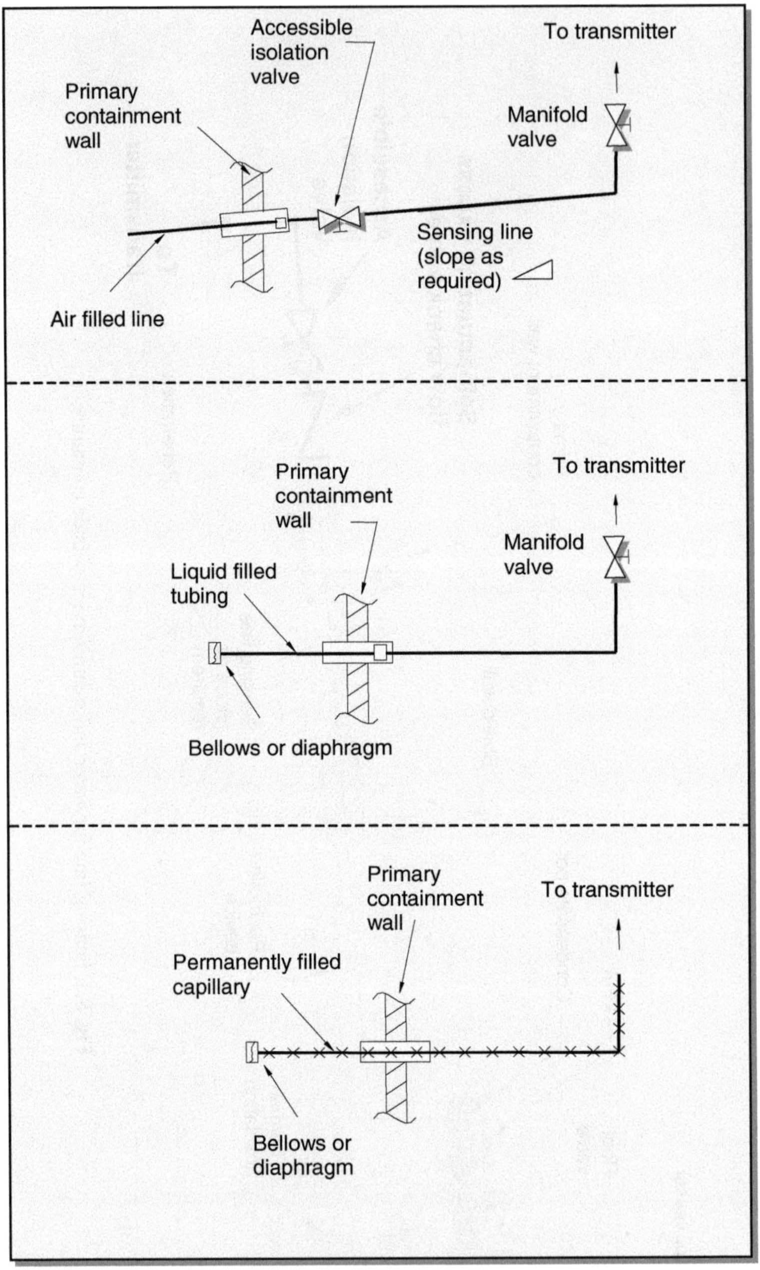

Fig. 8.4. Typical sensing-line installations for containment pressure transmitters

Any combination of these five problems can increase the pressure sensing system's response time or cause other problems. For example, the presence of air in the sensing line can cause not only increases in response time but also can cause resonances that produce pressure variations and false pressure indications. Although air may dissolve in the fluid at high pressures, there have been many cases where entrapped air has remained undissolved or has come out of solutions as soon as the plant depressurized below a threshold pressure. Aside from causing transient response problems, air in the sensing line can affect the accuracy of pressure indications.

Tables 8.1 and 8.2 present examples of LER and NPRDS reports on sensing-line problems experienced by the nuclear power industry. (The language used to describe the problem in Tables 8.1 and 8.2 comes from the LER and NPRDS summaries but has been edited by the author to reduce the wording. Also note that the NPRDS results are given without the name of the corresponding plant because the author was not authorized to release them.)

In Chap. 9, we describe how the noise analysis technique can be used during plant operation by way of on-line testing to detect blockages, voids, and leaks in nuclear plant pressure sensing systems.

8.4.2 BWR Level Measurement

It's important to accurately measure the water level in the pressure vessel of BWRs because reactor vessel water-level signals are used to actuate automatic safety systems and for guidance to operators during and after an accident. When the plant is undergoing a rapid depressurization, a problem can arise that can interfere with accurate measurement of the reactor water level in BWRs. This problem occurs because non-condensable gases may become dissolved in the reference leg of the sensing lines leading to level transmitters. These dissolved gases may reappear during rapid depressurization below about 30 bars, causing inaccurate level measurements. More specifically, the dissolved gases, which accumulate over time during normal operation, can rapidly come out of solution and displace water from the reference leg. This reduces the reference leg level and results in an erroneously high level indication. Fortunately, safety system actuation in BWRs occurs at higher pressures than 30 bars, at which pressure the dissolved gases are expelled. For this reason, the reactor is normally shut down before it reaches the pressure at which this level measurement problem begins. Nevertheless, the problem is important because operators use reactor water-level information during shutdown and other times while the pressure is below 30 bars.

8.4.3 Shared Sensing Lines

Redundant transmitters in nuclear power plants sometimes share a sensing line. The problem with shared sensing lines is that they can produce a common mode failure if there is a leak, blockage, or void in the common leg. Also, when a valve fails on the sensing line it can affect all the transmitters that share the line.

Table 8.1. Sample results of search of LER database on sensing-line problems in nuclear power plants

Nuclear Power Station	Problem	Cause	Resolution
	Sensing-line Blockages		
Indian Point	SG level transmitter tracking sluggishly	Partial blockage of impulse line due to sludge buildup	Line blown free
	SG level transmitter drifting high	Blockage in the impulse lines	Lines blown out
Browns Ferry	Several differential pressure transmitters failed downscale (non-conservative) on five separate occasions	Pulse-damping devices (snubbers) installed in sensing lines	Snubbers removed
Salem	Loss of indication of boric acid tank level	Sensing lines plugged with boric acid	Both sensing lines cleaned and blown down with pressurized nitrogen gas
H.B. Robinson	Reactor trip due to SG low-level indication	Partial blockage of reference leg	Blown down sensing lines
Ginna	Boric acid storage tank level reading lower than allowed by technical specifications	Level indication inaccuracy due to partial plugging of the sensing lines	Not stated
Farley	SG flow transmitter inoperable due to erroneously high indication	Obstruction in low-pressure sensing line	Sensing line cleared
Prairie Island	Plant tripped during startup on high steam generator level	Sluggish response of SG level transmitters; significant amount of magnetite in the line	Blowdown of variable and reference leg of SG level transmitters

Table 8.1. (continued)

Nuclear Power Station	Problem	Cause	Resolution
	Air or Void in Sensing Line		
North Anna	During startup, SG narrow-range level channel reading 10% below redundant channels and drifting low	Air pocket in SG level transmitter on low side of sensing line	Line blown down
Zion	Pressurizer level channel reading 6% lower than redundant channels due to a zero shift; six previous LERs with similar problems	Air pockets trapped in transmitter sensing lines	Reslope sensing lines
Brunswick	Reactor water cleanup system (RWCS) flow transmitter used for RWCS leak detection showing erroneous indication (leak Hi-Hi alarm)	Entrapped air in sensing line from procedural inadequacy in the RWCS high- flow response time test	Entrapped air removed and test procedure corrected
	Sensing-line Freezing		
Point Beach	SG pressure transmitter indicating higher than other channels; two previous LERs with the same problem	Frozen sensing lines due to incomplete insulation and cold weather	Improved insulation and turned on spare heat tracer
Sequoyah	SG pressure transmitter declared inoperable due to freezing sensing lines (also happened on refueling water storage tank-level transmitter and feed-water flow transmitter, causing high readings)	Sensing line freezing	Lines were defrosted and additional insulation and heaters used
	Leaking		
Arkansas Nuclear One	Upper-seal cavity pressure transmitter on reactor coolant pump low; similar LERs on three previous occasions	Leak in upper-seal pressure sensing line due to weld crack from vibration	Replaced faulty weld; resolved vibration-induced cracking in seal of sensing lines
Browns Ferry	A safety-related nitrogen level transmitter indicating greater than 100%	Leak in the sensing line resulting in false high reading	Sensing line repaired

Table 8.2. Sample results of search of NPRDS database on sensing-line problems in nuclear power plants

Problem	Causes	Resolution
	Blockages	
Level transmitter not working properly	Low-pressure tap found clogged with foreign material	Cleaned out (back flush) sensing lines
Control channel for service-water recirculation to discharge control valve indicates 2 bars instead of 8 bars	Sensing lines clogged with sediment	Sensing line blown out
SG pressure indicator remained pegged low during plant heatup	Sensing lines clogged	Cleared sensing lines
High-pressure safety injection loop transmitter very sluggish	Bourdon tube clogged with boric acid	Bourdon tube replaced
	Air or Voids in Sensing Lines	
Pressure transmitter failed high	Air in sensing line	Transmitter vented
Safety injection pump discharge pressure reading low	Air trapped in sensing line	Vented and transmitter calibrated
SG level indicator lower than other level channels	Air in transmitter reference loop due to improper setup	Filled pressure loop with water
Recirculation jet pump transmitter reading higher than other jet pump indicators	Excessive air in bellows assembly	Vented and calibrated

Aside from common mode problems, the dynamic response of redundant pressure transmitters that share a sensing line may be dominated by the response time of the most compliant transmitter on the common leg. The most compliant transmitter is normally the slowest-responding one. This could, therefore, cause all transmitters on the common sensing line to be almost as slow as the most compliant (i.e., the slowest) transmitter.

8.4.4 Use of Snubbers

Snubbers (also called *pulsation dampers*) are sometimes used in pressure sensing lines to reduce the effect of noise. The sources of such noise can be everything from process fluctuations, sensing-line vibration, acoustic resonances, and steam-line resonances to control system malfunctions and resonances caused by undissolved air pockets in liquidfilled sensing lines.

Snubbers reduce the effect of noise by increasing the dynamic response time of the pressure sensing system. They must therefore be used cautiously, especially in those cases where response time is important. An NRC information notice (included in Appendix D) describes an event in which a nuclear power plant had to be shut down because the use of snubbers on sensing lines produced unacceptable response times in the pressure sensing systems.

An alternative to using snubbers is electronic lowpass filters. These filters can provide any level of noise reduction, but they increase the response time of the system in the same way that snubbers do. One advantage of electronic filters is that they remove not only any mechanical or acoustic noise but also any electrical noise in the system. Another advantage is that they can be designed to have a precise rolloff frequency (i.e., with a known response time). The disadvantage of electronic filters is that, unlike snubbers, they do not protect the pressure transmitter's sensing element from mechanical fatigue resulting from vibration.

Several manufacturers provide pressure transmitters that have a built-in filter to dampen the noise. The damping adjustment in these transmitters must be used carefully to ensure that the dynamics of the pressure sensing system are not compromised.

8.5 Sensing-line Dynamics

A pressure sensing system may be represented by a spring-mass system (see Fig. 8.5). As the process pressure increases, the pressure surge is transmitted through the sensing line, producing a volume change (.ΔV_t.) in the transmitter cavity. For a pressure change of $\Delta.P_S$., the transmitter compliance (.C_t.) may be written as:

$$C_t = \frac{\Delta V_t}{\Delta P_S} \tag{8.1}$$

The dynamic behavior of the springmass system may be represented by a linear second-order model. In this case, the undamped natural frequency of the system (ω_n) may be written as:[18]

$$\omega_n = \frac{\pi U_a}{2L}\sqrt{\frac{V_{FS}}{\frac{\pi^2}{4}\left[BC_t + B\left(\frac{V_b}{\gamma P_b}\right) + V_t\right] + V_{FS}}} \tag{8.2}$$

where:

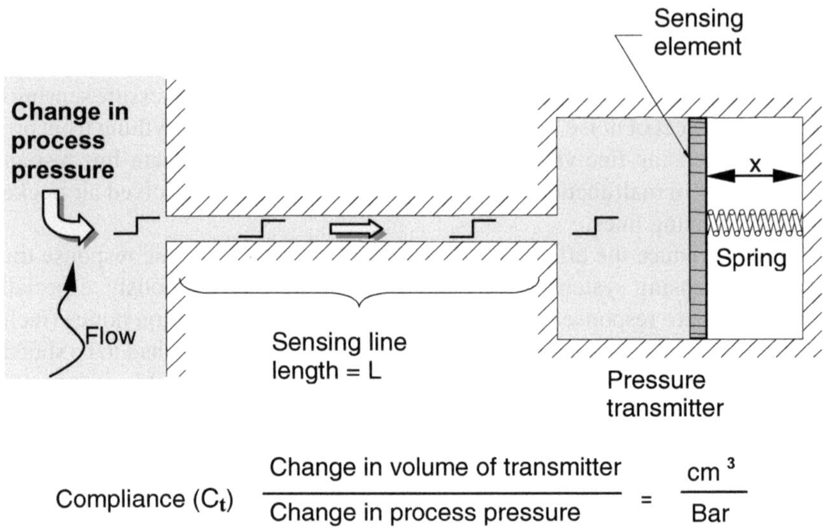

Fig. 8.5. Simplified model of a pressure sensing system and definition of compliance

U_a = acoustic velocity of fluid in the sensing line
L = length of the sensing line
V_{FS} = volume of fluid in the sensing line
V_t = volume of fluid in the transmitter
B = bulk modulus of the fluid
V_b = volume of any gas bubble present in the sensing line
γ = ratio of specific heat capacities of gas bubble at constant pressure and constant volume (C_p/C_v)
P_b = pressure applied to gas bubble

The damping ratio (ζ) of the system may be written as:

$$\zeta = \frac{16\nu}{\omega_n d_S^2} \tag{8.3}$$

where:
ν = kinematic viscosity of the fluid
d_S = inside diameter of the sensing line

Using the results from Equations 8.2 and 8.3, we can write the dynamic response, x(t), of the system for two input signals that are of typical interest in nuclear power plants: a step input and a ramp input. For a step input, the dynamic response is:

$$x(t) = K\left[1 - \frac{\omega_n}{\omega_d} e^{-\alpha t} \sin\left(\omega_d t + \arctan(\frac{\omega_d}{\alpha})\right)\right] \tag{8.4}$$

where:

K = system gain
ω_d = damped natural frequency ($\omega_n \sqrt{1-\zeta^2}$)
α = damping coefficient ($\omega_n \zeta$)
t = time in seconds

For a ramp input, the dynamic response is:

$$x(t) = Kr\left[t - \frac{2\alpha}{\omega_n^2} + \frac{1}{\omega_d} e^{-\alpha t} - \sin\left(\omega_d t + 2\arctan(\frac{\omega_d}{\alpha})\right)\right] \qquad (8.5)$$

where r is the ramp rate of the input signal.

Equations 8.4 and 8.5 represent the underdamped responses of the pressure sensing system, neglecting the dynamic response of such components as the transmitter's electronics, any mechanical linkages beyond the sensing element, and so on. As a result, the estimated response times these models generate will only be useful in demonstrating how sensing line length, blockages, and voids can affect the system response times in theory. In actual practice, the pressure sensing system's response time can only be identified experimentally by laboratory measurements or through field testing, as described in Chap. 9.

8.5.1 Effect of Length on Response Time

The pressure sensing system's response time increases as the length of the sensing line is increased. This is evident in the results given in Table 8.3 for three representative pressure transmitters of the types used in nuclear power plants. These results were calculated based on the theoretical models described in the previous section using compliance values obtained from the manufacturer's literature for each transmitter.

The response-time results in Table 8.3 correspond to onethird of the time it takes for the step response of the underdamped model to reach the first peak in its response to a step input (Fig. 8.6). These results are compared in Table 8.4 with corresponding values from laboratory measurements. The good agreement between the theoretical and experimental results testifies to the validity of the theoretical models and the equations used here to calculate the response-time values.

8.5.2 Effect of Blockages on Response Time

The effect of sensing-line blockages on the response times of representative pressure transmitters is shown in Table 8.5 and Fig. 8.7 for various sensing-line diameters. These results demonstrate that a pressure sensing system's dynamic response can be increased by simulating a sensing-line blockage by reducing the sensing line's diameter. These results were obtained theoretically based on the assumption that the blockage is rigid and extends the whole length of the sensing line. In reality, however, blockages are usually caused by a local obstruction and do not normally extend the

Table 8.3. Theoretical estimates of response time of pressure sensing lines as a function of sensing-line length and transmitter type

Sensing Line Length (meters)	Response Time (seconds)		
	Barton	Foxboro	Rosemount
	Sensing-line Inside Diameter = 6.35 mm		
15	0.22	0.03	0.11
30	0.31	0.04	0.15
60	0.44	0.06	0.22
90	0.54	0.07	0.27
120	0.63	0.09	0.31
150	0.71	0.11	0.35
	Sensing-line Inside Diameter = 9.53 mm		
15	0.14	0.02	0.07
30	0.20	0.03	0.10
60	0.29	0.04	0.15
90	0.35	0.06	0.18
120	0.41	0.07	0.21
150	0.46	0.09	0.24

The three transmitters are Barton Model 764, Foxboro (now Weed) Model E13DM, and Rosemount Model 1153 Range Code 3.

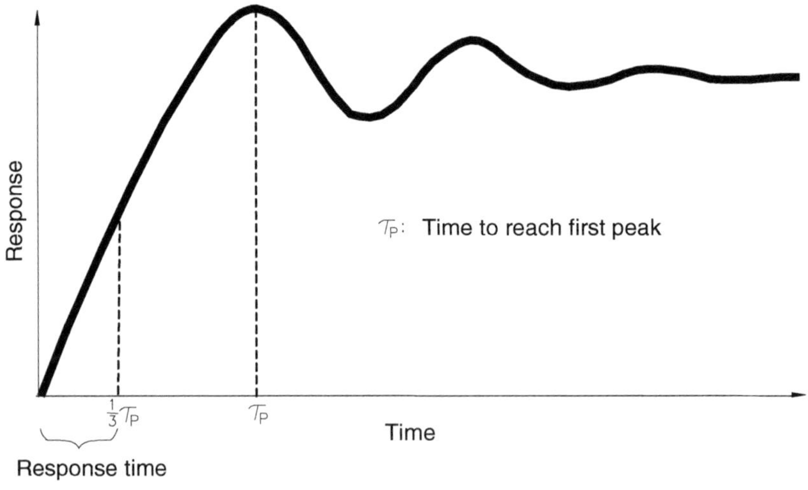

Fig. 8.6. Output of an underdamped system to a step input and calculation of system-response time

Table 8.4. Comparison of theoretical estimates and measured values of response times of pressure sensing lines as a function of sensing-line length and transmitter type

Sensing-line Length (meters)	Response Time (sec) Theoretical	Experimental
	Barton	
30	0.15	0.07
60	0.22	0.15
120	0.31	0.29
	Foxboro	
30	0.02	0.02
60	0.04	0.05
120	0.07	0.10
	Rosemount	
30	0.02	0.02
60	0.03	0.02
120	0.06	0.06

1. These results are for a sensing line with an inside diameter of 12.7 mm.
2. The experimental results in this table were obtained by performing the following two series of measurements and subtracting the results: 1) laboratory measurement of response times of a Barton, a Foxboro (Weed), and a Rosemount (Range Code 7) transmitter with sensing line lengths from 30 to 120 meters; and 2) laboratory measurement of response times of the same three transmitters with short (negligible) sensing line lengths.

line's entire length. For that reasons, the effects of actual blockages may be different than those simulated here.

Fig. 8.8 shows experimental results from the effect of sensing line blockages on the response time of pressure transmitters. This data confirms the theoretical estimates shown in Table 8.5 and Fig. 8.7.

8.5.3 Effect of Void on Response Time

A void in a pressure sensing line can affect both the accuracy and response time of a pressure sensing system. Table 8.6 shows theoretical response times of representative pressure transmitters as a function of void in the sensing lines. The results are shown for two pressures: 0.25 bars and 15 bars. Note that the effect of void on response time diminishes significantly with pressure.

The results given in Table 8.6 correspond to the following three cases: no air bubble, an air bubble with a length of 15 cm, and an air bubble with a length of 150 cm. The length of the sensing line producing these results is 7 meters, with an inside diameter of 9.5 mm.

Table 8.5. Theoretical effects of diameter (simulating blockage) on the response time of representative nuclear plant pressure transmitters at the end of a 15-meter sensing line

Sensing Line's Inside Diameter (cm)	Response Time (seconds)		
	Barton	Foxboro	Rosemount
16	0.086	0.012	0.044
13	0.108	0.014	0.054
10	0.143	0.018	0.072
5	0.216	0.026	0.108
3	0.637	0.050	0.232

The three transmitters are Barton Model 764, Foxboro (Weed) Model 13DM, and Rosemount Model 1153 Range Code 3.

The results in Table 8.6 were obtained using Equation 8.2. The value of V_b in this equation was changed as necessary to simulate the effect of the void on the response time of pressure transmitters with different compliances.

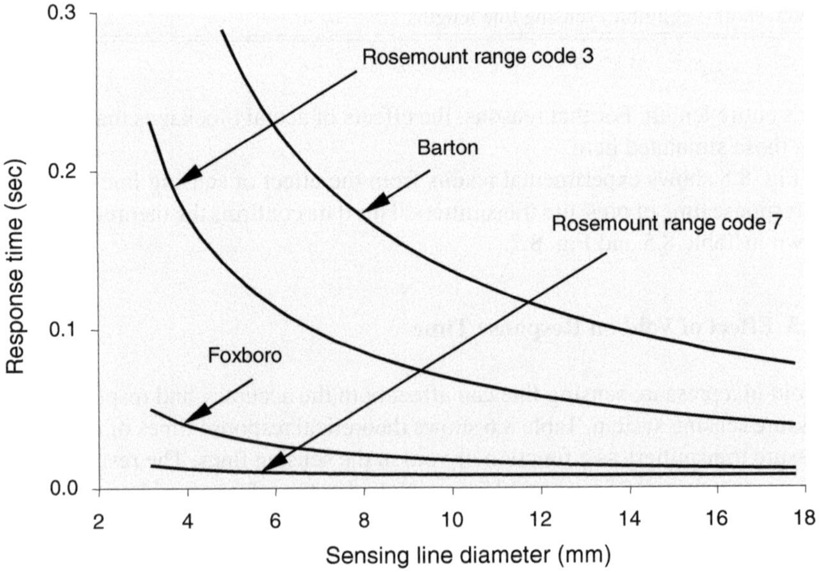

Fig. 8.7. Theoretical response time of representative pressure transmitters as a function of sensing line's inside diameter

8.6 Summary

In an operating nuclear plant, pressure transmitters are typically located away from the process to minimize temperature, vibration, radiation, and other effects and to make it easier for plant personnel to access the transmitter. Sensing lines, also referred to as *impulse lines* or *instrument lines*, are used to connect the process medium to a pressure transmitter. Typically, there are two sensing lines per pressure transmitter. At normal operation, there is no flow through the sensing lines.

Sensing lines are usually made of small-diameter, thick-wall, stainless steel tubing. They are designed to allow for thermal expansion and vibration without deformation, to ensure gravity-induced drainage, and to provide for self-venting. For fluid-filled sensing lines, self-venting is accomplished by sloping the sensing line downward so any gas or air in the line vents to the process.

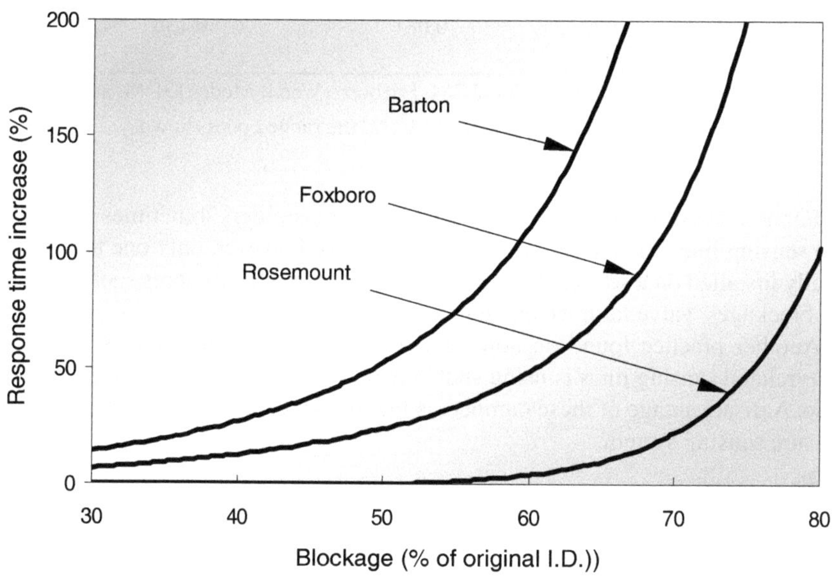

Fig. 8.8. Laboratory measurement results demonstrating the effect of sensing-line blockages on response time of representative pressure transmitters

Depending on the plant's physical layout, sensing lines range from about 10 meters to over 200 meters in length, with an average length of about 35 meters. The sensing lines' length is usually kept to a minimum to optimize response time. Sensing lines that are free of obstructions or voids do not noticeably delay the system's overall response time. However, numerous cases of blockages and voids in sensing lines that can cause significant dynamic delays have been reported in nuclear power plants. Some examples were reviewed in this chapter.

Table 8.6. Theoretical effect of sensing-line void on response time of representative nuclear plant pressure transmitters

	Response Time (sec)		
Manufacturer	**No Void**	**15 cm Void**	**150 cm Void**
		Pressure = 0.25 Bar	
Barton	0.143	0.307	0.880
Foxboro	0.018	0.272	0.868
Rosemount	0.008	0.271	0.868
RC 7	0.072	0.281	0.871
RC 3			
		Pressure = 15 Bar	
Barton	0.143	0.148	0.184
Foxboro	0.018	0.040	0.116
Rosemount	0.008	0.037	0.115
RC 7	0.072	0.081	0.136
RC 3			

The three transmitters are Barton Model 764, Foxboro (Weed) Model 13DM, and two Rosemount transmitters, both Model 1153, and the range codes shown.

For non-safety-related applications, multiple transmitters sometimes share a common sensing line. For safety system measurements, however, only one transmitter is usually installed on a sensing line to avoid common mode problems such as sensing-line blockages, valve failures, and the like.

Another practice found on non-safety-related sensing lines that is not used on safetyrelated sensing lines is using snubbers or pulsation dampers to reduce process noise. A disadvantage of these dampers is that they increase the response time of the pressure sensing system.

9

Measurement of Pressure Sensor and Sensing-Line Dynamics

Dynamic response measurements are made in nuclear power plants for one or more of at least four reasons:

1. comply with a plant's technical specifications and/or regulatory requirements for response time testing;
2. troubleshooting to identify sensor or sensing line problems including blockages, voids, and leaks;
3. manage component aging, estimate residual life, and assess reliability of pressure sensing systems; or
4. establish objective sensor replacement schedules.

The dynamic response of nuclear power plant pressure sensors and their associated sensing lines is measured using the noise analysis technique as described in this chapter. This technique can also be used to evaluate the dynamic characteristics of RTDs and thermocouples.

9.1 Noise Analysis Technique: Description

The noise analysis technique is based on analyzing the natural fluctuations that exist at the output of pressure transmitters while the plant is operating. These fluctuations (noise) are caused by turbulence that is induced by the flow of water in the system, by vibration, and by other naturally occurring phenomena.

The noise analysis technique provides a passive method for the dynamic testing of pressure sensing systems. It yields the response time for both a pressure transmitter and its sensing lines in the same test. The tests can be performed remotely while the plant is operating, do not require transmitters to be removed from service, do not interfere with plant operation, and can be performed on several transmitters simultaneously. The test involves three steps – data acquisition, data qualification, and data analysis – which are described in the following paragraphs.

9.1.1 Data Acquisition

A pressure transmitter's normal output is a DC signal on which the process noise (AC signal) is superimposed. The noise is extracted from the transmitter output by removing the signal's DC component and amplifying the AC component. This is accomplished simply by using commercial signal-conditioning equipment including amplifiers, filters, and other components. The AC signal is then digitized using a high sampling rate (e.g., 1 or 2 kHz) and stored for subsequent analysis. The analysis may be performed in real time as the data is collected or off line by retrieving the data from storage.

Fig. 9.1 shows a 50-second record of noise data from a pressure transmitter in a nuclear power plant. For each transmitter (or each group of transmitters), about an hour of such noise data is typically recorded for use in the analysis.

9.1.2 Data Qualification

The raw data must first be thoroughly scanned and screened before any analysis can resume. This is normally accomplished using data qualification algorithms embedded in software, which check for the stationary and linearity of the raw data and look for other abnormalities. For example, the raw data's amplitude probability density (APD) is plotted, as shown in Fig. 9.2, and examined for skewness. A skewed APD (lower plot of Fig. 9.2) could be caused by any number of anomalies in the data, including the non-linearity of the sensor from which the data is retrieved.

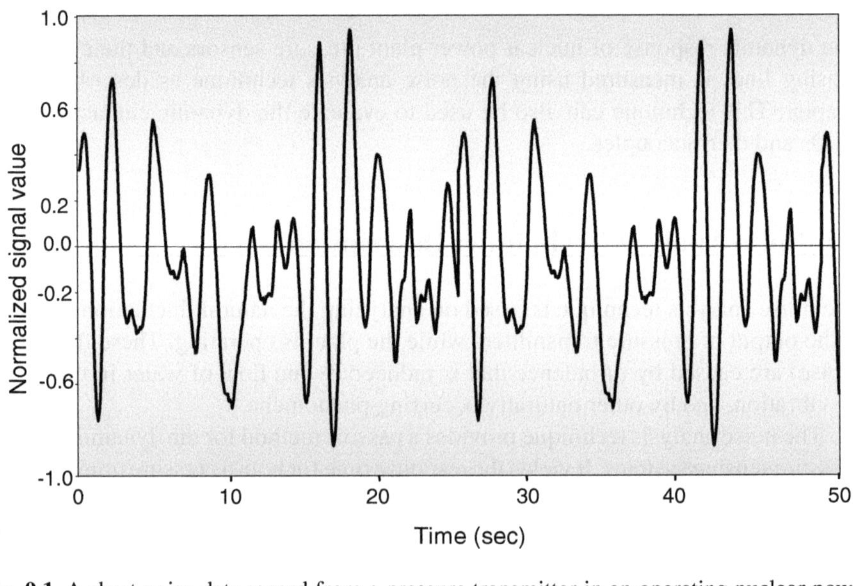

Fig. 9.1. A short noise data record from a pressure transmitter in an operating nuclear power plant

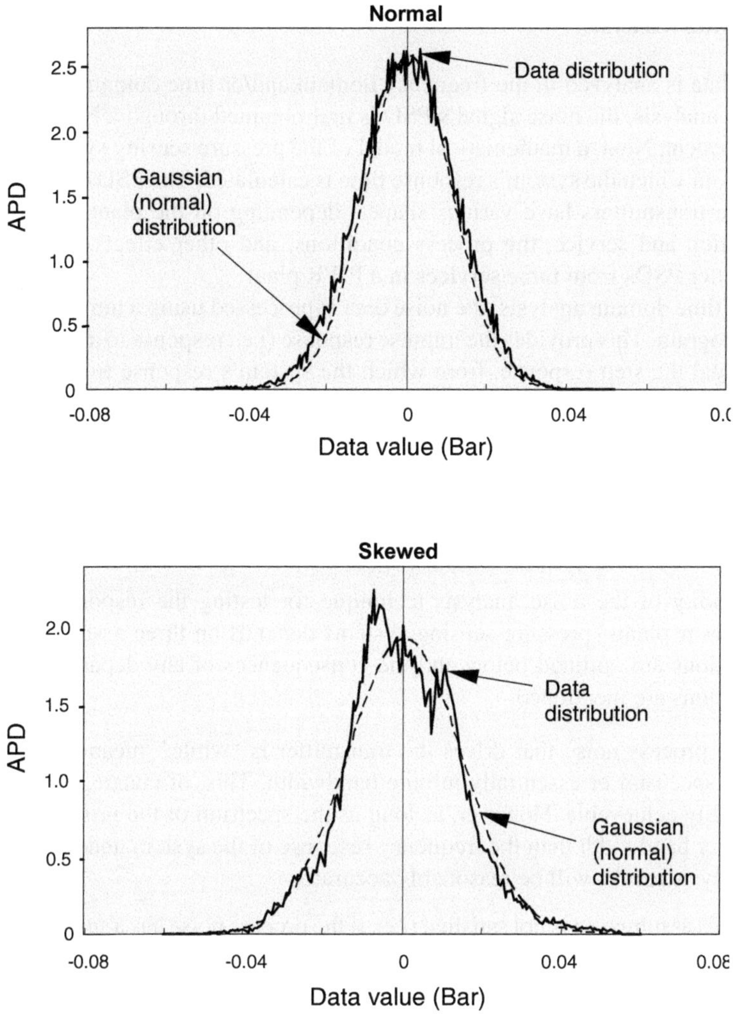

Fig. 9.2. Normal and skewed APDs of noise signals from nuclear plant pressure transmitters

The top APD in Fig. 9.2 is perfectly symmetrical about the mean value of the data and fits the Gaussian distribution (the bell-shaped curve) that is superimposed on the APD. A Gaussian distribution is also referred to as a *normal* distribution (i.e., the words *Gaussian* and *normal* are synonymous).

In addition to APD for noise data qualification, the mean, variance, skewness, and flatness of each block of raw data is calculated and scanned to verify that no saturated blocks, extraneous effects, missing data, or other undesirable characteristics are present. Any data block that has an anomaly is removed from the record before it is analyzed.

9.1.3 Data Analysis

Noise data is analyzed in the frequency domain and/or time domain. For frequency domain analysis, the noise signal's PSD is first obtained through a FFT algorithm or its equivalent. Next, a mathematical model of the pressure sensing system is fit to the PSD, from which the system's response time is calculated. The PSDs of nuclear plant pressure transmitters have various shapes, depending on the plant, the transmitter installation and service, the process conditions, and other effects. Fig. 9.3 shows transmitter PSDs from three services in a PWR plant.

For time domain analysis, the noise data is processed using a univariate AR modeling program. This provides the impulse response (i.e., response to a narrow pressure pulse) and the step response, from which the system's response time is calculated. Typically, the noise data is analyzed in both the frequency domain and time domain, and the results are averaged to obtain the system's response time.

9.2 Noise Analysis Technique: Assumptions

The validity of the noise analysis technique for testing the response time of nuclear power plants' pressure sensing systems depends on three assumptions. These assumptions are outlined below and the consequences of any departure from these assumptions are mentioned.

1. The process noise that drives the transmitter is "white," meaning that it has a flat spectrum or essentially infinite bandwidth. This, of course, is ideal but not readily achievable. However, as long as the spectrum of the process noise has a larger bandwidth than the frequency response of the system under test, the noise analysis results will be reasonably accurate.

   If this assumption is not satisfied (i.e., if the process noise has a smaller bandwidth than the system under test), then the noise analysis results will be dominated by the process bandwidth. Consequently, the response time results obtained from the noise analysis technique will be larger than the actual response time of the pressure sensing system. This is acceptable in nuclear power plants because it produces conservative results.

2. The process noise should not have large resonances that can shift the rolloff frequency of the noise spectrum to higher frequencies. If this assumption is not satisfied, corrective measures must be implemented when the data is being analyzed or the results being interpreted; or the response time values obtained from the noise analysis technique may be non-conservative.

3. The transmitter to be tested must be predominantly linear. If the transmitter is not linear, the noise analysis results will be valid if the response time of interest is one that can be measured with a small-amplitude test signal at a pressure setpoint that is close to the pressure at which the transmitter normally operates in the plant (plotting the APD of the raw data as described in Sect. 9.1.2 and checking

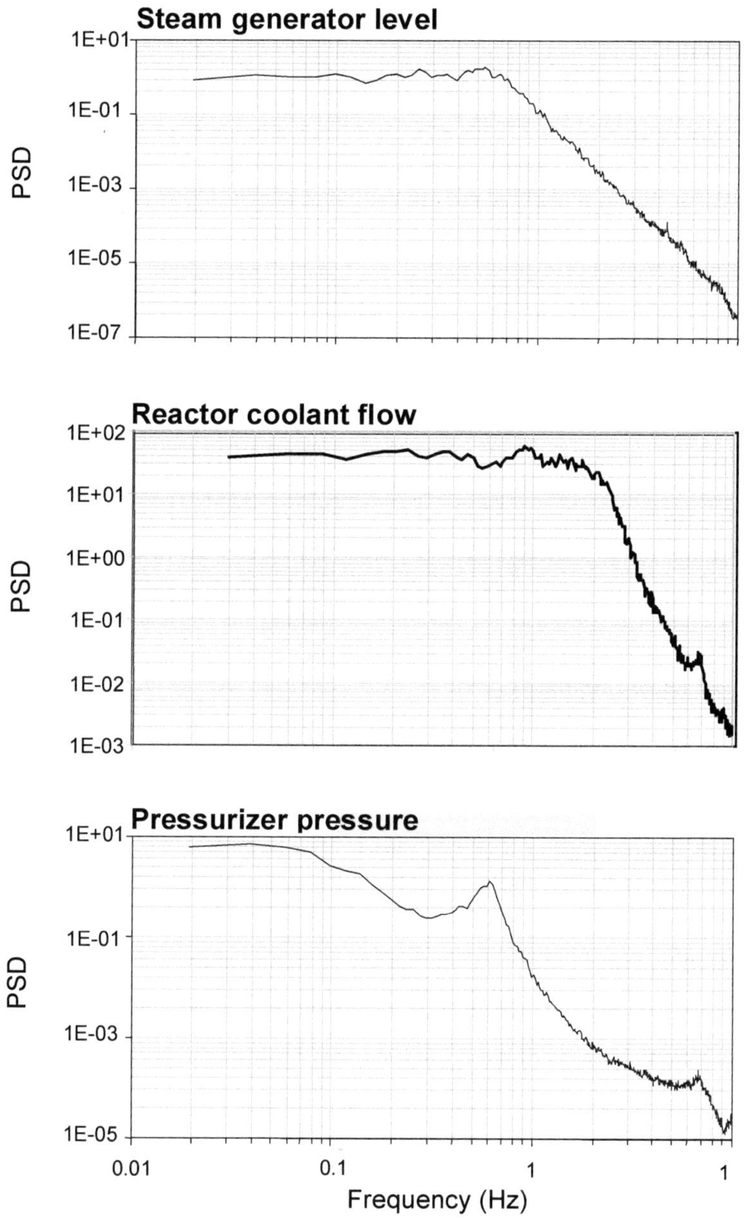

Fig. 9.3. Examples of PSDs of nuclear plant pressure transmitters

it for skewness is a good step toward verifying the linearity of a pressure sensing system).

Experience has shown that these three assumptions are normally met for nuclear plant pressure transmitters. The exceptions are containment pressure transmitters, water storage tank-level transmitters, and others whose process parameters fluctuate very little or not at all. For these transmitters, a method referred to as *pink noise* test has been developed to remotely measure response time. We describe this test in Sect. 9.4.

9.3 Noise Analysis Technique: Validation

The validity of the noise analysis technique for testing the response time of pressure transmitters in nuclear power plants has been established experimentally by both laboratory and in-plant measurements and by simulations.[18] We describe both validation methods in this section.

Since the ramp test method is the standard means for response time testing of nuclear power plants' pressure transmitters, the validation tests described here have used the ramp test results as the basis for validating the noise analysis technique.

9.3.1 Laboratory Validation

The procedure for laboratory validation of the noise analysis technique is as follows:

1. Measure the transmitter's response time using the ramp test method;
2. Install the transmitter in a laboratory flow loop that provides wideband pressure noise (monitor the bandwidth of the noise using a fastresponse reference transmitter);
3. Record the transmitter's AC output for up to one hour;
4. Qualify the data and then analyze it to obtain the transmitter's response time; and
5. Compare the results of the ramp tests with those of the noise analysis technique.

This procedure has been used on numerous pressure transmitters of the types used in nuclear power plants. Table 9.1 shows representative results in terms of response time values from the ramp and corresponding noise tests. Fig. 9.4 shows three examples of results in terms of frequency domain and time domain PSDs.

The difference between response time results from the ramp test and the noise analysis technique is also shown in Table 9.1 for each transmitter. Ideally, the response time values from the ramp and noise tests should be identical. However, inherent repeatability problems and other issues in both ramp and noise testing of pressure transmitters usually prevent obtaining identical response time results. The differences between the results of the ramp and noise tests as shown in Table 9.1 is less than ± 0.05 seconds, notwithstanding a few outliers. This is a fairly reasonable degree of

Table 9.1. Representative results of laboratory validation of noise analysis technique for nuclear-grade pressure transmitters

Response Time (sec)		
Ramp Test	**Noise Analysis**	**Difference**
	Barton	
0.05	0.09	0.04
0.17	0.20	0.03
0.17	0.25	0.08
0.12	0.15	0.03
0.12	0.20	0.08
0.11	0.15	0.04
0.12	0.18	0.06
	Foxboro	
0.13	0.16	0.03
0.21	0.18	−0.03
0.16	0.13	−0.03
0.09	0.12	0.03
0.29	0.30	0.01
0.25	0.15	−0.10
0.28	0.25	−0.03
	Rosemount	
0.05	0.06	0.01
0.32	0.28	−0.04
0.07	0.05	−0.02
0.10	0.07	−0.03
0.11	0.08	−0.03
0.09	0.08	−0.01
0.09	0.09	0.00
	Other Manufacturers	
0.15	0.15	0.00
0.21	0.18	−0.03
0.02	0.08	0.06
0.03	0.07	0.04
0.08	0.11	0.03
0.15	0.27	0.12
0.33	0.37	0.04

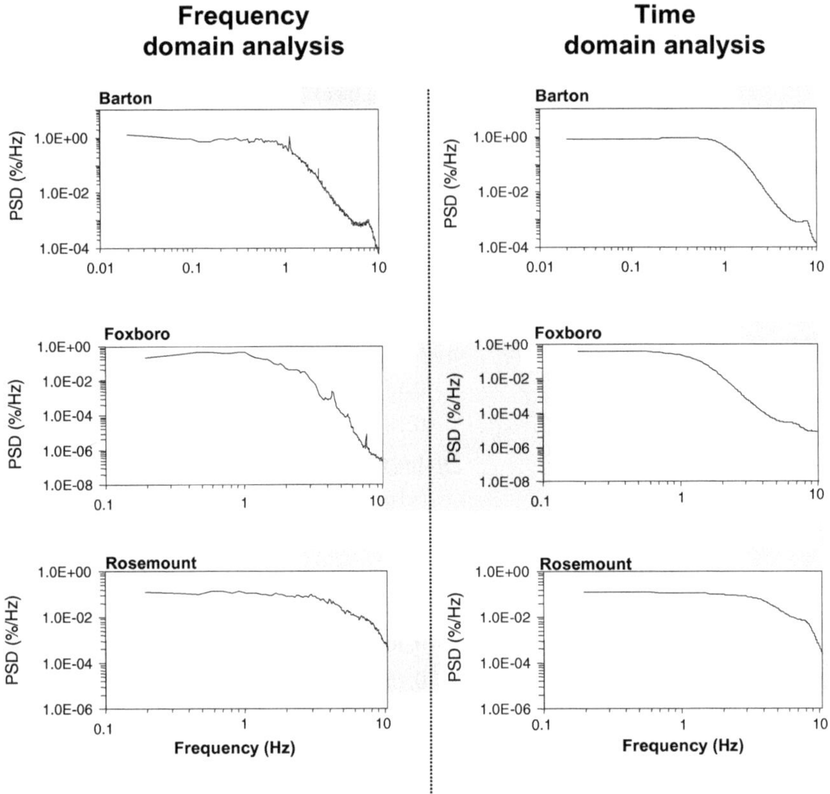

Fig. 9.4. PSDs from frequency and time domain analyses of laboratory noise data for representative nuclear-grade pressure transmitters

agreement considering all the factors that can affect the response time measurements using these two methods.

The results in Table 9.1 are based on tests of pressure transmitters configured as they are normally used. To further confirm the validity of the noise analysis technique, researchers artificially degraded several transmitters in order to increase their response time. This enabled them to verify that the noise analysis technique could successfully reveal response time degradation. The response time of these transmitters was then tested using both the ramp method and the noise analysis techniques. The results are presented in Table 9.2. For each transmitter shown, its response time was measured as the transmitter was incrementally degraded. Clearly, the degradation of the transmitters' response times is reflected in the noise analysis results.

9.3.2 In-Plant Validation

Several experiments have tested nuclear plant pressure transmitters on line using the noise analysis technique and then shortly afterward off line using the conventional ramp test method. These tests have generated in-plant validation results for the noise analysis technique (Table 9.3).

The reasonable agreement between the results of the noise analysis technique and the ramp test method is apparent in the data shown in Table 9.3, with the exception of one transmitter (Gould PT-505). This transmitter was found to have a skewed APD, as shown in Fig. 9.5, while the APDs of the other two Gould transmitters (PT-524 and PT-526) in the same plant are normal (Gaussian). The APD could be skewed because of non-linearity in the transmitter, problems in the data, noise, or other causes. What is important is that the erroneous response time result for PT-505 is correlated with its abnormal APD. Therefore, one should plot and examine the APD of any sensor whose response time is being measured using the noise analysis technique as the APD can provide clues as to whether or not the response time results are reliable.

Note in Table 9.3 that the differences between the results of the two methods are larger for the Barton transmitters. This is because the noise analysis results given in Table 9.3 include the effect of sensing lines which is typically larger for Barton transmitters.

9.3.3 Software Validation

Noise analysis software programs are usually validated using synthetic analog or digital noise data and theoretical models. The procedure is as follows:

1. Determine the dynamic model (theoretical equation) for the pressure sensing system of the type used in nuclear power plants.
2. Feed the model with synthetic digital or analog noise data. Synthetic digital data can be obtained using a random number generator and synthetic analog data from a wideband noise generator.
3. Using the software being validated, analyze the model's dynamic response to the synthetic input data and calculate the model's response time.
4. Compare the results of Step 3 with the actual response time of the model calculated from the model parameters.

Table 9.4 presents results of this four-step procedure for four theoretical models, which represent the normal and degraded response times of typical pressure transmitters. As expected, the differences between the two results are negligible (less than 5 percent), which testifies to the validity of the noise analysis software programs used in these cases.

9.3.4 Hardware Validation

For hardware validation, analog data from a simulator is sampled by a digital data acquisition system, screened by data qualification software, and then analyzed in the frequency domain and/or time domain. In this process, pressure transmitter simulators with known response times are used. The hardware validation procedure is as follows:

Table 9.2. Representative results of noise analysis validation for artificially degraded transmitters

Response Time (sec).		
Ramp Test		**Noise Analysis**
	Barton	
0.11		0.12
0.16		0.27
0.50		0.73
	Foxboro	
0.12		0.15
0.16		0.19
0.33		0.44
	Rosemount	
0.05		0.18
0.35		0.35
0.67		0.68
	Other Manufacturers	
0.15		0.15
0.19		0.19
0.30		0.35
0.02		0.02
0.04		0.03
0.42		0.50
0.08		0.11
0.12		0.20
0.25		0.35

1. Develop a simulator (e.g., an RC network) to mimic the dynamics of a pressure transmitter.

2. Measure the simulator's response time using a step (or ramp) input signal, as illustrated in Fig. 9.6.

3. Use a signal generator to feed wideband random noise to the simulator, as illustrated in Fig. 9.7.

4. Record the simulator's output data, and analyze it to obtain the simulator's response time.

Table 9.3. In-plant validation of noise analysis technique

Response Time (sec)		
Ramp Test	**Noise Analysis**	**Difference**
	Barton	
0.23	0.36	0.13
0.23	0.38	0.15
0.23	0.39	0.16
	Rosemount	
0.08	0.06	−0.02
0.11	0.13	0.02
0.21	0.33	0.12
0.10	0.11	0.01
	Gould	
0.13	0.16	0.03
0.12	0.18	0.06
0.16	0.78	0.62*
	Other Manufacturers	
0.04	0.05	0.01
0.21	0.21	0.00
0.02	0.02	0.00
0.04	0.03	−0.01

*Transmitter has a skewed APD (see Fig. 9.5). Footnote: The noise analysis results in this table include the contribution of sensing lines, while the ramp test results do not. However, the sonic delays and the effect of sensing line length were estimated for each case and added to the ramp test results.

5. Compare the results of Steps 2 and 4. The two results must be almost identical to ensure the validity of the process.

Table 9.5 presents results of this five-step procedure for four simulators. The differences between direct measurements of the simulators' response time and those of the noise analysis technique are negligible (less than 5 percent). A noise analysis data acquisition and data analysis system which passes the software and hardware validation just discussed is said to be qualified for transmitter response time testing in nuclear power plants.

9.4 Pink Noise Technique

As mentioned in Sect. 9.2, testing the response time of pressure sensing systems in nuclear power plants requires wideband process fluctuations or "white noise." Although, the term *white noise* is commonly used, process fluctuations do not normally have white noise characteristics. However, this does not pose a problem as long as the bandwidth of the process fluctuations is sufficiently greater than the expected bandwidth of the pressure sensing system being tested.

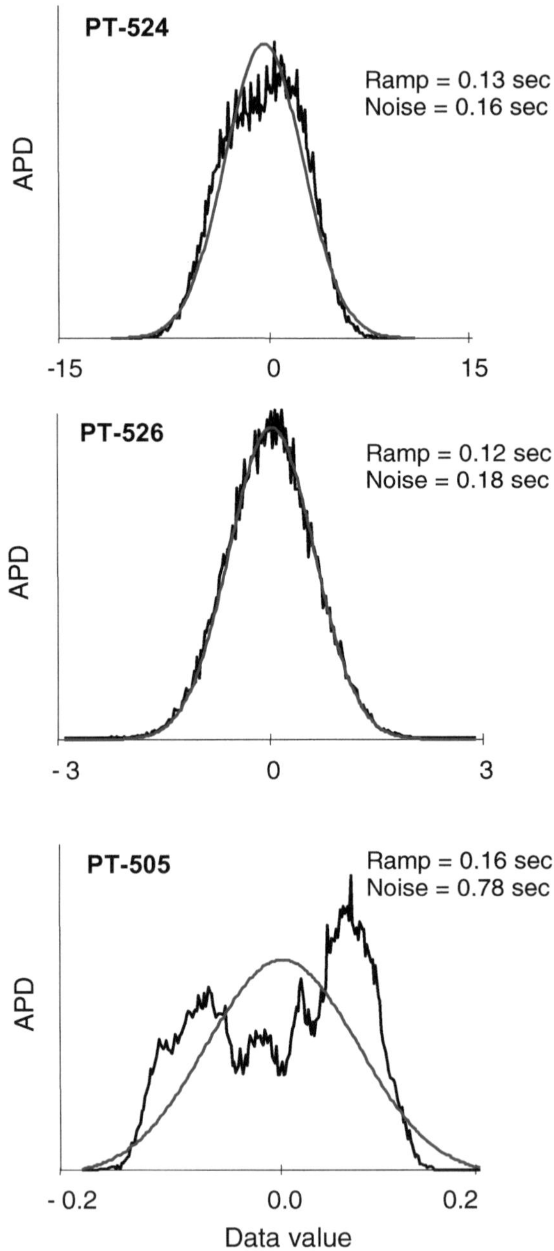

Fig. 9.5. APDs of Gould transmitters from in-plant testing at a PWR (*Transmitter is probably nonlinear)

Table 9.4. Representative results of validation of noise analysis software

	Response Time (sec)	
Theory	**Noise Analysis Software**	**Difference**
	Model 1	
0.05	0.05	0.00
0.10	0.11	0.01
0.80	0.80	0.00
3.18	3.18	0.00
	Model 2	
0.01	0.01	0.00
0.10	0.10	0.00
1.15	1.14	−0.01
2.32	2.35	0.03
	Model 3	
0.06	0.06	0.00
0.31	0.30	−0.01
0.61	0.64	0.03
2.02	2.04	0.02
	Model 4	
0.23	0.23	0.00
1.73	1.80	0.07
2.02	1.93	−0.09

For some pressure transmitters in nuclear power plants, such as containment pressure transmitters and water storage tank-level transmitters, process fluctuations do not normally exist or they are inadequate for using the noise analysis technique to test response time. As such, these transmitters' response times are tested using either the conventional ramp test method or by injecting artificial pressure noise into the transmitter. The artificial pressure noise is generated using a current-to-pressure (I-to-P) converter, which is driven by a random noise signal generator (Fig. 9.8).

The resulting signal is referred to as *pink noise*, and the test method is thus called the *pink noise technique*. The advantage of the pink noise technique is that it can be used to measure the response time of pressure transmitters remotely (e.g., from outside the containment). The pink noise is typically fed to the transmitter through existing lines, which are accessed from outside the containment.

The pink noise method has been validated for testing pressure transmitters' response times and has been used successfully in nuclear power plants. Representative results of laboratory validation of the pink noise technique are presented in Table 9.6.

9.5 Accuracy of Noise Analysis Technique

The accuracy of the noise analysis technique for testing the response time of pressure sensing systems has been established experimentally using pressure transmitters

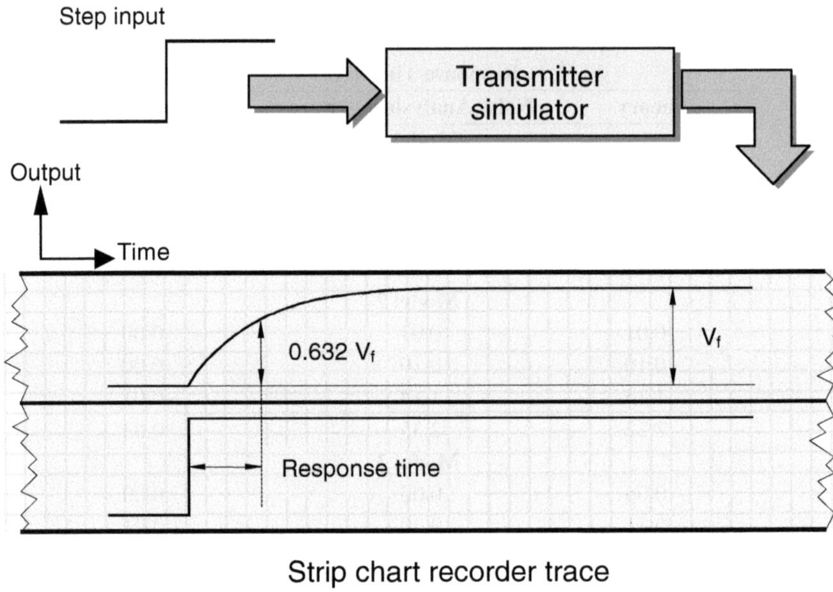

Fig. 9.6. Test setup to measure the response time of a pressure sensing system simulator

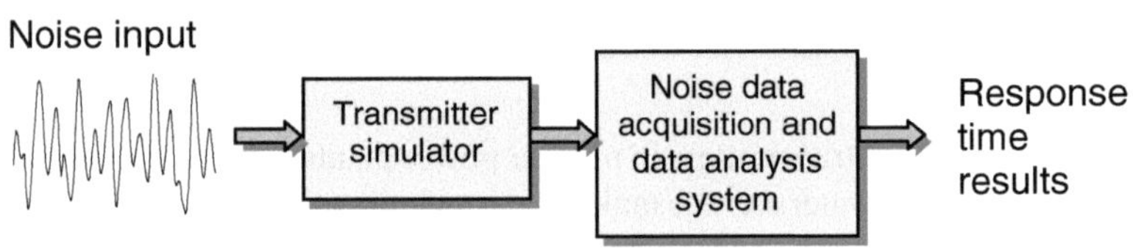

Fig. 9.7. Test setup for validating noise data acquisition hardware

like those used in nuclear power plants. For each transmitter, the response time was measured first using the ramp method and then using the noise analysis technique. In doing so, the reliability of the ramp test results was first established by laboratory measurements. Toward this end, two sets of measurements in particular were made. One set involved ramp testing pressure transmitters using a variety of ramp rates, and the other involved repeatability tests performed by three different engineers. Tables 9.7 and 9.8 present representative results of these measurements expressed in terms of response time values and in terms of the difference between the smallest and largest response time results for each transmitter. The results in Table 9.7 include measurements using three to eight ramp rates. The results in Table 9.8 are from repeatability tests performed by three engineers (identified in the table by their initials; MH, REF, and KMP). Except for a few outliers, the results in both Tables 9.7 and 9.8 are repeatable to better than 0.05 seconds, and there is little or no dependence on the ramp rate.

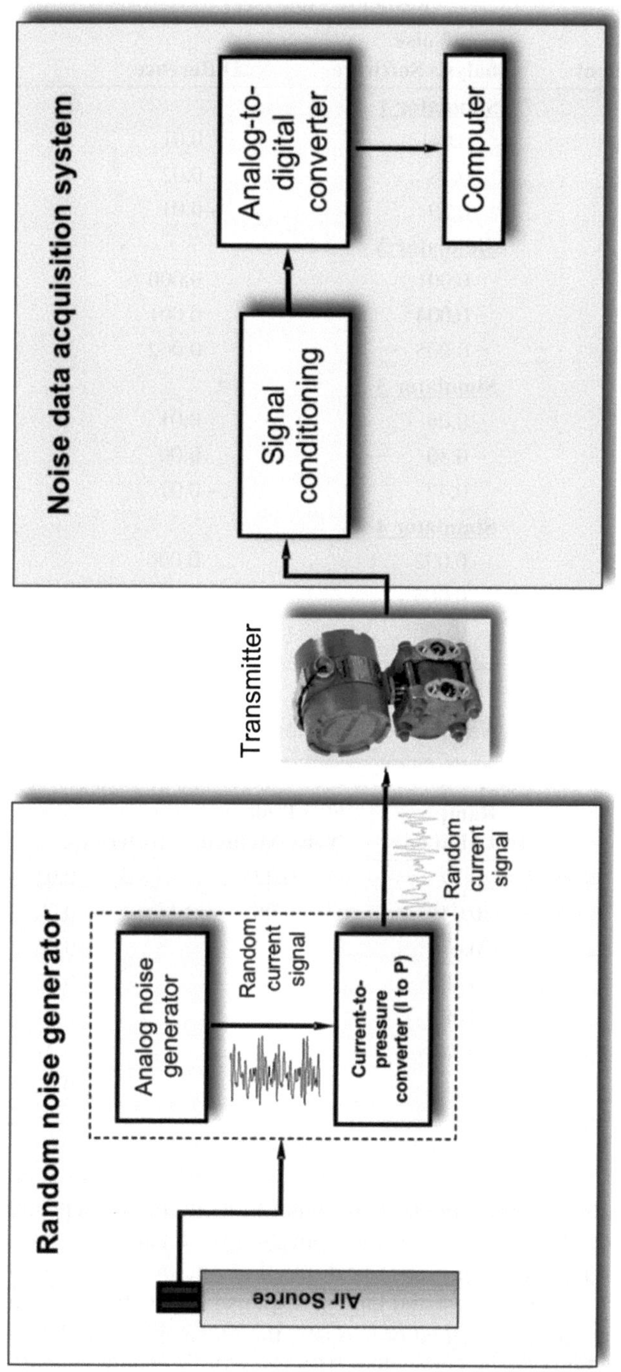

Fig. 9.8. Equipment setup for response time testing of containment pressure transmitters and other sensors using the pink noise technique

Table 9.5. Representative results of noise analysis hardware validation

Direct Measurement (Response Time, sec)	Noise Analysis Software (Response Time, sec)	Difference (Response Time, sec)
	Simulator 1	
0.03	0.04	0.01
0.26	0.28	0.02
0.30	0.29	−0.01
	Simulator 2	
0.001	0.001	0.000
0.003	0.004	0.001
0.003	0.005	0.002
	Simulator 3	
0.05	0.06	0.01
0.30	0.30	0.00
0.54	0.47	−0.07
	Simulator 4	
0.002	0.002	0.000
0.006	0.005	−0.001
0.006	0.006	0.000

Table 9.6. Representative results of validation of pink noise analysis technique

Item	Ramp Test Method (Response Time, sec)	Pink Noise Method (Response Time, sec)	Difference (Response Time, sec)
1	0.19	0.22	0.03
2	0.04	0.06	0.02
3	0.47	0.42	−0.05
4	0.48	0.46	−0.02
5	0.08	0.09	0.01
6	0.34	0.33	−0.01
7	0.31	0.33	0.02

Next, laboratory measurements were made to examine the repeatability of the noise analysis results. Table 9.9 shows representative results in terms of response time values from repeated noise tests performed at one or two different times. The differences between the smallest and largest response time values are also shown. As in the case of ramp tests described previously, the repeatability of the noise analysis results shown in Table 9.9 is better than 0.05 seconds, a couple of outliers notwith-

Table 9.7. Examples of results of laboratory response-time measurements versus ramp rate

Transmitter	Ramp Rate (bar/sec)	Response Time (sec)	Difference (sec)
1	0.3, 0.7, 1, 1.5	0.13, 0.13, 0.14, 0.13	0.01
2	0.3, 0.7, 1.5	0.23, 0.21, 0.20	0.03
3	0.2, 0.3, 0.5, 0.7	0.10, 0.13, 0.11, 0.10	0.03
4	0.7, 1, 1.5	0.14, 0.18, 0.17	0.04
5	0.6, 1.2, 2, 2.5	0.08, 0.08, 0.09, 0.09	0.01
6	0.2, 0.4, 0.5, 0.6	0.13, 0.12, 0.12, 0.12	0.01
7	0.1, 0.2, 0.4, 0.7, 1.7, 2	0.04, 0.04, 0.05, 0.05, 0.04, 0.04	0.01
8	1.3, 2.2, 4.2, 6.7, 7.7, 9.1, 10, 120	0.14, 0.21, 0.21, 0.18, 0.18, 0.21, 0.17, 0.18	0.07
9	0.4, 0.6, 0.7	0.15, 0.15, 0.14	0.01
10	2, 2.1, 4, 2.2, 6, 6.7, 7.7, 9.1	0.17, 0.19, 0.19, 0.20, 0.19, 0.19, 0.20, 0.20	0.03
11	0.1, 0.2, 0.3, 0.4	0.32, 0.30, 0.30, 0.30	0.02
12	4, 8.7, 127, 16.7	< 0.01, < 0.01, < 0.01, < 0.01	0.00
13	9, 27, 35	< 0.01, < 0.01, < 0.01	0.00
14	0.3, 1.3, 1.9, 2	0.10, 0.07, 0.07, 0.07	0.03
15	0.2, 0.4, 0.5, 0.6, 0.7	0.20, 0.16, 0.15, 0.13, 0.12	0.08
16	17, 47, 60, 45, 16, 44, 56	< 0.01, < 0.01, < 0.01, < 0.01, < 0.01, < 0.01	0.00
17	0.1, 0.5, 0.7, 0.8	0.05, 0.07, 0.08, 0.08	0.03
18	0.1, 0.3, 0.4	0.20, 0.17, 0.17	0.03
19	10, 20, 21, 30, 38	0.07, 0.07, 0.07, 0.08, 0.08	0.01
20	0.1, 0.2, 0.3, 0.5	0.39, 0.39, 0.27, 0.27	0.12

standing. This is reasonable considering the potential effects that can influence the noise analysis results.

In addition to the above, the results presented earlier in Tables 9.1, 9.2, and 9.3 help to determine the accuracy of the noise analysis technique. In particular, the results in Tables 9.1 through 9.3 came from a large number of tests, which led to the following conclusions:[18]

- 79 percent of response time results from the noise analysis technique fall within $\pm$ 0.05 seconds of ramp test results performed on the same transmitters under the same conditions.

- 16 percent of response time results from the noise analysis technique fall between 0.05 and 0.10 seconds ($\pm$) of ramp test results performed on the same transmitters under the same conditions.

- 5 percent of response time results from the noise analysis technique fall within $\pm$ 0.10 seconds of ramp test results performed on the same transmitters under the same conditions.

Table 9.8. Representative results of laboratory testing for repeatability of ramp test method

Transmitter	Test Engineer	Response Time (sec)	Difference (sec)
1	MH	0.15, 0.16	0.04
	REF	0.13, 0.13, 0.13, 0.13	
	KMP	0.14, 0.12, 0.15, 0.13	
2	MH	0.23, 0.22	0.04
	REF	0.23, 0.21, 0.20, 0.20	
	KMP	0.19, 0.19, 0.19, 0.19	
3	MH	0.18, 0.16, 0.16	0.08
	REF	0.10, 0.13, 0.11, 0.10	
	KMP	0.12, 0.12, 0.13, 0.11	
4	MH	0.16, 0.16, 0.16	0.04
	REF	0.12, 0.14, 0.14, 0.12	
	KMP	0.14, 0.12, 0.12, 0.12	
5	MH	0.04, 0.04, 0.04	0.01
	REF	0.05, 0.05, 0.04, 0.04	
	KMP	0.05, 0.05, 0.04, 0.04	
6	MH	0.32, 0.32, 0.32	0.03
	REF	0.29, 0.30, 0.32	
	KMP	0.29, 0.30, 0.31	
7	REF	0.08, 0.06, 0.09, 0.09	0.01
8	MH	0.28, 0.28, 0.30, 0.30	0.02
9	KMP	0.28, 0.26, 0.26, 0.23	0.05
10	MH	< 0.01, < 0.01, < 0.01	0.00
11	KMP	< 0.01, < 0.01, < 0.01	0.00
12	REF	0.04, 0.03, 0.04, 0.05, 0.05, 0.05, 0.04, 0.05	0.02

Based on all the foregoing data, the nuclear power industry has concluded that the noise analysis technique provides the response time of pressure sensing systems with an accuracy of better than 0.10 seconds.

9.6 Experience from Testing in Nuclear Power Plants

Response time testing using the noise analysis technique has been performed on nuclear power plant pressure transmitters since the early 1980s. As a result, a database of response time values and records of raw data, PSD plots, and interesting observations has accumulated since then. A few examples are reviewed in this section.

Fig. 9.9 shows noise-test PSDs for two-loop, three-loop, and four-loop PWRs and a BWR plant. Three PSDs are shown in each case for a pressure, a level, and a flow transmitter in each plant type.

In several plants, noise testing has been performed on more than one occasion, making it possible to examine the repeatability of the results. Fig. 9.10 shows two PSDs for a steam generator level transmitter in a three-loop PWR plant. The tests were performed approximately three years apart. The results are essentially identical, which indicates that the noise tests for this transmitter is very repeatable and that the transmitter has experienced no response time changes over this three-year period.

On another occasion, two redundant transmitters measuring the same steam generator level signal were tested at the same time in a four-loop PWR. The PSD results are shown in Fig. 9.11. It is apparent that one of these transmitters is significantly faster than the other (by about an order of magnitude). This is unusual because the response times of redundant transmitters are normally expected to be comparable. In this particular case, the two transmitters are from two different manufacturers, and they were probably installed without considering that the two transmitters might have vastly difference response times. This type of difference in response time is also seen in redundant transmitters when there is blockage in the sensing line. However, in the case shown in Fig. 9.11, the difference is not the result of sensing line blockage.

9.7 Oil Loss in Nuclear Plant Pressure Transmitters

9.7.1 Problem Description

In the late 1980s, some Rosemount pressure transmitters in nuclear power plants were found to be leaking silicon oil from their sensing cell. The silicon oil is used to transfer pressure signals from the isolation diaphragm to the sensing diaphragm at the center of the sensing cell. Thus, if the oil leaks, both the transmitter's steady-state (calibration) and dynamic response are affected. The oil loss problem in Rosemount pressure transmitters caused significant concern in the nuclear power industry and a number of documents were issued by Rosemount and the NRC. The Rosemount documents were provided in terms of technical bulletins that were widely distributed in the nuclear industry to help with resolution of the oil loss problem. The NRC documents provided the regulatory recommendations as to how the issue shall be addressed. Appendix E contains two examples of NRC documents published on the oil loss problem.

Fig. 9.12 shows the responses of two Rosemount flow transmitters at the Millstone nuclear power station Unit 3 (a U.S. PWR plant) after a reactor coolant pump trip.

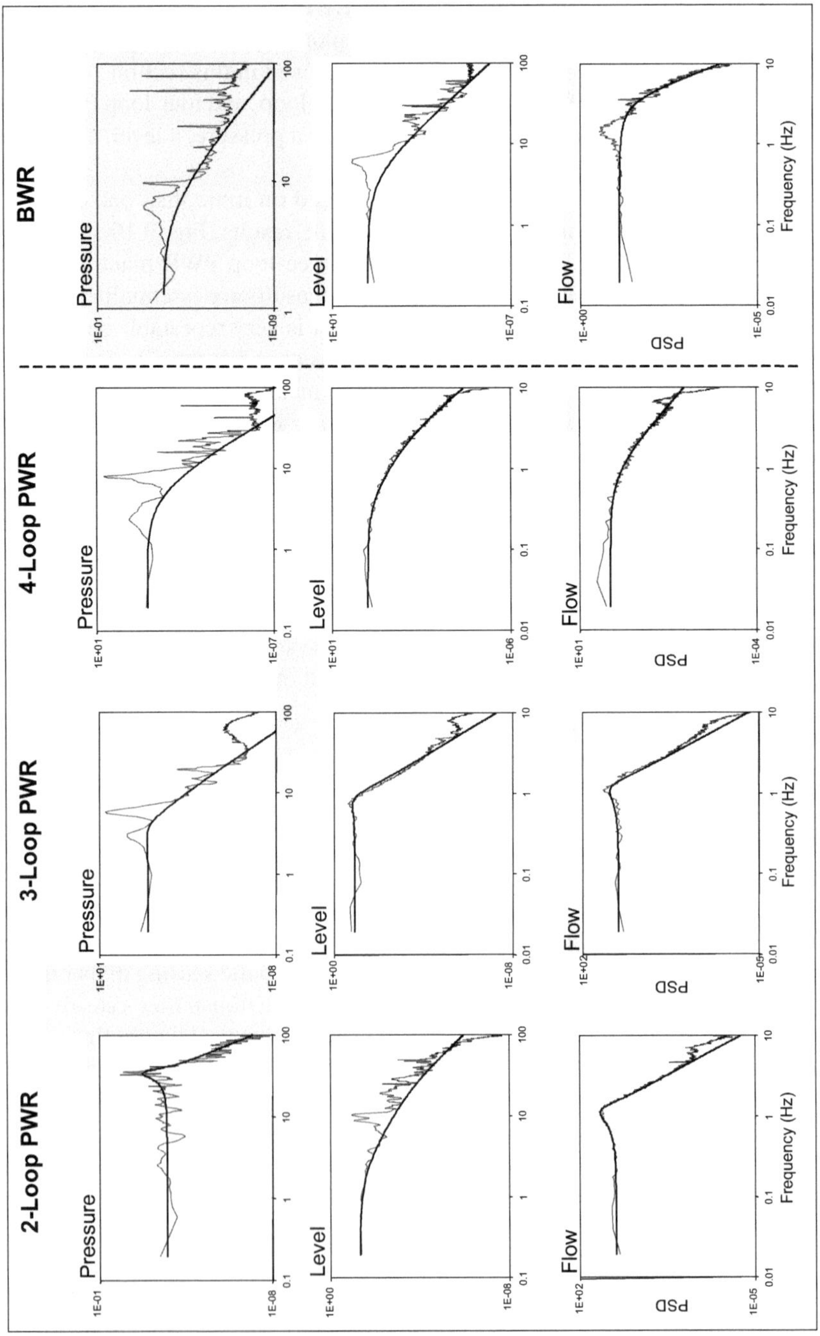

Fig. 9.9. Examples of typical PSDs of pressure, level, and flow transmitters in PWRs and BWRs

Table 9.9. Repeatability of noise analysis results in laboratory tests

Transmitter	Date of Test	Measured Response Time (sec)	Difference (sec)
1	Week 1	0.11, 0.12, 0.16, 0.16	0.06
	Week 3	0.17, 0.16, 0.17, 0.17	
2	Week 1	0.16, 0.16, 0.21, 0.23	0.07
	Week 3	0.16, 0.16, 0.17, 0.17	
3	Week 1	0.15, 0.17, 0.14	0.02
	Week 3	0.14, 0.14	
4	Week 1	0.13, 0.13, 0.13, 0.13	0.02
	Week 3	0.14, 0.12, 0.12, 0.14	
5	Week 1	0.32, 0.27, 0.28, 0.28	0.10
	Week 3	0.23, 0.34, 0.33, 0.24	
6	Week 1	0.05, 0.05, 0.06	0.03
	Week 3	0.03, 0.04, 0.06, 0.04	
7	Week 1	0.07, 0.07	0.01
	Week 3	0.07, 0.08	
8	Week 1	0.21, 0.19, 0.21, 0.21	0.05
	Week 3	0.22, 0.22, 0.24, 0.24	
9	Week 1	0.26, 0.20, 0.25	0.06
	Week 3	0.26, 0.25, 0.26	
10	Week 1	0.10, 0.11, 0.12, 0.11	0.02
11	Week 1	0.17, 0.17, 0.18, 0.18	0.01
12	Week 1	0.09, 0.09, 0.10, 0.08	0.02
13	Week 1	0.23, 0.22, 0.22, 0.22	0.01
14	Week 1	0.33, 0.35, 0.36, 0.38	0.05

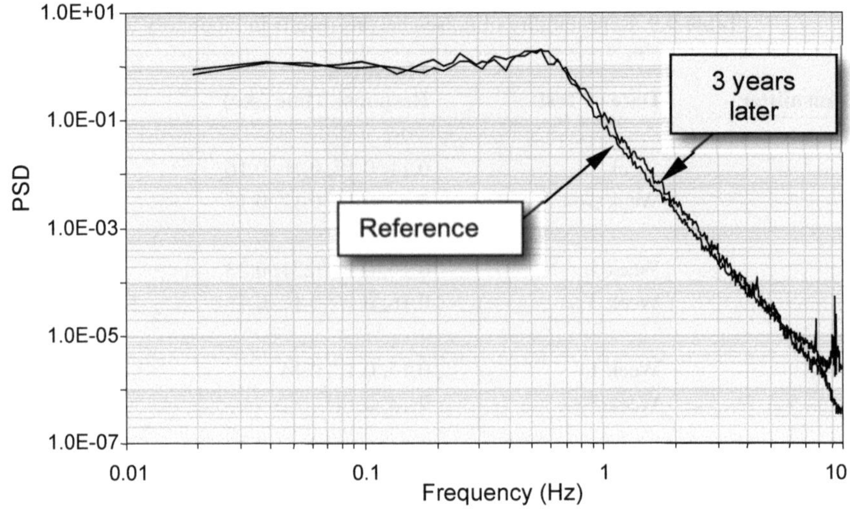

Fig. 9.10. PSDs of a nuclear plant pressure transmitter measured three years apart

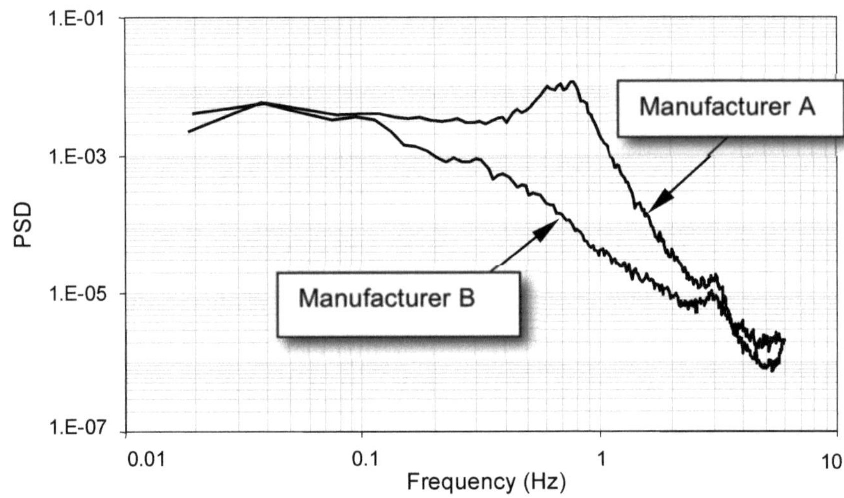

Fig. 9.11. PSDs of two redundant steam generator level transmitters in a four-loop PWR plant

Note that one transmitter (FT-444) responds quickly to flow reduction as expected, but the other transmitter (FT-445) is extremely sluggish. The FT-445 was later confirmed as having suffered from the oil loss problem. In fact, it was the data shown in Fig. 9.12 which led to the discovery of the oil loss problem in Rosemount pressure transmitters.

Fig. 9.13 shows raw noise data for a normal and a failed (from oil loss) Rosemount Model 1153 transmitter, both used in the same service in an operating nuclear power plant. As expected, the amplitude of the noise signal from the failed transmitter is much smaller than the normal transmitter.

Fig. 9.14 shows diagrams of the sensing cell of Rosemount transmitters in normal, over-pressurized, and oil-loss conditions. The oil is said to leak out of the sensing cell at the intersection of the glass and metal, as shown in Fig. 9.15.

9.8 Oil Loss Diagnostics

Upon discovery of the oil loss problem in Rosemount transmitters in the late 1980s, the author and his colleagues at AMS developed noise diagnostics for detecting oil loss in Rosemount transmitters. This involved calculating the second to the fifth moments of the noise data as well as the ratio of these moments from noise records above and below the signal's mean value. The first moment of noise data is its mean value, its second moment is the variance, the third moment is skewness, and so on. Table 9.10 is an example of noise diagnostic descriptors for four steam generator level transmitters, designated as LT 518, 528, 538, and 548, in a PWR plant. The descriptors' normal values are also shown in the table. Note that the values of descriptors for LT 528 are much different than for the other transmitters. This transmitter was later removed from the plant and sent to Rosemount where it was determined that the problem was due to oil loss in the transmitter's sensing module. This and other work in this area concluded that the noise analysis technique can provide a useful means for oil loss diagnostics. It should be pointed out here that the root cause of the oil loss problem in Rosemount transmitters was fortunately identified and resolved by the manufacturer very quickly. Therefore, the nuclear industry did not suffer any adverse consequences. Also, since the problem was successfully resolved early, the noise diagnostics for detection of oil loss did not become a routine activity in nuclear power plants.

Table 9.10. Example of oil-loss diagnostic results

		Measured Values of Diagnostic Descriptors			
Diagnostic Descriptor	Normal Value	LT518	**LT528**	LT538	LT548
Skewness	0.0	0.02	**0.23**	0.08	0.05
5th moment	0.0	0.07	**2.12**	0.70	0.36
Variance ratio	1.0	1.03	**1.25**	1.09	1.06
Skewness ratio	1.0	1.00	**1.06**	1.01	1.02
5th moment ratio	1.0	1.00	**1.21**	1.04	1.06

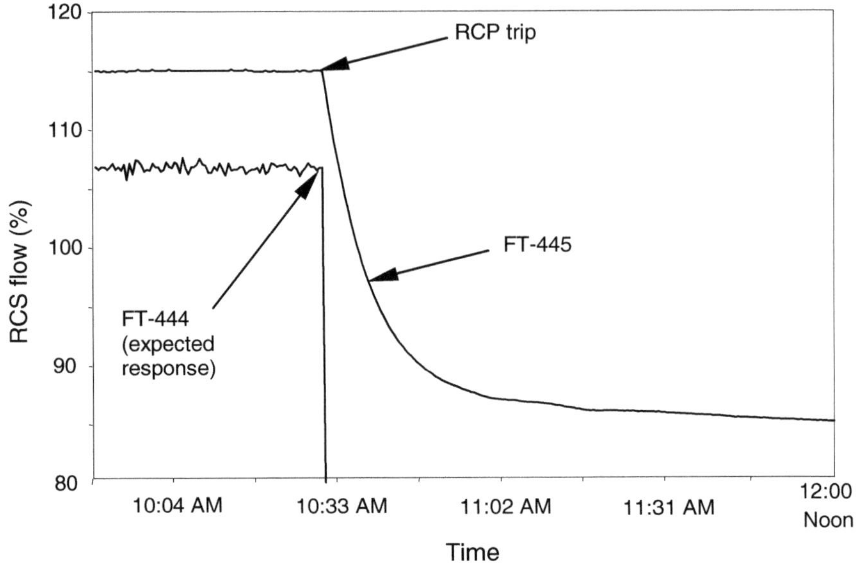

Fig. 9.12. Dynamic response of two Rosemount transmitters during the shutdown of Millstone nuclear power station Unit 3

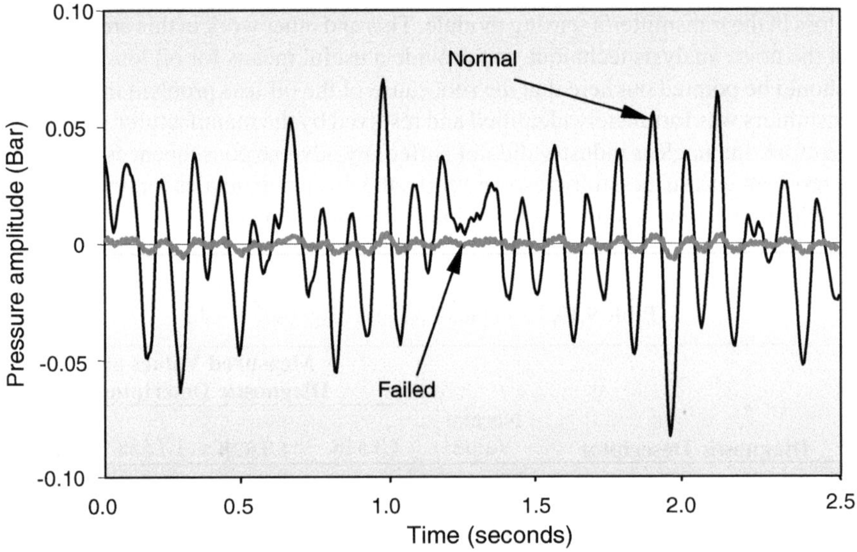

Fig. 9.13. Noise output of a normal and a failed Rosemount transmitter from testing in an operating nuclear power plant

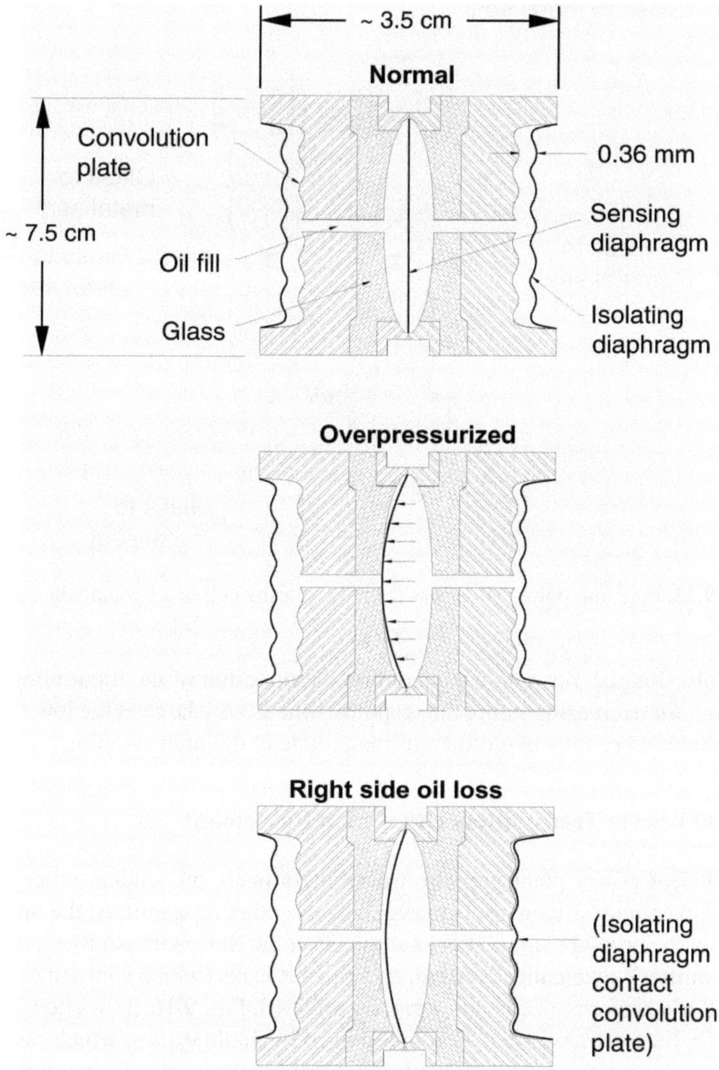

Fig. 9.14. Sensing cell of Rosemount transmitters under normal and oil-loss conditions

9.8.1 Effect of Oil Loss on Transmitter Linearity

Oil loss affects not only a transmitter's static and dynamic response but can also cause the transmitter to become non-linear. Table 9.11 shows ramp test results for a normal and a failed (due to oil loss) Rosemount Model 1153 transmitter. The ramp tests were performed with both increasing and decreasing ramp signals and at three setpoint pressures. The normal transmitter is unaffected by the direction of the ramp test signal and the setpoint pressure, while the response time of the failed transmitter

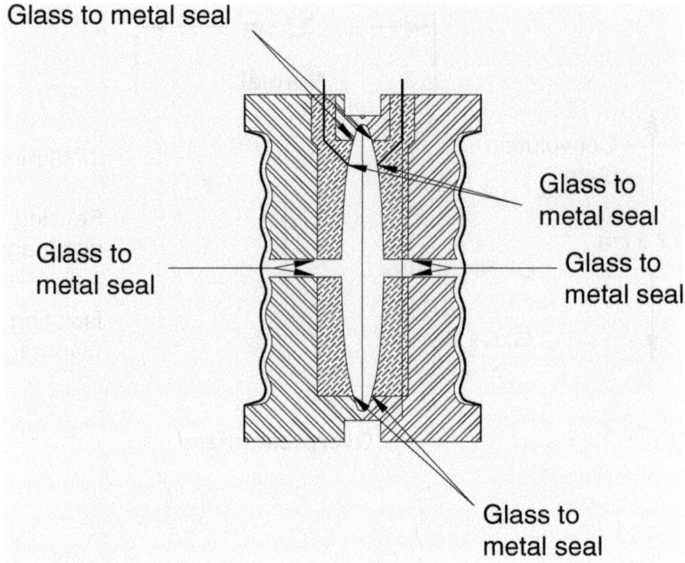

Fig. 9.15. Potential points of oil loss from the sensing cell in a Rosemount transmitter

is not only sluggish but also dependent on the direction of the input ramp signal. In particular, for decreasing ramp, the response time is very large at the low setpoint and decreases by more than two orders of magnitude at the high setpoint.

9.8.2 Oil Loss in Transmitters Other than Rosemount

Some nuclear power plant pressure transmitters made by vendors other than Rosemount also contain silicon oil. However, in these other transmitters, the oil is not used to transfer the pressure signal. For example, in some Barton transmitters, oil is used in the transmitter's mechanical system, and when it leaks (which sometimes occur), the transmitter's performance is not normally affected. Fig. 9.16 shows how oil loss can occur in a Barton transmitter. This is followed by Table 9.12 in which response time results are shown for a Barton Model 764 transmitter with varying amounts of oil loss. The oil in this transmitter was intentionally removed and laboratory tests were performed to verify that oil loss in this type of transmitter do not necessarily lead to a significant loss of performance or linearity. Also shown in Table 9.12 are response time results for a Foxboro and a Tobar transmitter with and without oil loss.

9.9 Response Time Degradation

The response time of nuclear plant pressure transmitters does degrade, but it is not as prevalent a problem for pressure transmitters as it is for RTDs. Conversely, calibration drift is more of a problem in pressure transmitters than it is in RTDs. Fig. 9.17

Table 9.11. Results of response-time measurements made to demonstrate the effect of oil loss on transmitter linearity

Setpoint Pressure	Response Time (sec) Increasing Ramp	Decreasing Ramp
	Normal Transmitter	
Low	0.12	0.13
Medium	0.12	0.13
High	0.15	0.13
	Failed Transmitter	
Low	0.23	171.0
Medium	0.25	19.0
High	0.25	1.1

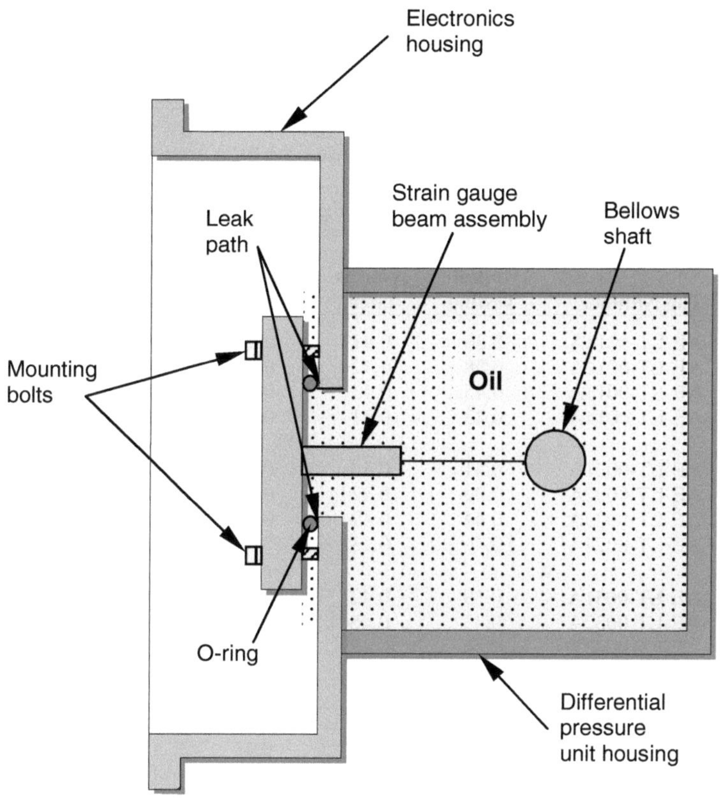

Fig. 9.16. Sensing module of a Barton transmitter and O-ring where oil loss can occur

Table 9.12. Laboratory response time testing results for a Barton Module 764 transmitter with and without oil loss

	Response Time (sec)	
Amount of Oil Loss	**Increasing Ramp**	**Decreasing Ramp**
	Barton 764	
0%	0.19	0.19
50%	0.16	0.16
75%	0.12	0.12
100%	0.10	0.11
	Foxboro E13DM	
0%	0.17	0.12
100%	0.12	0.08
	Tobar 32DP	
0%	0.17	0.18
100%	0.11	0.12

Table 9.13. Typical results of trending of response time for a group of nuclear plant pressure transmitters

	Response Time (sec)				
Tag Number	**Initial Testing**	**18 months later**	**36 months later**	**48 months later**	**60 months later**
AE-LT-0011A	0.36	0.41	0.43	0.44	0.44
AE-LT-0012A	0.38	0.42	0.43	0.43	0.43
AE-LT-0013A	0.45	0.43	0.45	0.47	0.41
AE-LT-0014A	0.43	0.41	0.44	0.47	0.43
AE-LT-0021A	0.41	0.45	0.43	0.43	0.42
AE-LT-0022A	0.39	0.42	0.42	0.43	0.42
AE-LT-0023A	0.44	0.49	0.47	0.46	0.43
AE-LT-0024A	0.46	0.48	0.44	**0.66**	0.41**
AE-LT-0031A	0.39	0.42	0.41	0.41	0.40
AE-LT-0032A	0.43	0.46	0.44	0.48	0.42
AE-LT-0033A	0.45	0.48	0.44	0.46	0.44
AE-LT-0034A	0.45	0.47	0.42	0.45	0.41
AE-LT-0041A	0.38	0.44	0.40	0.41	0.44
AE-LT-0042A	0.44	0.42	0.43	0.45	0.41
AE-LT-0043A	0.43	0.44	0.42	0.41	0.40
AE-LT-0044A	0.45	0.44	0.41	0.42	0.40

**Sensor's response time degraded between 36 and 48 months of service. The problem was corrected during an outage at 48 months.

Table 9.14. Examples of results of search of NPRDS database on problems with pressure transmitters in nuclear power plants

Problem	Cause	Resolution
During plant depressurization, pressure transmitter tracked very slowly and indicating 50 bar too high	Defective oscillator, force motor, and detector coil	Replaced force motor, detector coil, and oscillator
High end of flow transmitter's output sluggish	Unknown	Replace transmitter
Main steam line differential transmitter failed response time testing	Bad amplified card in transmitter	Amplified card replaced
Transmitter for post accident loss of coolant level indicating high and would not respond to different inputs	Differential transmitter cell failed due to leakage of material in cell	Replaced transmitter
Reactor coolant flow transmitter sluggish	Diaphragm ruptured due to wear out	Replaced transmitter
Transmitter for post loss of coolant accident level indication reading erroneously	Differential pressure transmitter cell failed due to loss of oil in cell	Replaced transmitter
Reactor coolant flow transmitter have a leak at the back of its sensing block	Unknown	Replaced transmitter's bellows unit
Silicon oil leaking into the level transmitter's electronics housing from the instruments differential pressure unit	O-ring groove was too deep	Replaced transmitter; all transmitters of this type inspected for oil leakage
Reactor vessel upper level transmitter found to have silicon oil leaking into the instrument's electronics housing from instrument's differential pressure unit	Silicon oil mistakenly left in the electronic housing during assembly	Transmitter replaced

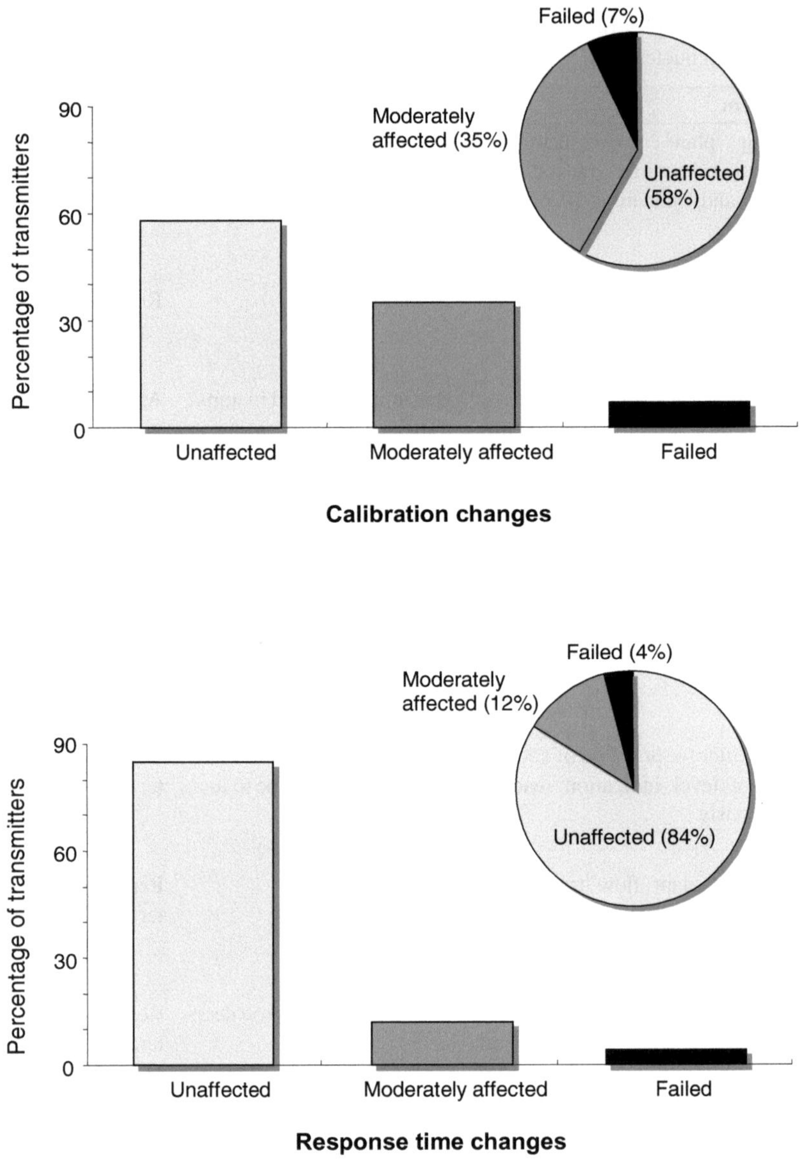

Fig. 9.17. Summary of results of experimental aging research on performance of nuclear plant pressure transmitters

summarizes the results of an aging research project to quantify the effects of normal aging on the calibration and response time of a sample of nuclear-grade pressure transmitters.[18] It is clear that aging affects the calibration of pressure transmitters more so than their response time.

Table 9.13 shows response time results from noise analysis testing performed on 16 transmitters over five years. The measurements were made using the noise analysis technique. Only one transmitter suffered response time degradation of about 30 percent after 36 months of service and this was later determined not to be due to the transmitter but due to a sensing line blockage. This is consistent with the nuclear industry's experience that response time degradation in pressure transmitters are more often due to sensing line blockages than degradation in the transmitter itself.

Table 9.14 provides a few examples from a search of the NPRDS database on failure of nuclear plant pressure transmitters. It is evident that response time degradation, although not prevalent, has been responsible for some of the failures that the nuclear industry has experienced over the years.

10

On-line Detection of Sensing Line Problems

We mentioned several times in Chap. 9 that the noise analysis technique provides the dynamic response of not only a pressure transmitter but also its sensing line. That is, when the response time of pressure transmitters are tested using the noise analysis technique, the results include any delays caused by the sensing line's length; and the effect of any blockages or voids in the sensing lines are automatically accounted for in the results.[20] These points are further verified by the examples presented in this chapter.

10.1 Sensing Line Blockages

Pressure sensing lines can become blocked for any number of reasons, including crud buildup, boron solidification, and isolation and equalizing valves that have been improperly lined up or incorrectly seated. These effects are accounted for when the noise analysis technique is used to measure response time of pressure transmitters. This is demonstrated by the results of laboratory measurements presented in Table 10.1 involving ramp and noise tests performed under the same conditions. The test setup and a photograph of the facilities used for these tests are shown in Figs. 10.1 and 10.2, respectively. The tests were performed according to the following procedure:

1. The response times of two fast-response reference transmitters were measured against one another. As expected, both the ramp and noise tests indicated a response time of essentially zero for the reference transmitter.

2. One of the two reference transmitters was moved away from the other using 35 meters of sensing lines. Its response time was then measured against the other reference transmitter. The response time did not increase significantly in spite of the 35 meters of sensing line. This is because the compliance of the reference transmitter is very small, and the effect of sensing line length on the response time is therefore insignificant.

3. The long sensing lines were removed, a snubber was installed in the sensing line leading to one of the reference transmitters, and the ramp and noise measurements were repeated. The snubber added about 0.3 seconds to the response time. This increase is indicated in the results of both the ramp and noise tests. It is apparent from the results in Table 10.1 that the noise analysis technique identifies the effect of the snubber (simulating a blockage).
4. The reference transmitter was replaced with a Barton transmitter and the previous three tests were repeated. As indicated by the results in Table 10.1, the effect of length and blockage is very significant for the Barton transmitter as this transmitter has a large compliance. More specifically, the 35 meters of sensing line added about 0.3 seconds to the response time of the Barton transmitter, and the snubber added about 3.0 seconds. In all three cases, the noise analysis technique provided comparable results to that of the ramp test.

Table 10.1. Experimental results on detection of sensing line blockages using the noise analysis technique

	Response Time (sec)	
Test Configuration	**Ramp Test**	**Noise Analysis**
	Reference Transmitter	
Reference alone	0.00	0.00
Reference & 35 meters	0.01	0.00
Reference & snubber	0.34	0.27
	Barton Transmitter	
Barton alone	0.12	0.17
Barton & 35 meters	0.27	0.28
Barton & snubber	3.00	2.94

Tests were performed with a 6.35 cm-diameter sensing line.

Based on numerous laboratory tests, such as the example just described as well as the author's observations of in-plant tests over 20 years, the following remarks can be made about the effect that sensing lines have on the dynamic response of a pressure sensing system:

- Long sensing lines and blockages (simulated here by a snubber) increase the response time of a pressure sensing system.
- Increases in response times caused by sensing line length and blockages depend on the compliance of the transmitter. The response time of transmitters with larger compliances (e.g., Barton transmitters) is more sensitive to sensing line length and blockages than that of transmitters with small compliances.

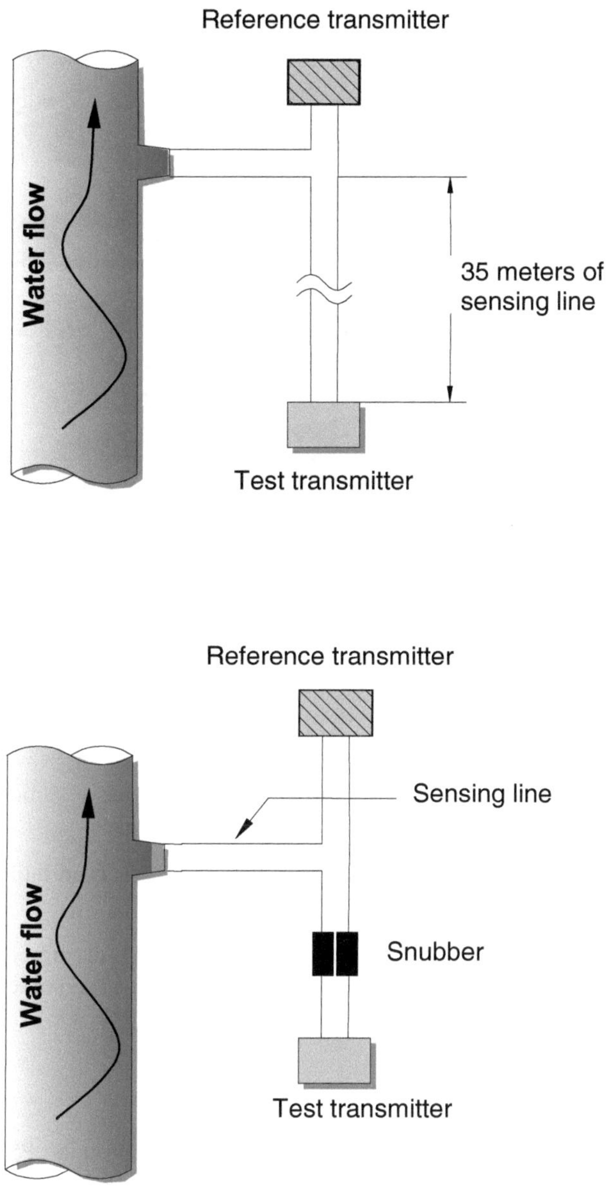

Fig. 10.1. Laboratory test setup to measure the effects of sensing line length and blockages on the response times of pressure sensing systems

Fig. 10.2. A portion of a laboratory test loop used to develop noise diagnostics for pressure sensing lines

- Response time results obtained using the noise analysis technique include the effects of long sensing lines and any significant blockages.
- Sensing line blockages occur because of valve failures, crud buildup, boron solidification, freezing, and the like.
- When noise analysis identifies a sensing line blockage, the line must be purged (flushed) and noise tests repeated to ensure the problem is resolved.

10.2 Air in Sensing Lines

Air or void in pressure sensing lines usually manifest itself on the PSDs of pressure noise signals. Fig. 10.3 shows theoretical results that demonstrate the effect of air on the dynamics of an underdamped pressure sensing system. The figure illustrates how air causes the resonance to move to lower frequencies, and how the response time increases as the amount of air in the sensing line is increased.

Fig. 10.4 shows PSDs of a Rosemount pressure transmitter that was tested in a laboratory with and without a large air pocket in the system. This experiment was performed using a 25-meter long, 6.35-mm-diameter steel pipe at a pressure of 5 bars. As expected, the air pocket creates a resonance on the PSD and reduces the transmitter's dynamic response.

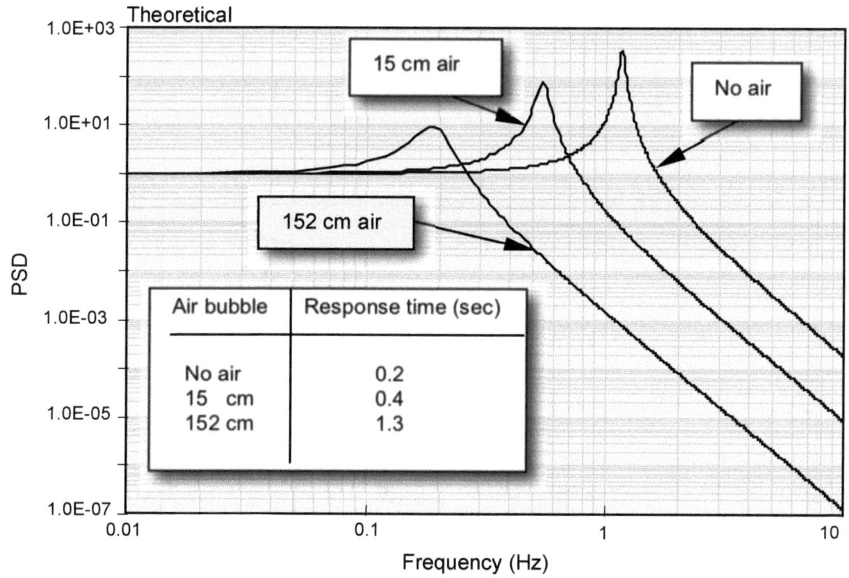

Fig. 10.3. Theoretical PSDs demonstrating the effect of air on dynamics of a pressure sensing system (sensing line inside diameter = 9.5 mm, at a pressure of 0.3 bar)

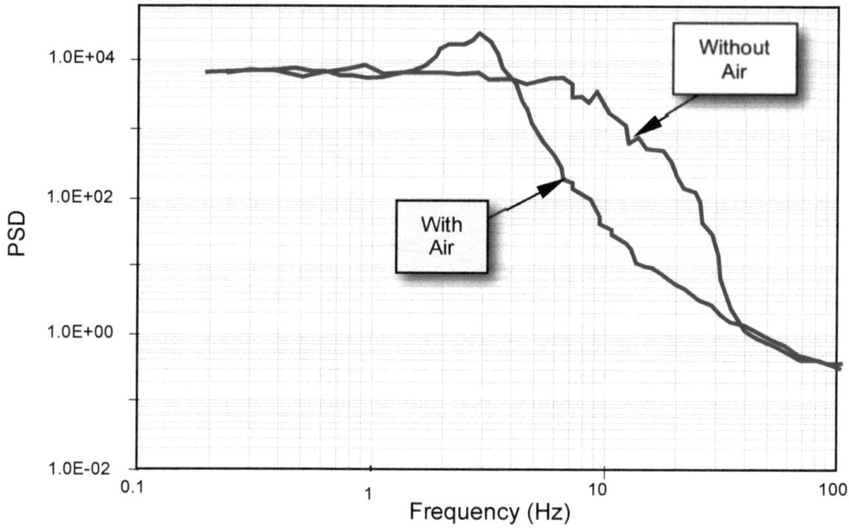

Fig. 10.4. Effect of air pocket on the shape and bandwidth of PSD of a pressure transmitter

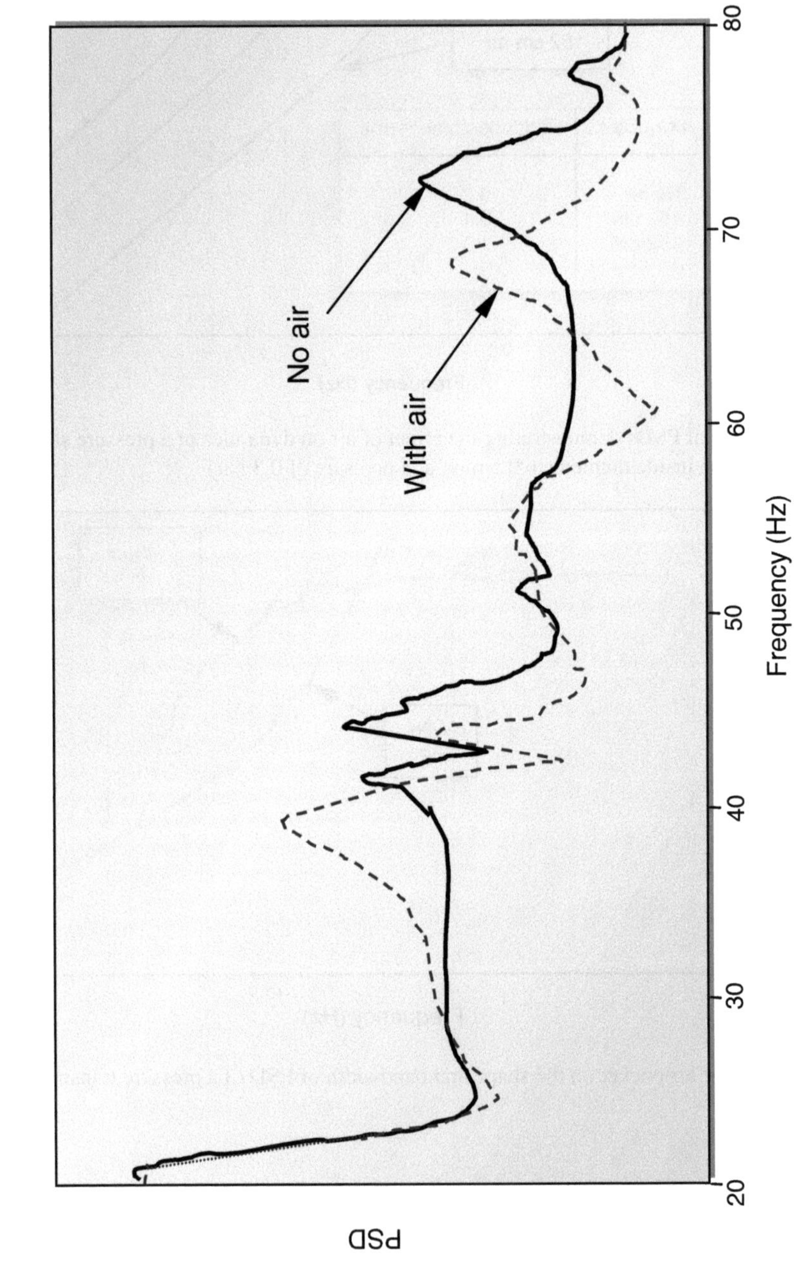

Fig. 10.5. Effect of void on PSD of noise signal for a pressure transmitter

Fig. 10.5 shows another demonstration of the effect of air on the dynamic response of a pressure sensing system. The resonances on this spectrum all shift to lower frequencies when air is injected into the system.

10.3 Detecting Sensing Line Leaks

Leaks in pressure sensing lines are common and usually cause drift at the output of the affected transmitters. The leak can sometimes be detected by monitoring the amplitude of the process noise at the pressure transmitter's output. Fig. 10.6 compares the normal noise output of a pressure transmitter with that of a transmitter that has a leak in its sensing line. It is apparent that the leak reduces the amplitude of the noise

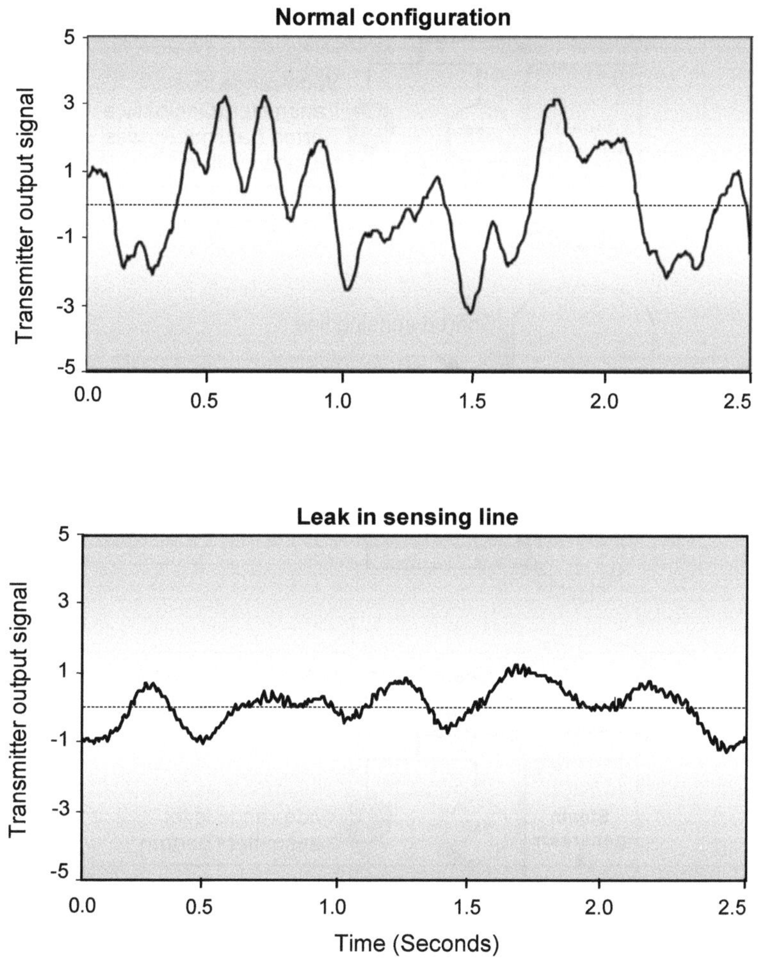

Fig. 10.6. Noise output of pressure transmitters with and without a leak in their sensing line

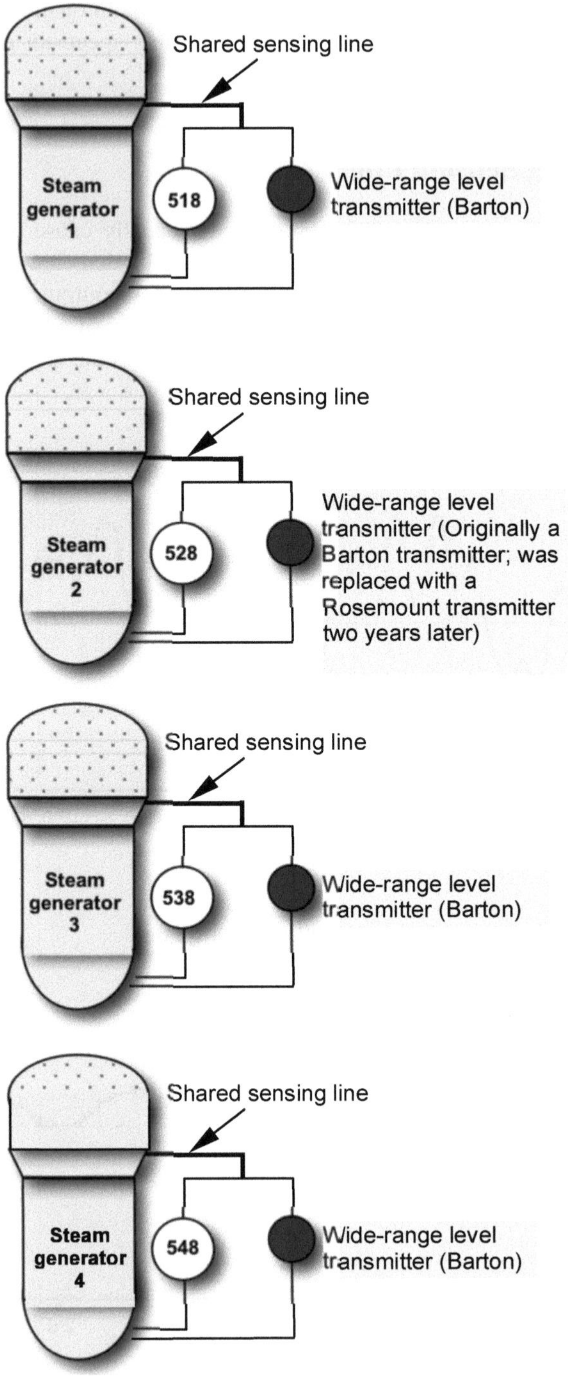

Fig. 10.7. Example of shared sensing line arrangement in a nuclear power plant

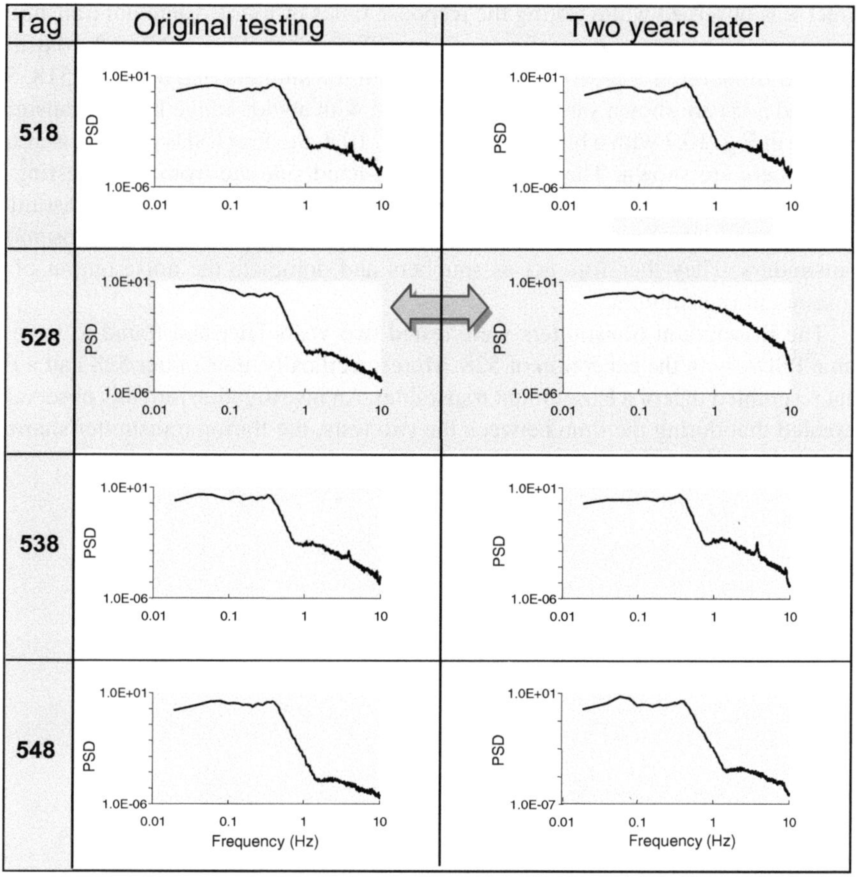

Fig. 10.8. PSDs of transmitters with shared sensing lines

signal. Therefore, if baseline noise data is available, sensing line leaks can be detected by using the noise analysis technique. If baseline data is not available, then the output of redundant transmitters may be intercompared to identify the leaking sensing line. To provide automated leak diagnostics, measure and track the RMS value of the noise data.

10.4 Problems with Shared Sensing Lines

In some plants, redundant transmitters may share a sensing line. In these plants, any blockages, voids, or leaks would affect all the transmitters that share the sensing line.

Shared sensing lines could also be a concern when transmitters with different compliances are installed on the same sensing lines. In these situations, each transmitter's response time is dominated by the transmitter that has the highest compliance. This effect was observed while testing the response times of four Rosemount transmitters used to measure steam generator level in a PWR plant. Figs. 10.7 and 10.8 illustrate the situation. In Fig. 10.7, four Rosemount transmitters (tag numbers 518, 528, 538, and 548) are shown sharing a sensing line with a wide-range Barton transmitter (shown in Fig. 10.7 with a black circle). In Fig. 10.8, the four PSDs of the Rosemount transmitters are shown. The PSDs on the left-hand side are from noise testing the Rosemount transmitters, but their shapes correspond to that of Barton transmitters. This is because the Barton transmitters have larger compliances than the Rosemount transmitters. They therefore act as snubbers and dominate the noise output of the Rosemount transmitters.

The Rosemount transmitters were tested two years later and found to have the same PSDs, with the exception of 528. More specifically, transmitter 528 had a PSD that resembled that of a Rosemount transmitter. An investigation into this observation revealed that during the time between the two tests, the Barton transmitter sharing a sensing line with 528 was replaced with a Rosemount transmitter.

About the Author

H.M. Hashemian is the founder and President of Analysis Measurement Services Corporation (AMS), a nuclear engineering company established in Knoxville, Tennessee in 1977. Headquartered in the United States with representatives in Europe and Asia, AMS specializes in testing the instrumentation and control systems of nuclear power plants, reactor diagnostics, and automated test equipment and software development for process and power industries. As a full-time employee of AMS for nearly thirty years, Mr. Hashemian has worked for numerous utilities around the world and a variety of organizations such as the U.S. Department of Energy, U.S. Nuclear Regulatory Commission, National Aeronautics and Space Administration, U.S. Air Force, International Atomic Energy Agency (IAEA), and International Electrotechnical Commission (IEC). He is a Fellow of the Instrumentation, Systems, and Automation Society (ISA), present or past member of the American Nuclear Society, Institute of Electrical and Electronics Engineers, American Society for Testing and Materials, and European Nuclear Society. He is the author or co-author of nearly 150 publications including books and book chapters, journal articles, conference papers, and government reports. Additionally, he has been directly responsible for developing a number of ISA and IEC standards on the performance of nuclear plant instrumentation and control equipment.

Mr. Hashemian was born in Tehran, Iran and came to the United States in 1974 after receiving his Bachelor of Science degree in physics from the National University of Iran. He entered the University of Tennessee in 1975 where he received his Master of Science degree in nuclear engineering in 1977. He continued his education in the University of Tennessee's doctoral program in nuclear engineering, but left in the early 1980s to attend full time to his company.

Acknowledgement

The effort of all technical and administrative staff of Analysis and Measurement Services Corporation who helped with the production of this book is gratefully acknowledged. In particular, Ms. Elain Crumley is appreciated for her role in managing the administrative aspects of this book and working with the publisher, Mr. Darrell Mitchell is recognized for directing the production of the figures, drawings, photographs and coordinating the effort of technical staff who helped the author. Also, Mr. Paul Bodine, an independent technical book editor, is acknowledged for his role in proofreading and editing this book.

Acronyms and Abbreviations

AMS	Analysis and Measurement Services Corporation
ANO	Arkansas Nuclear One
ANSI	American National Standards Institute
APD	amplitude probability density
AR	autoregressive (refers to autoregressive modeling)
ASTM	American Society for Testing and Materials
B&W	Babcock and Wilcox
BTP-13	Branch Technical Position 13
BWR	Boiling Water Reactor
C	capacitance
CANDU	Canadian deuterium reactor (Canadian heavy water reactor)
CEA	Commissariat à l'Energie Atomique
CF	correction factor
CFR	Code of Federal Regulations
CRBR	Clinch River Breeder Reactor
CRDM	control rod drive mechanism
DBE	design basis event
DOD	U.S. Department of Defense
DOE	U.S. Department of Energy
DPU	differential pressure unit
EdF	Electricité de France
EMF	electromotive force
EMI/RFI	electromagnetic/radio frequency interference
EPRI	Electric Power Research Institute

FFT	fast Fourier transform
HELB	high-energy line break
HFIR	high flux isotope reactor
HRP	Halden Reactor Project
HTGR	high-temperature gas-cooled reactor
I&C	instrumentation and control
IAEA	International Atomic Energy Agency
IEC	International Electrotechnical Commission
IEEE	Institute of Electrical and Electronics Engineers
INPO	Institute of Nuclear Power Operations
IR	insulation resistance
ISA	Instrumentation, Systems, and Automation Society (formerly Instrument Society of America)
L	inductance
LCR	inductance, capacitance, and resistance
LCSR	loop current step response (refers to the method for in-situ response time testing of RTDs and thermocouples)
LER	licensee event report
LMFBR	liquid metal fast breeder reactor
LOCA	loss-of-coolant accident
LPRM	local power range monitor
mA	Milliampere
mW	Milliwatt
NASA	National Aeronautics and Space Administration
NBS	National Bureau of Standards (now National Institute of Standards and Technology – NIST)
NI	neutron instrumentation
NII	Nuclear Installation Inspectorate (UK)
NIST	National Institute of Standards and Technology (formerly the National Bureau of Standards – NBS)
NMAC	Nuclear Maintenance Assistance Center
NPAR	Nuclear Plant Aging Research Program
NPRDS	Nuclear Plant Reliability Data System
NRC	Nuclear Regulatory Commission (U.S.)
NRR	Office of Nuclear Regulatory Regulation
NTIS	National Technical Information Service
ORNL	Oak Ridge National Laboratory

PRT	platinum resistance thermometer
PSD	power spectral density
PWR	pressurized water reactor
QA	quality assurance
R	resistance
R&D	research and development
RCP	reactor coolant pump
RES	NRC Office of Research
RMS	root mean square
RSS	root sum squared
RTD	resistance temperature detector
SBIR	Small Business Innovation Research
SER	safety evaluation report
SG	steam generator
SHI	self-heating index
SPRT	standard platinum resistance thermometers
TDR	time domain reflectometry
TECDOC	IAEA technical report
TMI	Three Mile Island (refers to Three Mile Island nuclear power plant)
TVA	Tennessee Valley Authority
UT	University of Tennessee in Knoxville
VVER	Russian PWR

References

1. Hashemian, H.M. (2005) Sensor Performance and Reliability. ISA - Instrumentation, Systems, and Automation Society, Research Triangle Park, North Carolina.
2. Analysis and Measurement Services Corporation (1979) Response Time Qualification of Resistance Thermometers in Nuclear Power Plant Safety System. Topical Report submitted to the NRC under Millstone 2 Docket No. 50–336, AMS, Knoxville, Tennessee.
3. Electric Power Research Institute (2000) On-line Monitoring of Instrument Channel Performance. EPRI Topical Report No TR-104965-RI, NRC SER, Palo Alto, California.
4. Electric Power Research Institute (1997) Rod Control System Maintenance Guide for Westinghouse Pressurized Water Reactors. Nuclear Maintenance Application Center (NMAC) Report No TR-108152, Palo Alto, California.
5. U.S. Nuclear Regulatory Commission (1998) Advanced Instrumentation and Maintenance Technologies for Nuclear Power Plants. NUREG/CR-5501, Washington, DC.
6. Tew, WL, SW, Benz, SP, Dresselhaus, P, and Hashemian, HM (2005) New Technologies for Noise Thermometry with Applications in Harsh and-or Remote Operating Environments. 51st International Instrumentation Symposium of Instrumentation, Systems and Automation Society (ISA), Knoxville, Tennessee.
7. U.S. Nuclear Regulatory Commission (1990) Aging of Nuclear Plant Resistance Temperature Detectors. NUREG/CR-5560, Washington, DC.
8. U.S. Nuclear Regulatory Commission (1995) On-Line Testing of Calibration of Process Instrumentation Channels in Nuclear Power Plants. NUREG/CR-6343, Washington, DC.
9. Hashemian, HM (2003) Instrument Calibration. Instrument Engineers' Handbook, Fourth Edition, Chapter 1.8, Process Measurement and Analysis, Volume 1, CRC Press.
10. U.S. Nuclear Regulatory Commission (1997) Standard Review Plan for the Review of Safety Analysis Reports for Nuclear Power Plants. NUREG-0800, Washington, DC.
11. U.S. Nuclear Regulatory Commission (1981) Safety Evaluation Report Review of Resistance Temperature Detector Time Response Characteristics. NUREG-0809, Washington, DC.
12. Electric Power Research Institute (1978) In-Situ Response Time Testing of Platinum Resistance Thermometers. EPRI Report No NP-834, Vol 1, Palo Alto, California.
13. Electric Power Research Institute (1980) Temperature Sensor Response and Characterization. EPRI Report No NP-1486, Palo Alto, California.
14. Electric Power Research Institute (1977) In Situ Response Time Testing of Platinum Resistance Thermometers. EPRI Report Number NP-459, Project 503-3, Palo Alto, California.
15. Hashemian, HM (2003) Response Time and Drift Testing. Instrument Engineers' Handbook, Fourth Edition, Chapter 1.9, Process Measurement and Analysis, Vol 1, CRC Press.
16. Hashemian, HM (2002) Safety Instrumentation and Justification of Its Cost. Instrument Engineers' Handbook, Third Edition, Chapter 2.11, Process Software and Digital Networks, CRC Press.
17. Hashemian, HM (2002) Optimized Maintenance and Management of Aging of Critical Equipment in Nuclear Power Plants, Power Plant Surveillance and Diagnostics. Chapter 3, Applied Research with Artificial Intelligence, Springer-Verlag, New York.
18. U.S. Nuclear Regulatory Commission (1993) Long Term Performance and Aging Characteristics of Nuclear Plant Pressure Transmitters. NUREG/CR-5851, Washington, DC.
19. U.S. Nuclear Regulatory Commission (1995) Assessment of Fiber Optic Pressure Sensors. NUREG/CR-6312, Washington, DC.
20. International Atomic Energy Agency (2000) Management of Ageing of I&C Equipment in Nuclear Power Plants. IAEA Publication TECDOC-1147, Vienna, Austria.

Appendix A
Bibliography

Appendix A: Bibliography

This appendix contains a comprehensive listing of publications by the author and his colleagues at AMS and elsewhere on topics that are related directly or indirectly to the material covered in this book. Most of the publications contained in this appendix are available from one or more of the following resources. Note that the attached list includes patents awarded or pending to author and his colleagues.

U.S. National Technical Information Services (NTIS) www.ntis.gov
U.S. Government Printing Office www.gpo.gov
IEC www.iec.ch
IAEA www.iaea.org
NRC Public Document Room www.nrc.gov
AMS www.ams-corp.com
ISA www.isa.org
ASTM www.astm.org
United States Patent and Trademark Office www.uspto.gov

Addresses for organizations just mentioned are:

- National Technical Information Service, 5285 Port Royal Road, Springfield, VA 22161 USA
- U.S. Government Printing Office, 732 North Capitol St. NW, Washington, DC 20401 USA
- IEC Central Office, 3, rue de Varembé, P.O. Box 131, CH - 1211 GENEVA 20, Switzerland
- International Atomic Energy Agency, P.O. Box 100, Wagramer Strasse 5, A-1400 Vienna, Austria
- AMS, 9111 Cross Park Drive, Building A-100, Knoxville, TN 37923 USA
- ISA, 67 Alexander Drive, PO Box 12277, Research Triangle Park, NC 27709 USA
- ASTM International, 100 Barr Harbor Drive, PO Box C700, West Conshohocken, PA, 19428-2959 USA
- Office of Public Affairs, U.S. Patent and Trademark Office, P. O. Box 1450, Alexandria, VA 22313-1450 USA

U.S. Government Publications

G1. Hashemian, H.M., Holbert, K.E., Kerlin, T.W., Upadhyaya, B.R., "A Low Power Fourier Transform Processor." NASA Goddard Space Flight Center, Contract Number NAS5-28635 (July 1985).

G2. Hashemian, H.M., "Determination of Installed Thermocouple Response." U.S. Air Force, Arnold Engineering Development Center, Report Number AEDC-TR-86-46 (December 1986).

G3. Hashemian, H.M., Holbert, K.E., Thie, J.A., Upadhyaya, B.R., Kerlin, T.W., Petersen, K.M., Beck, J.R., "Sensor Surveillance Using Noise Analysis." U.S. Department of Energy, Contract Number DE-AC05-86ER80405 (March 1987).

G4. Hashemian, H.M., et al., "Degradation of Nuclear Plant Temperature Sensors." U.S. Nuclear Regulatory Commission, Report Number NUREG/CR-4928 (June 1987).

G5. Hashemian, H.M., et al., "Effect of Aging on Response Time of Nuclear Plant Pressure Sensors." U.S. Nuclear Regulatory Commission, NUREG/CR-5383 (June 1989).

G6. Hashemian, H.M., et al., "Aging of Nuclear Plant Resistance Temperature Detectors." U.S. Nuclear Regulatory Commission, Report Number NUREG/CR-5560 (June 1990).

G7. Hashemian, H.M., "New Technology for Remote Testing of Response Time of Installed Thermocouples." United States Air Force, Arnold Engineering Development Center, Report Number AEDC-TR-91-26, Volume 1 - Background and General Details (January 1992).

G8. Hashemian, H.M., and Mitchell, D.W., "New Technology for Remote Testing of Response Time of Installed Thermocouples." United States Air Force, Arnold Engineering Development Center, Report Number AEDC-TR-91-26, Volume 2 - Determination of Installed Thermocouple Response Research Data (January 1992).

G9. Hashemian, H.M., et al., "New Technology for Remote Testing of Response Time of Installed Thermocouples." United States Air Force, Arnold Engineering Development Center, Report Number AEDC-TR-91-26, Volume 3 - Thermocouple Response Time Test Instrumentation (January 1992).

G10. Hashemian, H.M., et al., "Validation of Smart Sensor Technologies for Instrument Calibration Reduction in Nuclear Power Plants." U.S. Nuclear Regulatory Commission, NUREG/CR-5903 (January 1993).

G11. Hashemian, H.M., "Long Term Performance and Aging Characteristics of Nuclear Plant Pressure Transmitters." U.S. Nuclear Regulatory Commission, NUREG/CR-5851 (March 1993).

G12. Hashemian, H.M., et al., "Improved Temperature Measurement in Composite Material for Aerospace Applications." National Aeronautics and Space Administration, Marshall Space Flight Center, Contract Number NAS8-39814, MSFC, AL (July 1993).

G13. Hashemian, H.M., et al., "Assessment of Fiber Optic Pressure Sensors." U.S. Nuclear Regulatory Commission, NUREG/CR-6312 (April 1995).

G14. Hashemian, H.M., et al., "New Sensor for Measurement of Low Air Flow Velocity." U.S. Nuclear Regulatory Commission, NUREG/CR-6334 (August 1995).

G15. Hashemian, H.M., "On-Line Testing of Calibration of Process Instrumentation Channels in Nuclear Power Plants." U.S. Nuclear Regulatory Commission, NUREG/CR-6343 (November 1995).

G16. Hashemian, H.M., Shell, C.S., and C.N. Jones, “New Instrumentation Technologies for Testing the Bonding of Sensors to Solid Materials." National Aeronautics and Space Administration, Marshall Space Flight Center, NASA/CR-4744 (May 1996).
G17. Hashemian, H.M., et al., “Advanced Instrumentation and Maintenance Technologies for Nuclear Power Plants." U.S. Nuclear Regulatory Commission, NUREG/CR-5501 (August 1998).

Journal Articles

J1. Kerlin, T.W., Miller, L.F., and Hashemian, H.M., “In-Situ Response Time Testing of Platinum Resistance Thermometers." ISA Transactions, Volume 17, No. 4, page 71-88 (1978).
J2. Kerlin, T.W., Hashemian, H.M., and Petersen, K.M., “Time Response Testing of Temperature Sensors." ISA Transactions, Volume 20, No. 1 (1981).
J3. Hashemian, H.M., Thie, J.A., and Upadhyaya, B.R., “Reactor Sensor Surveillance Using Noise Analysis." Nuclear Science and Engineering, Vol. 98, Number 2, pp. 96-102 (February 1988).
J4. Hashemian, H.M., et al., “In Situ Response Time Testing of Thermocouples." ISA Transactions, Volume 29, Number 4, pp. 97-104 (1990).
J5. Hashemian, H.M., and Petersen, K.M., “Measurement of Performance of Installed Thermocouples." ISA Test Measurement Division Newsletter, Volume 29, Number 2 (April 1992).
J6. Hashemian, H.M., “Effects of Normal Aging on Calibration and Response Time of Nuclear Plant Resistance Temperature Detectors and Pressure Sensors." Nuclear Safety Technical Progress Journal, Volume 35, Number 2, pp. 223-234 (1994).

Magazine Articles

M1. Hashemian, H.M., and Petersen, K.M., “Boosting Accuracy of Industrial Temperature Measurement." Power Magazine, Vol. 133, No. 3, pp. 74-76 (May 1989).
M2. Hashemian, H.M., and Petersen, K.M., “Assuring Accurate Temperature Measurement." Intech Magazine, Vol. 36, No. 10, pp. 45-48 (October 1989).
M3. Hashemian, H.M., et al., “Measuring Critical Process Temperatures." Control Magazine, Vol. III, No. 4, pp. 62-69 (April 1990).
M4. Riner, J.L., “In-Situ Testing of Industrial Sensors." Control Magazine, Vol. 38, Number 15, pp. 175-176 (November 1991).
M5. Petersen, K.M., “Testing Sensors For Accuracy and Speed." Chemical Engineering Magazine, pp. 131-134 (February 1992).
M6. Hashemian, H.M., Petersen, K.M., “In-Situ Tests Gauge Thermocouple Performance, Part 1." Intech, Vol. 40, No. 1, pp. 30-32 (January 1993).
M7. Hashemian, H.M., Petersen, K.M., “In-Situ Tests Gauge Thermocouple Performance, Part 2." Intech, Vol. 40, No. 6, pp. 31-33 (June 1993).
M8. Hashemian, H.M., “Performance Testing of Nuclear Plant Temperature and Pressure Sensors." Nuclear Plant Journal, Vol. 12, No. 7, pp. 46 (November-December 1994).
M9. Hashemian, H.M., Riner, J.L., K.M. Petersen, D.D. Beverly, “Online Testing Assures Integrity of Sensors." Chemical Engineering, Vol. 102, No. 3, pp. 84-88 (March 1995).

M10. Hashemian, H.M., "Catching the Drift." Pressure, a supplement to InTech Magazine, an ISA publication (August 2003).
M11. ISA Editor (for H.M. Hashemian's book), "Impulse Lines are Vital Intelligence Link." Intech Magazine, an ISA publication (November 2005).

Technical Reports

T1. Hashemian, H.M., Jacquot, J.P., and Guerin, B., "Preliminary Report: Response Time Testing of Resistance Temperature Detectors (RTDs) in EDF Test Loop." Electricité de France, EDF Report Number HP/236/79/04 (January 1978).
T2. Kerlin, T.W., Miller, L.F., Hashemian, H.M., and Poore, W.P., "In-Situ Response Time Testing of Platinum Resistance Thermometers." Electric Power Research Institute, EPRI Report Number NP-834, Vol. 1 (July 1978).
T3. Kerlin, T.W., Miller, L.F., Hashemian, H.M., Poore, W.P., Shorska, M., Upadhyaya, B.R., Cormault, P., Jacquot, J.P., "Temperature Sensor Response Characterization." Electric Power Research Institute, EPRI Report Number NP-1486 (August 1980).
T4. Beverly, D., Fain, R., Mitchell, D., Shell, C., Hashemian, H.M., Dressler, R., "Rod Control System Maintenance Guide for Westinghouse Pressurized Water Reactors." Electric Power Research Institute, EPRI Report Number TR-108152 (April 1997).

IAEA Publications (following are publications in which the author has been directly involved)

IA1. "Management of Ageing of I&C Equipment in Nuclear Power Plants." IAEA Publication TECDOC-1147, Vienna, Austria (June 2000).
IA2. "Harmonization of the Licensing Process for Digital Instrumentation and Control Systems in Nuclear Power Plants." IAEA Publication TECDOC-1327, Vienna, Austria (December 2002).
IA3. "Management of Life Cycle and Aging at Nuclear Power Plants: Improved I&C Maintenance,"IAEA Publication TECDOC-1402, Vienna, Austria (August 2004).
IA4. "On-Line Monitoring for Nuclear Power Plants, Part 2: Instrumentation Channel Performance Monitoring," New IAEA document due for publication in 2006.
IA5. "On-Line Monitoring for Nuclear Power Plants Part 2: Process and Component Condition Monitoring and Diagnostics," New IAEA documented due for publication in 2007.

Books or Book Chapters

B1. Hashemian, H.M., " Safety Instrumentation and Justification of Its Cost." Instrument Engineers' Handbook, Third Edition, Chapter 2.11, Process Software and Digital Networks, CRC Press, 2002.
B2. Hashemian, H.M., " Instrument Calibration." Instrument Engineers' Handbook, Fourth Edition, Chapter 1.8, Process Measurement and Analysis, Volume 1, CRC Press, 2003.

B3. Hashemian, H.M., " Response Time and Drift Testing." Instrument Engineers' Handbook, Fourth Edition, Chapter 1.9, Process Measurement and Analysis, Volume 1, CRC Press, 2003.
B4. Lipták, Béla and Hashemian, H.M., " Cross-Correlation Flow Metering." Instrument Engineers' Handbook, Fourth Edition, Chapter 2.5, Process Measurement and Analysis, Volume 1, CRC Press, 2003.
B5. Hashemian, H.M., "Optimized Maintenance and Management of Aging of Critical Equipment in Nuclear Power Plants," Power Plant Surveillance and Diagnostics, Chapter 3, Applied Research with Artificial Intelligence, Springer-Verlay, 2002.
B6. Hashemian, H.M., "Sensor Performance and Reliability." Book published by ISA—The Instrumentation, Systems, and Automation Society, © 2005.

International Publications

I1. Kerlin, T.W., Hashemian, H.M., and Petersen, K.M., "Response Characteristics of Temperature Sensors Installed in Processes." Vol. V/I, International Measurement Confederation (IMEKO), 9th World Congress, Berlin, Germany (May 1982).
I2. Hashemian, H.M., Thie, J. A., and Upadhyaya, B. R., Holbert, K.E., "Sensor Response Time Monitoring Using Noise Analysis." Proceedings of the Fifth Specialists Meeting on Reactor Noise, Progress in Nuclear Energy, Pergamon Press, Vol. 21, pp. 583–592, Munich, FRG (October 1987).
I3. Hashemian, H.M., Petersen, K.M., et al., "Aging Effects on Calibration and Response Time of Temperature Sensors in PWRs." Proceedings of 1989 Conference on Operability of Nuclear Systems in Normal and Adverse Environments (OPERA '89), Vol. 1, pp. 275–282, Lyon, France (September 1989).
I4. Hashemian, H.M., Petersen, K.M., and Miller, L.F., "Dynamic Testing of Pressure Sensing Systems in Nuclear Power Plants." Proceedings of 1989 Conference on Operability of Nuclear Systems in Normal and Adverse Environments (OPERA '89), Vol. 2, pp. 935–942, Lyon, France (September 1989).
I5. Hashemian, H.M., and Petersen, K.M., "On-Line Testing of Performance of Nuclear Plant Pressure Transmitters Using Noise Analysis." IMORN-22 Conference, Delft University of Technology, Interfaculty Reactor Institute, The Netherlands (June 1990).
I6. Hashemian, H.M., and Petersen, K.M., "Achievable Accuracy and Stability of Industrial RTDs." Published by the American Institute of Physics, Seventh International Symposium on Temperature, Volume Six, pp. 427-432, Toronto, Canada (May 1992).
I7. Hashemian, H.M., and Petersen, K.M., "Loop Current Step Response Method For In-Place Measurement of Response Time of Installed RTDs and Thermocouples." Published by American Institute of Physics, Seventh International Symposium on Temperature, Volume Six, pp. 1151–1156, Toronto, Canada (May 1992).
I8. Hashemian, H.M., "Measurement of Static and Dynamic Performance of Temperature and Pressure Instrumentation in Nuclear Power Plants." Presented at the 1992 COMADEM International Congress, CETIM, Paris, France (July 1992).
I9. Hashemian, H.M., et al., "Aging of Nuclear Plant RTDs and Pressure Transmitters." Proceedings of PLEX '93 International Conference and Exhibition, pp. 85–99, Zurich, Switzerland (November 29-December 1, 1993).
I10. Hashemian, H.M., Riner, J.L., "On-Line Monitoring of Calibration Drift of Process Instrumentation Channels in Nuclear Power Plants." Proceedings of PLEX '93 International Conference and Exhibition, pp. 488–491, Zurich, Switzerland (November 29 -December 1, 1993).

I11. Hashemian, H.M., Riner, J.L., "In-Situ Response Time Testing of RTDs and Pressure Transmitters in Pressurized Water Reactors." Proceedings of PLEX '93 International Conference and Exhibition, pp. 476–479, Zurich, Switzerland (November 29 - December 1, 1993).

I12. Hashemian, H.M., "Aging of Nuclear Plant Instrumentation." Proceedings of 1st International Symposium on Nuclear Energy (AREN), pp. 129-152, Bucharest, Romania (December 1993).

I13. Hashemian, H.M., "Precision Calibration of RTDs for Nuclear Power Plants." Proceedings of 1st International Symposium on Nuclear Energy (AREN), pp. 153–169, Bucharest, Romania (December 1993).

I14. Hashemian, H.M., "Response Time Testing of Temperature and Pressure Sensors in Nuclear Power Plants." Proceedings of 1st International Symposium on Nuclear Energy (AREN), pp. 170–194, Bucharest, Romania (December 1993).

I15. Hashemian, H.M., "Aging Characteristics of Nuclear Plant RTDs and Pressure Transmitters." Proceedings of the 4th International Topical Meeting on Nuclear Thermal Hydraulics, Operations and Safety, Paper #32-C, Vol 2, pp 32-C-1 - 32-C-6, Taipei, Taiwan (April 1994).

I16. Hashemian, H.M., and Mitchell, D.W., "On-Line Testing of Calibration of Process Instrumentation Channels in Nuclear Power Plants." Proceedings of the Fourth International Topical Meeting on Nuclear Thermal Hydraulics, Operations and Safety, Paper #33-B, Vol 2, pp. 33-B-1 - 33-B-6, Taipei, Taiwan (April 1994).

I17. Hashemian, H.M., "On-Line Measurement of Response Time of Temperature and Pressure Sensors in PWRs." Proceedings of the Fourth International Topical Meeting on Nuclear Thermal Hydraulics, Operations and Safety, Paper #33-C, Vol 2, pp. 33-C-1 - 33-C-6, Taipei, Taiwan (April 1994).

I18. Hashemian, H.M., Jakubenko, I., Forejt, V., "Response Time Testing and Calibration of Temperature and Pressure Sensors in Nuclear Power Plants." Proceedings of the Specialist's Meeting on Instrumentation and Control of WWER Type Nuclear Power Plants, Prague/Rez, Czech Republic (September 1994).

I19. Hashemian, H.M., Jakubenko, I., Forejt, V., "Measurement of Core Barrel Vibration and Testing for Core Flow Anomalies in Pressurized Water Reactors." Proceedings of the Specialist's Meeting on Instrumentation and Control of WWER Type Nuclear Power Plants, Prague/Rez, Czech Republic (September 1994).

I20. Hashemian, H.M., Jakubenko, I., Forejt, V., "Measurement of Drop Time of Control and Shutdown Rods and Testing of CRDMs in Nuclear Power Plants." Proceedings of the Specialist's Meeting on Instrumentation and Control of WWER Type Nuclear Power Plants, pp. 61–76, Prague/Rez, Czech Republic (September 1994).

I21. Hashemian, H.M., "In-Situ Response Time Testing of Temperature and Pressure Sensors in Nuclear Power Plants." Presented at the ENC '94 ENS-ANS-FORATOM World Exhibition, Lyon, France (October 1994).

I22. Hashemian, H.M., "New Methods for On-Line Testing of Calibration of Temperature and Pressure Sensors in Nuclear Power Plants." Presented at the ENC '94 ENS-ANS-FORATOM World Exhibition, Lyon, France (October 1994).

I23. Hashemian, H.M., "On-Line Response Time and Calibration Testing of Instrumentation in Nuclear Power Plants." Presented at the INEC 2nd International Conference on Control & Instrumentation in Nuclear Installations, London, England (April 1995).

I24. Hashemian, H.M., "On-Line Testing of Response Time and Calibration of Temperature and Pressure Sensors in Nuclear Power Plants." Proceedings of the Third International Conference on Nuclear Engineering – ICONE-3, pp. 1501–1506, Kyoto, Japan (April 1995).

I25. Hashemian, H.M., "Experience with a PC-Based System for Noise and DC Signal Analysis in PWRs." Presented at SMORN VII, Avignon-Palais des Papes, France (June 1995).
I26. Hashemian, H.M., "Results of Aging Research on Nuclear Plant Instrumentation." Presented at the IAEA Specialists Meeting, San Carlos de Bariloche, Argentina (October 1995).
I27. Hashemian, H.M., "Automated Testing of Critical Nuclear Plant Equipment." Presented at the IAEA Specialists Meeting, San Carlos de Bariloche, Argentina (October 1995).
I28. Hashemian, H.M., "PC-Based Equipment and Techniques for Testing the Performance of Critical Equipment in Nuclear Power Plants." Presented at the PLEM + PLEX 95 Nuclear Plant Life Management and Extension Conference, Nice, France (November 1995).
I29. Hashemian, H.M., "New Methods for Monitoring the Performance of Critical Equipment in Nuclear Power Plants." Invited paper for the IAEA Specialists Meeting, Session 2, San Carlo de Bariloche, Argentina (October 17–19, 1995).
I30. Mediavilla,F, Hashemian, H.M., "Medida Automatica Del Tiempo De Caida De Barras De Control Y De Parada y Pruebas De Actuacion Del Mecanismo De Accionamiento De Barras." Proceedings of 23rd Annual Meeting of the Spanish Nuclear Society, pp. 286–288, La Coruna, Spain (November 5–7, 1997).
I31. Hashemian, H.M., "History of On-Line Calibration Monitoring Developments in Nuclear Power Plants." Presented at the Technical Meeting on Increasing Instrument Calibration Interval Through On-Line Calibration Technology, International Atomic Energy Agency (IAEA), Halden, Norway (September 27–29, 2004).
I32. Hashemian, H.M., "Implementation of On-Line Monitoring to Increase the Calibration Interval of Pressure Transmitters." Presented at the Technical Meeting on Increasing Instrument Calibration Interval Through On-Line Calibration Technology, International Atomic Energy Agency (IAEA), Halden, Norway (September 27–29, 2004).
I33. Hashemian, H.M., "Equipment Life Cycle Management Through On-Line Condition Monitoring." Presented at the Technical Meeting on On-Line Condition Monitoring of Equipment and Processes in Nuclear Power Plants Using Advanced Diagnostic Systems, International Atomic Energy Agency (IAEA), Knoxville, TN (June 27–30, 2005).
I34. Hashemian, H.M., "Aging Management Through On-Line Condition Monitoring." Presented at the PLIM + PLEX 2006 Conference, Paris, France (April 10–11, 2006).

National Publications

N1. Kerlin, T.W., Mott, J.E., Warner, DC, Hashemian, H.M., Arendt, J.S., Gentry, T.S., and Cain, D.G., "Progress in Development of a Practical Method for In-Situ Response Time Testing of Platinum Resistance Thermometers." Transactions of American Nuclear Society, Vol. 23, 428 (June 1976).
N2. Kerlin, T.W., Miller, L.F., and Hashemian, H.M., "In-Situ Response Time Testing of Temperature Sensors." Transactions of American Nuclear Society, Volume 27, page 679–680 (1977).
N3. Hashemian, H.M., and Kerlin, T.W., "Response Time Testing of Platinum-Resistance Thermometers at St. Lucie Nuclear Station." Transactions of American Nuclear Society, page 532 (June 1978).
N4. Hashemian, H.M., and Kerlin, T.W., "Validation of Techniques for Response Time Testing of Temperature Sensors in PWRs." Transactions of American Nuclear Society, page 736 (June 1980).

N5. Kerlin, T.W., Hashemian, H.M., Petersen, K.M., Thomas, J., Snodgrass, D., Haynes, H., and Elder, R.T., "Response Time Testing of Resistance Thermometers in Operating Pressurized Water Reactors." Transactions of American Nuclear Society, pp. 324–325 (1980).

N6. Kerlin, T.W., Hashemian, H.M., and Petersen, K.M., "Time Response of Temperature Sensors." Paper C.I. 80-674, Proceedings of the ISA '80 International Conference and Exhibit, Houston, TX (October 1980).

N7. Hashemian, H.M., et al., "Resistance Thermometer Response Time Characterization." SECON'82, "Instrumenting for a Better World." 28th Southeastern Conference and Exhibit, Sponsored by the Instrument Society of America, Richmond, VA (May 1982).

N8. Kerlin, T.W., and Hashemian, H.M., "New Methods for Response Time Qualification of Temperature Sensors." Conference Proceedings Volume 2, Sensors and Systems 82 Conference, Chemical Sensors/Various Sensors Application, Chicago, IL (June 1982).

N9. Kerlin, T.W., Shepard, R.L., Hashemian, H.M., and Petersen, K.M., "Response of Installed Temperature Sensors." Temperature its Measurement and Control in Science and Industry, Volume 5, pp. 1357–1366, American Institute of Physics, Washington, DC (1982).

N10. Hashemian, H.M., and Kerlin, T.W., "Experience with RTD Response Time Testing in Nuclear Power Plants." Proceedings of the Industrial Temperature Measurement Symposium, pp. 14.01-14.23, Knoxville, TN (September 1984).

N11. Hashemian, H.M., Kerlin, T.W., and Petersen, K.M., "In-Situ Response Time Testing of Temperature and Pressure Sensors." Presented at IEEE Winter Power Meeting, New York, NY (February 1986).

N12. Hashemian, H.M., Thie, J.A., and Upadhyaya, B.R., "Reactor Sensor Surveillance Using Noise Analysis." Proceedings of the Topical Meeting on Reactor Physics and Safety, NUREG/CP-0080, Vol. 2, Saratoga Springs, NY (September 1986).

N13. Hashemian, H.M., et al., "New Methods for Response Time Testing of Industrial Temperature and Pressure Sensors." Proceedings of Annual Meeting of American Society of Mechanical Engineers (ASME), pp. 79–85, Anaheim, CA (December 1986).

N14. Hashemian, H.M., "Response Time of Nuclear Power Plant RTDs." Transactions of the American Nuclear Society, 1987 Annual Meeting, Dallas, TX (June 1987).

N15. Hashemian, H.M., and Petersen, K.M., "Performance of Nuclear Plant RTDs." Proceedings of the 13th Biennial Conference on Reactor Operating Experience, International Meeting on Nuclear Power Plant Operation, ANS Transactions, Suppl. #1, Vol. 54, pp. 138–139, Chicago, IL (August-September 1987).

N16. Hashemian, H.M., and Petersen, K.M., "Aging Degradation of Primary System RTDs." American Nuclear Society Transactions 1987 Winter Meeting, Vol. 55, pp. 518–520, Los Angeles, CA (November 1987).

N17. Hashemian, H.M., and Petersen, K.M., "New Methods for In-Situ Response Time Testing of Pressure Sensors in Nuclear Power Plants." Proceedings of the International Nuclear Power Plant Aging Symposium, U.S. Nuclear Regulatory Commission, NUREG/CP-0100, Bethesda, MD (August 1988).

N18. Hashemian, H.M., and Petersen, K.M., "Effect of Aging on Performance of Nuclear Plant RTDs." Proceedings of the International Nuclear Power Plant Aging Symposium, U.S. Nuclear Regulatory Commission, NUREG/CP-0100, Bethesda, MD (August 1988).

N19. Hashemian, H.M., and Petersen, K.M., "Calibration and Response Time Testing of Industrial RTDs." Proceedings of the 34th International Instrumentation Symposium, Test Measurement Division of the Instrument Society of America, Albuquerque, NM (May 1988).

N20. Hashemian, H.M., and Petersen, K.M., "New Methods for Response Time Testing of Pressure Transmitters." Proceedings of the 34th International Instrumentation Symposium, Instrument Society of America, Albuquerque, NM (May 1988).

N21. Thie, J.A., and Hashemian, H.M., "BWR Stability Measurements Using Neutron Noise Analysis." Presented at the 1988 Informal Meeting on Reactor Diagnostics, Sponsored by Duke Power Company, Orlando, FL (June 1988).

N22. Hashemian, H.M., Thie, J.A., and Petersen, K.M., "Validation of Noise Analysis for Response Time Testing of Pressure Sensors in Nuclear Power Plants." Presented at the 1988 Informal Meeting on Reactor Diagnostics, Sponsored by Duke Power Company, Orlando, FL (June 1988).

N23. Kerlin, T.W., and Hashemian, H.M., "Uncertainty in Temperature Measurements with Industrial Platinum Resistance Thermometers." Proceedings of the 11th Triennial World Congress of the International Measurement Confederation (IMEKO), Vol. Sensors, pp. 593–601, Houston, TX (October 1988).

N24. Hashemian, H.M., and Petersen, K.M., "Accuracy of Industrial Temperature Measurement." Proceedings of the ISA '88 International Conference and Exhibit, Houston, TX (October 1988).

N25. Hashemian, H.M., et al., "Application of Noise Analysis for Response Time Testing of Pressure Sensors in Nuclear Power Plants." Proceedings of 7th Power Plant Dynamics, Control and Testing Symposium, Vol. 2, pp. 53.01–53.23, Knoxville, TN (May 1989).

N26. Miller, L.F., and Hashemian, H.M., et al., "In-Situ Response Time Testing of Force-Balance Pressure Transmitters in Nuclear Power Plants." Proceedings of 7th Power Plant Dynamics, Control and Testing Symposium, Vol. 2, pp. 49.01–49.18, Knoxville, TN (May 1989).

N27. Hashemian, H.M., et al., "In-Situ Response Time Testing of Thermocouples." Proceedings of the 35th International Instrumentation Symposium, Instrument Society of America, Orlando, FL (May 1989).

N28. Hashemian, H.M., et al., "Accurate and Timely Temperature Measurements Using RTDs." Proceedings of the 36th International Instrumentation Symposium, Instrument Society of America, pp. 465–470, Denver, CO (May 1990).

N29. Hashemian, H.M., "Effects of Aging on Calibration and Response Time of Nuclear Plant RTDs and Pressure Transmitters." Proceedings of the 18th Water Reactor Safety Information Meeting, U.S. Nuclear Regulatory Commission, NUREG/CP-0114, Volume 3, pp. 547–570, Rockville, MD (October 1990).

N30. Hashemian, H.M., "Advanced Methods for Management of Aging of Nuclear Plant Instrumentation." Proceedings of the American Nuclear Society, Nuclear Power Plant and Facility Maintenance International Meeting, Volume 2, pp. 473–489, Salt Lake City, UT (April 1991).

N31. Hashemian, H.M., and Petersen, K.M., "Measurement of Performance of Installed Thermocouples." Proceedings of the Aerospace Industries and Test Measurement Divisions of The Instrument Society of America, 37th International Instrumentation Symposium, pp. 913–926, ISA Paper #91–113, San Diego, CA (May 1991).

N32. Hashemian, H.M., and Petersen, K.M., "Experience With On-Line Measurement of Response Time of Pressure Transmitters Using Noise Analysis." Proceedings of the Sixth Symposium on Nuclear Reactor Surveillance and Diagnostics, SMORN VI, pp. 68.01–68.12, Volume 2, Gatlinburg, TN (May 1991).

N33. Hashemian, H.M., and Petersen, K.M., " Response Time Testing of Pressure Transmitters in Nuclear Power Plants." Proceedings of the Instrument Society of America, First Annual Joint ISA/EPRI Power Instrumentation Symposium, Volume 34, pp. 275–289, ISA Paper #91-720, St. Petersburg, FL (June 1991).

N34. Hashemian, H.M., and Petersen, K.M., "Response Time Testing and Calibration of RTDs in Conjunction with By-Pass Manifold Elimination Projects." Proceedings of the Instrument Society of America, First Annual Joint ISA/EPRI Power Instrumentation Symposium, Volume 34, pp. 233–265, ISA Paper #91–717, St. Petersburg, FL (June 1991).

N35. Hashemian, H.M., "On-Line Measurement of Response Time and Calibration of Temperature and Pressure Sensors in Nuclear Power Plants." Proceedings of the 15th Biennial Reactor Operations Division Topical Meeting on Reactor Operating Experience, American Nuclear Society, pp. 127–134, Bellevue, WA (August 1991).

N36. Hashemian, H. M., "Effects of Aging on Calibration and Response Time of Nuclear Plant Pressure Transmitters." Proceedings of the 19th Water Reactor Safety Information Meeting, U.S. Nuclear Regulatory Commission, NUREG/CP-0119, Bethesda, MD (October 1991).

N37. Hashemian, H.M., "Aging Evaluation of Nuclear Plant RTDs and Pressure Transmitters." Aging Research Information Conference, U.S. Nuclear Regulatory Agency, NUREG/CP-0121, Rockville, MD (March 1992).

N38. Hashemian, H.M., et al., "Detection of Core Flow Anomalies in Pressurized Water Reactors." Published by Instrument Society of America, ISA/EPRI Power Instrumentation Symposium, Kansas City, MO, Volume 35, pp. 253-272, ISA Paper #92–0621 (June 1992).

N39. Hashemian, H.M., and Mitchell, D.W., "On-Line Detection of Clogging of Pressure Sensing Lines in Nuclear Power Plants." Presented at the EPRI 5th Incipient Failure Detection Conference, Knoxville, TN (September 21–23, 1992).

N40. Hashemian, H.M., "On-Line Testing of Calibration of Pressure Transmitters in Nuclear Power Plants." Published by IEEE, 1992 IEEE Nuclear Science Symposium and Medical Imaging Conference, Volume 2, pp. 773–774, Orlando, FL (October 1992).

N41. Hashemian, H.M. and Petersen, K.M., "Loss of Fill Fluid in Nuclear Plant Pressure Transmitters." Presented at the 20th Water Reactor Safety Information Meeting, Bethesda, MD (October 1992).

N42. Hashemian, H.M. and Mitchell, D.W., "Evaluation of Smart Sensor Technologies for Instrument Calibration Reduction in Nuclear Power Plants." Presented at 20th Water Reactor Safety Information Meeting, Bethesda, MD (October 1992).

N43. Hashemian, H.M., "Effect of Sensing Line Blockages on Response Time of Nuclear Plant Pressure Transmitters." Presented at the 20th Water Reactor Safety Information Meeting, Bethesda, MD (October 1992).

N44. Hashemian, H.M., et al., "On-Line Testing of Calibration of Process Instrumentation Channels in Nuclear Power Plants." Proceedings of the 2nd ASME/JSME International Conference on Nuclear Engineering ICONE-2, pp. 767–774, San Francisco, CA (March 1993).

N45. Mitchell, D.W., Hashemian, H.M., and Shell, C.S., "On-Line Detection of Blockages in Pressure Sensing Systems." Proceedings of the 2nd ASME/JSME International Conference on Nuclear Engineering ICONE-2, pp. 775–781, San Francisco, CA (March 1993).

N46. Hashemian, H.M., and Stansberry, D.V., "Validation of Instrument Calibration Reduction Techniques for Nuclear Power Plants." Proceedings of the American Nuclear Society Topical Meeting, pp. 315–322, Oak Ridge, TN (April 1993).

N47. Hashemian, H.M., Mitchell, D.W., and Antonescu, C.E., "Validation of On-Line Monitoring Techniques for In-Situ Testing of Calibration Drift of Process Instrumentation Channels in Nuclear Power Plants." Proceedings of the International Atomic Energy Agency (IAEA) Specialist Meeting, NUREG/CP-0134, pp. 209–233, Rockville, MD (May 1993).

N48. Hashemian, H.M., and Riner, J.L., "Effects of Normal Aging on Calibration and Response Time of Nuclear Plants RTDs and Pressure Sensors." Proceedings of the International Atomic Energy Agency (IAEA) Specialist Meeting, NUREG/CP-0134, pp. 157–178, Rockville, MD (May 1993).

N49. Hashemian, H.M., and Riner, J.L., "In-Situ Measurement of Response Time of RTDs and Pressure Transmitters in Nuclear Power Plants." Proceedings of the International Atomic Energy Agency (IAEA) Specialist Meeting, NUREG/CP-0134, pp. 89–113, Rockville, MD (May 1993).

N50. Hashemian, H.M., et al., "RTD Cross Calibration in Pressurized Water Reactors." Proceedings of the 3rd International Joint ISA POWID/EPRI Controls and Instrumentation Conference, Vol. 36, pp. 81–102, Phoenix, AZ (June 1993).

N51. Hashemian, H.M., and Beverly, D.D., "Precision Calibration of Industrial RTDs for Critical Temperature Measurements." Proceedings of the 3rd International Joint ISA POWID/EPRI Controls and Instrumentation Conference, Vol. 36, pp. 391–412, Phoenix, AZ (June 1993).

N52. Hashemian, H.M., and Riner, J.L., "On-Line Testing of Calibration of Pressure Transmitters in Nuclear Power Plants." Proceedings of the 3rd International Joint ISA POWID/EPRI Controls and Instrumentation Conference, Vol. 36, pp. 439–450, Phoenix, AZ (June 1993).

N53. Hashemian, H.M., Mitchell, D.W., "On-Line Calibration Monitoring for Instrumentation Channels in Nuclear Power Plants." Proceedings of the 21st Water Reactor Safety Meeting, NUREG/CP-0133, pp. 191–206, Bethesda, MD (October 1993).

N54. Riner, J.L., "On-Line Response Time Testing Methods for Process Instrumentation and the Effects of Normal Aging on Their Dynamic Characteristics." Presented at the AIChE 1994 Spring National Meeting, Atlanta, GA (April 1994).

N55. Riner, J.L., et al., "Validation of On-Line Monitoring Techniques for the Detection of Drift in Process Sensors." Presented at the AIChE 1994 Spring National Meeting, Atlanta, GA (April 1994).

N56. Hashemian, H.M., "Instrument Calibration Reduction System for Nuclear Power Plants." Proceedings of the American Power Conference, Vol. 56-II, pp. 1246–1250, Chicago, IL (April 1994).

N57. Hashemian, H.M., "New Methods for Remote Testing of Accuracy and Response Time of Installed Temperature and Pressure Sensors." Proceedings of the American Power Conference, Vol. 56-I, pp. 409–414, Chicago, IL (April 1994).

N58. Mitchell, D.W., and Hashemian, H.M, "New Technology for Testing the Installation Integrity of Thermocouples in Composite Materials for SRM Nozzles." Proceedings of the Instrument Society of America 40th International Instrumentation Symposium, ISA Paper #94-6257, pp. 233–242, Baltimore, MD (May 1994).

N59. Mitchell, D.W., and Hashemian, H.M., "Effect of Sensing Line Length and Blockages on Dynamic Performance of Pressure Sensing Systems." Proceedings of the 1994 POWID/EPRI Symposium, ISA Paper #94–440, pp. 201–210, Orlando, FL (June 1994).

N60. Fain, R.E., Petersen, K.M., Hashemian, H.M., "New Equipment for Rod Drop and Control Rod Drive Mechanism Timing Tests in PWRs." Presented at the 1994 POWID/EPRI Symposium, Orlando, FL (June 1994).

N61. Farmer, J.P., Hashemian, H.M., "Minimizing the Cost of Instrument Calibrations in Nuclear Power Plants." Presented at the 1994 POWID/EPRI Symposium, Orlando, FL (June 1994).

N62. Hashemian, H.M., Farmer, J.P., "A Versatile Data Acquisition System For Nuclear Power Plants." Proceedings of the EPRI Nuclear Plant Performance Improvement Seminar, pp. 1–16, Charleston, SC (August 1994).

N63. Hashemian, H.M., Riner, J.L., "Using On-Line Performance Testing Methods in Developing a Mechanical Integrity Program for Process Sensors." Presented at the American Chemical Industries Week '94, Philadelphia, PA (October 1994).

N64. Hashemian, H.M., Mitchell, D.W., Petersen, K.M., "Nondestructive Evaluation of the Attachment of Sensors in Solid Materials." Published in Conference Proceedings of JANNAF Nondestructive Evaluation Subcommittee Meeting, Ogden Air Logistics Center, Hill Air Force Base, UT (October 1994).

N65. Hashemian, H.M., Mitchell, D.W., Petersen, K.M., "New Technology for Testing the Attachment of Sensors in Composite Materials for SRM Nozzles." Presented at the JANNAF Rocket Nozzle Test Subcommittee Meeting, at the Boeing Defense and Space Group, Seattle, WA (November 1994).

N66. Hashemian, H.M., Black, C.L., "Fiber Optic Pressure Sensors for Nuclear Power Plants." Published in the proceedings of the 22nd Water Reactor Safety Meeting, Bethesda, MD (October 1994).

N67. Hashemian, H.M., Farmer, J.P., "On-Line Calibration of Process Instrumentation Channels in Nuclear Power Plants." Proceedings of the 22nd Water Reactor Safety Meeting, Bethesda, MD (October 1994).

N68. Hashemian, H.M., Jones, C.N., Shell, C.S., Harkelroad, J.D., "New Technology for Testing the Attachment of Sensors to Solid Materials." Presented at the 41st International Instrumentation Symposium, Aurora, CO (May 1995).

N69. Hashemian, H.M., "Measurement of Primary Coolant Temperatures in Conjunction with Steam Generator Replacement." Presented at the 9th Power Plant Dynamics, Control & Testing Symposium, Knoxville, TN (May 1995).

N70. Hashemian, H.M., "Application of Neural Networks for On-Line Testing of Calibration of Process Instrumentation in Nuclear Power Plants." Presented at the 9th Power Plant Dynamics, Control & Testing Symposium, Knoxville, TN (May 1995).

N71. Hashemian, H.M., Fain, R.E., Riner, J.L., Fidler, C.R., "Development of an Automated System for Diesel Generator Performance Monitoring." Presented at the Fifth International Joint ISA POWID/EPRI Controls and Instrumentation Conference, La Jolla, CA (June 1995).

N72. Hashemian, H.M., Fain, R.E., "Experience With Measurement of Drop Time of Control and Shutdown Rods and Testing of CRDMs in Nuclear Power Plants." Presented at the Fifth Intl. Joint Controls and Instrumentation Conference, La Jolla, CA (June 1995).

N73. Hashemian, H.M., Riner, J.L., Fain, R.E, Fidler, C.R., "Monitoring the Performance of Turbines to Improve Diagnostic Capabilities and Reduce Maintenance Costs." Presented at the Fourth EPRI Turbine/Generator Conference, Milwaukee, WI (August 1995).

N74. Hashemian, H.M., "Assessment of Fiber Optic Sensors and Other Advanced Sensing Technologies for Nuclear Power Plants." Presented at the 23rd Water Reactor Safety Meeting, NUREG/CP-0149, Vol. 2, pg. 61, Bethesda, MD (October 1995).

N75. Hashemian, H.M., "In-Situ Testing of the Attachment of Thermocouples, Strain Gages and RTDs to Solid Materials." Presented at the 1995 JANNAF Propulsion and Subcommittee Joint Meetings, Tampa, FL (November 1995).

N76. Hashemian, H.M., Fain, R.E., Riner, J.L., "Automated Diesel Generator Performance Monitoring for the Reduction of Plant Operating and Maintenance Costs." Presented at the NMAC Diesel Engine Analysis and Monitoring Workshop and Exhibition, Orlando, FL (December 4–5, 1995).

N77. Hashemian, H.M., Fain, R.E., "Automated Rod Drop Time Testing and Diagnostics in Soviet Designed RBMK and VVER Reactors." Published in NPIC & HMIT 96 Conference Proceedings, Vol. 2, pp. 1461–1468 (May 1996).

N78. Hashemian, H.M., Fain, R.E., "Reducing Outage Time Through Automated Tests To Meet Technical Specification Requirements." Published in NPIC & HMIT 96 Conference Proceedings, Vol 2., pp. 1289–1296 (May 1996).

N79. Hashemian, H.M., " On-Line Monitoring Techniques for Preventive Maintenance, Performance Measurements, and Aging Management in Nuclear Power Plants." Presented at the ANS/ENS International Meeting, Washington, DC (November 1996).

N80. Hashemian, H.M., "On-Line Monitoring Techniques for Preventive Maintenance, Performance Measurements, and Aging Management In Nuclear Power Plants." Presented at the 7th ISA POWID/EPRI Controls and Instrumentation Conference, Knoxville, TN (June 1997).

N81. Hashemian, H.M., "Cross Calibration of Primary Coolant RTDs in Compliance with Recent NRC Position." Presented at the 7th ISA POWID/EPRI Controls and Instrumentation Conference, Knoxville, TN (June 1997).

N82. Hashemian, H.M., Beverly, D.D., "Maintenance Guide for Control Rod Drive Systems in Westinghouse PWRs." Presented at the 8th ISA POWID/EPRI Controls and Instrumentation Conference, Scottsdale, AZ (June 1998).

N83. Hashemian, H.M., "IAEA Guidance on Aging of Nuclear Power Plant Instrumentation and Control Equipment, Periodical Testing and Maintenance Strategies." Presented at the 8th ISA POWID/EPRI Controls and Instrumentation Conference, Scottsdale, AZ (June 1998).

N84. Hashemian, H.M., "Experience with Use of LCSR Test for Measurement of Response Time of Temperature Sensors in Industrial Processes." Presented for the ASTM E20 Committee on Temperature Measurement, Seattle, WA (May 17, 1999).

N85. Hashemian, H.M., "Advanced Sensor and New I&C Maintenance Technologies for Nuclear Power Plants." Presented at the 7th International Joint ISA POWID/EPRI Controls and Instrumentation Conference, St. Petersburg, FL (June 1999).

N86. Hashemian, H.M., Morton, G.W., "Automated System for Pressure Transmitter Calibration." Presented at the 7th International Joint ISA POWID/EPRI Controls and Instrumentation Conference, St. Petersburg, FL (June 1999).

N87. Hashemian, H.M., Mitchell, D.W., "Examples of Instrumentation Problems in Power Industries." Presented at the 43rd Annual ISA POWID Conference 2000, San Antonio, TX (June 4–9, 2000).

N88. Hashemian, H.M., et al., "Experience with Rod Drop and CRDM Testing in Nuclear Power Plants." Presented at the 43rd Annual ISA POWID Conference 2000, San Antonio, TX (June 4–9, 2000).

N89. Hashemian, H.M., "Review of Advanced Instrumentation and Maintenance Technologies for Nuclear Power Plants." Presented at the 43rd Annual ISA POWID Conference 2000, San Antonio, TX (June 4–9, 2000).

N90. Hashemian, H.M., "Latest Trends in Electric Power Production." Presented at the 43rd Annual ISA POWID Conference 2000, San Antonio, TX (June 4–9, 2000).

N91. Hashemian, H.M., "Optimized Maintenance and Management of Ageing of Critical Equipment in Support of Plant Life Extension." Presented at the 2000 ANS/ENS International Meeting, Washington, DC (November 12–16, 2000).

N92. Hashemian, H.M., "Power Uprating in PWR Plants By Better Measurement of Reactor Coolant Flow." Presented at the 2000 ANS/ENS International Meeting, Washington, DC (November 12–16, 2000).

N93. Hashemian, H.M., "Increasing Instrument Calibration Intervals." Presented at the 44th Annual ISA POWID Conference 2001, Orlando, Florida (July 7–13, 2001).

N94. . Hashemian, H.M., "Implementation of On-Line Calibration Monitoring in Nuclear Power Plants." Presented at MARCON, Maintenance and Reliability Conference, Knoxville, TN (May 5–8, 2002).

N95. Hashemian, H.M., "Verifying the Performance of RTDs in Nuclear Power Plants." Presented at the 8th Temperature Symposium, ISA—The Instrumentation, Systems, and Automation Society, Chicago, IL (October 21–24, 2002).

N96. Hashemian, H.M., "Comparison of RTDs and Thermocouples for Industrial Temperature Measurements." Presented at the 8th Temperature Symposium, ISA—The Instrumentation, Systems, and Automation Society, Chicago, IL (October 21–24, 2002).

N97. Hashemian, H.M., "LCSR Method to Verify the Attachment of Temperature Sensors and Strain Gauges to Solid Material." Presented at the 8th Temperature Symposium, ISA—The Instrumentation, Systems, and Automation Society, Chicago, IL (October 21–24, 2002).

N98. Hashemian, H.M., "Extending the Calibration Interval of Pressure Transmitters in Nuclear Power Plants." Presented at the 13th Annual Joint ISA POWID/EPRI Control and Instrumentation Conference, Williamsburg, VA (June 15–20, 2003).

N99. Hashemian, H.M., "Instrument Calibration Reduction Through On-Line Monitoring." Presented at the 2004 Utility Working Conference & Vendor Technology Expo, American Nuclear Society, Amelia Island, FL (August 8–11, 2004).

N100. Hashemian, H.M., et al., "Calibration Reduction System Implementation at the Sizewell B Nuclear Power Plant." Presented at the 4th International Topical Meeting on Nuclear Plant Instrumentation, Control and Human Machine Interface Technology, American Nuclear Society, Columbus, OH (September 19–22, 2004).

N101. Hashemian, H.M., "Temperature Sensor Diagnostics." Proceedings of the 51st International Instrumentation Symposium, ISA—The Instrumentation, Systems, and Automation Society, Knoxville, TN (May 8–12, 2005).

N102. Tew, W.L., Nam, S.W., Benz, S.P. and Dresselhaus, P., and Hashemian, H.M., "New Technologies for Noise Thermometry with Applications in Harsh and-or Remote Operating Environments." Proceedings of the 51st International Instrumentation Symposium, ISA—The Instrumentation, Systems, and Automation Society, Knoxville, TN (May 8–12, 2005).

N103. Hashemian, H.M., " Maintenance of Cables in Industrial Processes." Presented at MARCON 2005, Maintenance and Reliability Conference, Knoxville, TN (May 3–6, 2005).

N104. Hashemian, H.M., "Sensor Performance and Reliability." Presented at the 15th Joint ISA/POWID/EPRI Controls and Instrumentation Conference, Nashville, TN (June 2005).

N105. Hashemian, H.M., Shumaker, B.D. "On-Line Condition Monitoring Applications in Nuclear Power Plants." Submitted for American Nuclear Society (ANS) Topical Meeting on Nuclear Power Plant Instrumentation and Control and Human Machine Interface Technologies (NPIC&HMIT), to be held concurrently with the ANS National Winter Meeting, Albuquerque, New Mexico, USA (November 2006).

Patents (awarded and pending)

P1. Morton, G.W., Shumaker, B.D., Hashemian, H.M., “Cross Calibration of Plant Instruments with Computer Data.” Publication No. 2005/0187730 A1, (August 2005) Patent Pending.

P2. Hashemian, H.M., “Testing of Wire Systems and End Devices Installed in Industrial Processes.” Publication No. US 2005/0182581 A1 (August 2005) Patent Pending.

P3. Hashemian, H.M., “Integrated System for Verifying the Performance and Health of Instruments and Processes.” Patent No. US 6,915,237 B2 (July 2005).

P4. Morton, G.M, Sexton, C.D., Beverly, D.D., Hashemian, H.M., “Nuclear Reactor Rod Drop Time Testing Method.” Patent No. US 6,404,835 B1 (July 2002)

P5. Hashemian, H.M., “Apparatus for Measuring the Degradation of a Sensor Time Constant." Patent No. US 4, 295, 128 (October 1981)

P6. Hashemian, H.M., “Instrument and Process Performance and Reliability Verification System.” Patent No. US 6,973,413 B2 (December 2005).

Appendix B

NRC Position on RTD Cross Calibration in Nuclear Power Plants

Appendix B: NRC Position on RTD Cross Calibration in Nuclear Power Plants

The NRC position on RTD cross calibration in nuclear power plants is provided in a public document, a copy of which is attached in this appendix. The document is referred to as the I&C Branch Technical Position 13 (BTP-13). BTP-13 is an Appendix to Chapter 7 of NRC's NUREG-0800. The NUREG-0800 is also referred to as the Standard Review Plan (SRP).

Note: Due to the consistency requirements, this appendix was retyped verbatim from the NRC version and formatted as closely as possible to the original NRC document.

NEW

Appendix 7-A

NUREG-0800

Branch Technical Position HICB-13

Guidance on Cross-Calibration of Protection System

Resistance Temperature Detectors

A. Background

The purpose of this branch technical position (BTP) is to identify the information and methods acceptable to the Staff for using cross-calibration techniques for surveying the performance of resistance temperature detectors (RTDs). These guidelines are based on experience in the detailed reviews of applicant/licensee submittals describing the application of in-situ cross-calibration procedures for reactor coolant RTDs, as well as NRC research activities. In addition, the Staff has completed reviews of applicant/licensee submittals and found that they met the requirements of the regulations identified.

Other methods, such as using a diverse parameter to provide a cross-correlation reference, can be used if adequate justification is provided.

1. Regulatory Basis

10 CFR 50.55a(h) requires in part that protection systems satisfy the criteria of ANSI/IEEE Std 279, "Criteria for Protection Systems for Nuclear Power Generating Stations," including the following:

- Section 3(9) regarding the bases for minimum performance requirements, including response times and accuracies.
- Section 4.9, "Capability for Sensor Checks."
- Section 4.10, "Capability for Test and Calibration."

10 CFR 50 Appendix A, General Design Criterion (GDC) 13, "Instrumentation and Control" requires in part that instrumentation be provided to monitor variables and systems, and that controls be provided to maintain these variables and systems within prescribed operating ranges.

10 CFR 50 Appendix A, GDC 20, "Protection System Functions," requires in part that the protection system be designed to initiate operation of appropriate systems to ensure that specified acceptable fuel design limits are not exceeded.

10 CFR 50 Appendix A, GDC 21, "Protection System Reliability and Testability," requires in part that the protection system be designed for high functional reliability and in-service testability commensurate with the safety functions to be performed.

10 CFR 50 Appendix A, GDC 24, "Separation of Protection and Control Systems," requires in part that the protection system be separated from the control systems to the extent that failure of any single control system component or channel, or failure or removal from service of any single protection system component or channel that is common to the protection system, leaves intact a system satisfying all reliability, redundancy, and independence requirements of the protection system.

10 CFR 50 Appendix A, GDC 29, "Protection against Anticipated Operational Occurrences," requires in part that protection and reactivity control systems be designed to ensure an extremely high probability of accomplishing their safety function in the event of an anticipated operational occurrence.

2. Relevant Guidance

Reg. Guide 1.153, "Criteria for Power, Instrumentation, and Control Portions of Safety Systems," endorses IEEE Std. 603, "IEEE Standard Criteria for Safety Systems for Nuclear Power Generating Stations" as an alternative to ANSI/IEEE Std. 279. IEEE Std 603 requires in part that the safety system design basis include the following:

- The increment allotted for inaccuracies, calibration uncertainties, and errors.
- The overall response times of the safety system used in establishing the setpoint allowable value.
- The basis to demonstrate that the assumed values used for instrumentation inaccuracy, calibration uncertainties and error, and time response are acceptable and reasonable.

Performance of an RTD is characterized by its accuracy and response time. Accuracy is a measure of how well the RTD indicates a static temperature, and response time indicates how quickly the RTD can sense a temperature change. NUREG/CR-5560, "Aging of Nuclear Plant Resistance Temperature Detectors," asserts that the calibration and response time of RTDs are affected by aging even within design conditions, but that the aging is manageable by periodic tests performed at each refueling interval. EPRI TR-106453-3925, "Temperature Sensor Evaluation," provides additional information on RTD performance.

3. Purpose

The purpose of this BTP is to provide guidance for NRC reviewers to verify that the previously cited regulatory bases and standards are met by an applicant's submittal. This BTP has two objectives:

- Confirm that calibration inaccuracies, uncertainties, and errors associated with a proposed cross-calibration method are consistent with design basis and setpoint analysis assumptions, and
- Confirm that a proposed cross-calibration method is adequate to confirm that RTD response times are consistent with accident analysis assumptions.

B. Branch Technical Position

1. Introduction

To ensure adequate performance of the RTD, its accuracy and response time should be verified at appropriate intervals. For reactor coolant system (RCS) RTD sensors, practical considerations may limit the extent and methods prudent for in-situ calibration and testing. Periodic removal and re-installation of RTDs solely to support verification of calibration or response time could potentially introduce errors due to installation and increasing personnel exposure. In addition, it may not be feasible or prudent to achieve the range of isothermal conditions in the RCS for in-situ verification of the complete calibration range of the RTDs. Nevertheless, the applicant licensee should provide assurance that the calibration and response time for each RTD has not significantly changed due to aging or degradation of the sensor and its installation.

One method acceptable to the Staff is to periodically provide an installed reference RTD that has been recently calibrated and response-time tested. The remaining "similar" RTDs may be cross-correlated to the reference RTD to identify any significant degradation in performance. The "similar" RTDs are those which can be shown to be subject to sufficiently similar temperature and flow conditions in the RCS. While this method does not provide for complete calibration verification of each RTD over its range, the Staff has found the method adequate for timely detection of drift or degradation of RTDs, provided that the guidance herein is applied. This guidance addresses the following topics:

- Traceability of the installed reference RTD to laboratory calibration data.
- Acceptable methods for in-situ testing of RTDs.
- Response time testing.
- "As-found" and "as-left" surveillance data.
- Control/protection interaction or common-mode failure during in-situ testing.

2. Information to be Reviewed

The information to be reviewed consists of specifications, drawings, and analyses of the proposed RTD cross-calibration program.

3. Acceptance Criteria

Supporting Analysis

Analyses, and information on the instrument maintenance and calibration program should be provided to support the adequacy of the cross-calibration program. The analysis should, as a minimum, address the following topics.

- Justification that the cross-calibration program is consistent with the characteristics of the RTD sensors, including RTD specifications, range, accuracy, repeatability, dynamic response, installed configuration, environmental qualification, calibration reference, calibration history, and calibration intervals.
- The specific methods or analyses used for signal conditioning or processing (for example, averaging, biasing, failure detection, data quality determination, and error compensation).
- The planned process for cross-calibration and response time determination.
- Justification that the performance requirements and failure criteria assumed in the plant accident/event analyses are satisfied by the cross-calibration process and testing results.
- The technical basis for the acceptance criteria and values of cross-calibration points monitored in-situ throughout the RTD range, to ensure that the data are adequate for detecting degradation or systematic drift.

Traceability of the Installed Reference RTD to Laboratory Calibration Data

Laboratory calibration involves measuring the RTD's resistance at several known temperatures. The data are then used to provide a calibration curve for the device.

In addition, the RTD response time can be determined under laboratory conditions using controlled temperature baths and a methodology to calculate the RTD response time over the measuring temperature range.

The installation of a calibrated RTD should include a test procedure to demonstrate the response time applicability of the laboratory test results. Loop current step response (LCSR) testing is an acceptable way to verify that the conditions of the installed RTD are adequately correlated to the laboratory test data.

Response time testing of the installed RTDs using LCSR should use an analytical technique such as the LCSR transformation identified in NUREG-0809, "Review of Resistance Temperature Detector Time Response Characteristics," to correlate the in-situ results with the results of a laboratory-type temperature test.

Acceptable Methods for In-Situ Testing

Verification of RTD calibrations should be accomplished by installing a newly calibrated reference RTD sensor and then cross-correlating with the measurements of the other RTDs subject to the same temperature and flow environment. A critical element in this approach is providing assurance that all sensor elements are subject to sufficiently similar temperature and flow environments. Other methods, such as using a diverse parameter to provide a cross-correlation reference, can be used if adequate justification is provided.

Before installing a reference or new RTD, the sensor should either be calibrated in a laboratory or, if the manufacturer's calibration data are to be used, the applicant/licensee should perform an analysis or test to verify the RTD has retained its calibration. The application temperatures should be within the manufacturer's highest calibration range.

All data should be taken at isothermal plant conditions and all loops (hot legs and cold legs) should be at similar temperatures. If this condition can not be assured then the applicant/licensee should provide for removal of one or more of the RTDs at each representative location and for replacement with a newly calibrated RTD.

The applicant/licensee should provide an analysis which states the limits of acceptable calibration, response times, and in-situ testing of the RTDs. Test procedures, with acceptance criteria, should state the limits of the calibration, particularly the dependency of the data on uniform coolant temperature and flow.

Correction factors or bias values should be established to compensate for non-isothermal conditions. Because plant temperatures cannot be perfectly controlled, fluctuations and drift in the primary coolant temperature might occur during in-situ testing. The test data should be corrected for the fluctuations and drift in the coolant temperature. If during the testing incomplete mixing of the reactor coolant should occur, the test data should be corrected for the temperature differences. Reactor coolant temperatures should be stable and uniform. In the event this is not the case the data should be corrected to account for these effects.

Equipment used in the test should be accurate to within the necessary tolerance and have stable performance. See BTP HICB-12 for guidance on determining plant instrumentation tolerances.

Response Time Testing

Even though response time testing is independent from the cross-calibration test, it should be performed for the existing and the newly installed reference sensors to account for installation effects and to identify degradation.

The resulting test data and analysis should support correlation of each of the existing sensors in the common flow path to its laboratory response time test data, and also to the laboratory response time test data for the reference sensor. Correlation between LCSR test results for the existing sensors and LCSR test results for the reference sensor may be used to establish the correlation with the reference RTD laboratory test data.

As-Found/As-Left Surveillance Data

The applicant/licensee should maintain a database of the “as-left" and “as-found" calibration and response time tests for each sensor.

To monitor systematic drift or degradation, at each refueling cycle a newly calibrated RTD or a new RTD with recent calibration data should be installed at representative location(s) detonated by analysis. The cross-correlation to the reference RTD(s) should be monitored using “as found" and “as left" data records.

Test data and analysis should identify and account for differences in isothermal conditions and demonstrate that the drift is random and is within an acceptable band as determined by setpoint analyses, and that systematic drift is not exhibited. If historical data reveals potential drift problems which would exceed the allowable values of temperature drift in testing for any sensor then the applicant/licensee should verify the calibration of the deviating sensor(s) and identify appropriate corrective action. Analysis to project RTD drift should be available for all RTDs within the protection system.

Control/Protection Interaction and Common-Mode Failure During In-Situ Testing

If the applicant/licensee uses test equipment common to redundant channels, qualified isolation should be provided to preclude single-failure effects on redundant channels or unacceptable protection/control interactions.

4. Review Procedures

The protection system design basis should be examined to identify the requirements for RTD accuracy and time response.

The cross-calibration method and calibration and response time data should be examined to identify calibration inaccuracies, uncertainties, and errors, and to confirm that the cross-calibration method is adequate.

The programmatic documentation of the cross-calibration process should be reviewed with respect to the acceptance criteria above. This review should confirm that the calibration process is consistent with all setpoint analysis assumptions and design basis requirements.

C. References

ANSI/IEEE Std 279-1971. "Criteria for Protection Systems for Nuclear Power Generating Stations."

EPRI Topical Report TR-106453-3925. "Temperature Sensor Evaluation.." Electric Power Research Institute, June 1996.

IEEE Std 603-1991. "IEEE Standard Criteria for Safety Systems for Nuclear Power Generating Stations."

NUREG-0809. "Review of Resistance Temperature Detector Time Response Characteristics." August 1981.

NUREG/CR-5560. "Aging of Nuclear Plant Resistance Temperature Detectors." June 1990.

Regulatory Guide 1.153. "Criteria for Power, Instrumentation, and Control Portions of Safety Systems." Office of Nuclear Regulatory Research, U.S. Nuclear Regulatory Commission, 1996.

Appendix C

Regulatory Guide 1.118

Periodic Testing of Electric Power and Protection Systems

Appendix C: Regulatory Guide 1.118 Periodic Testing of Electric Power and Protection Systems

The attached document is a reference for the material covered in Chapter 6 of this book. The document in this appendix was retrieved from the NRC's website.

Regulatory Guide 1.118 - Periodic Testing of Electric Power and Protection Systems

Revision 3
April 1995

A. Introduction

Section 50.55a, "Codes and Standards," of 10 CFR Part 50, "Domestic Licensing of Production and Utilization Facilities," requires in paragraph (h), "Protection Systems," that protection systems meet the requirements set forth in Institute of Electrical and Electronics Engineers Standard 279,[1] "Criteria for Protection Systems for Nuclear Power Generating Stations." Section 4.9 of IEEE Std. 279-1971 requires, in part, that means be provided for checking the operational availability of each protection system input sensor during reactor operation and includes examples of how this can be accomplished. Section 4.10 of IEEE Std. 279-1971 requires, in part, that capability be provided for testing and calibrating protection system equipment other than sensors and indicates when such equipment must be tested during reactor operation.

General Design Criterion 21, "Protection System Reliability and Testability," of Appendix A, "General Design Criteria for Nuclear Power Plants," to 10 CFR Part 50 requires, in part, that the protection system be designed to permit its periodic testing during reactor operation, including a capability to test channels independently to determine failures and losses of redundancy that may have occurred. General Design Criterion 18, "Inspection and Testing of Electric Power Systems," requires, in part, that electric power systems important to safety be designed to permit periodic testing, including periodic testing of the performance of the components of the system and the system as a whole. The testing should be carried out under conditions as close to design as practical and should involve the full operational sequence, including operation of portions of the protection system, as well as the transfer of power among the nuclear power unit, the offsite power system, and the onsite power system. Criterion XI, "Test Control," of Appendix B, "Quality Assurance Criteria for Nuclear Power Plants and Fuel Reprocessing Plants," to 10 CFR Part 50 requires, in part, that a test program be established to ensure that all testing, including operational testing required to demonstrate that systems and components will perform satisfactorily in service, is identified and performed in accordance with written test procedures.

This regulatory guide describes a method acceptable to the NRC staff for complying with the Commission's regulations with respect to the periodic testing of the electric power and protection systems.

The Advisory Committee on Reactor Safeguards has been consulted concerning this guide and has concurred in the regulatory position.

Any information collection activities mentioned in this regulatory guide are contained as requirements in 10 CFR Part 50, which provides the regulatory basis for this guide.

The information collection requirements in 10 CFR Part 50 have been approved by the Office of Management and Budget, Approval No. 3150-0011.

B. Discussion

IEEE Std. 338-1987,[(1)] “Criteria for the Periodic Surveillance Testing of Nuclear Power Generating Station Safety Systems," was prepared by Working Group 3.0 of Subcommittee 3, “Operations, Surveillance and Testing," of the IEEE Nuclear Power Engineering Committee and was approved by the IEEE Standards Board on September 10, 1987 (reaffirmed in 1993). The standard provides design and operational criteria for the performance of periodic testing as part of the surveillance program of nuclear power plant safety systems. The periodic testing consists of functional tests and checks, calibration verification, and time response measurements, as required, to verify that the safety system performs to meet its defined safety functions. The system status, associated system documentation, test intervals, and test procedures during operation are also addressed.

C. Regulatory Position

Conformance with the requirements of IEEE Std. 338-1987, “Criteria for the Periodic Surveillance Testing of Nuclear Power Generating Station Safety Systems," provides a method acceptable to the NRC staff for satisfying the Commission's regulations with respect to periodic testing of electric power and protection systems if the following exceptions are complied with:

1. The definitions of “safety systems," “safety function," and “safety group" in IEEE Std. 603-1991,[1] “Criteria for Safety Systems for Nuclear Power Generating Stations," are used instead of the definitions in IEEE Std. 338-1987.
2. Both Sections 5(15) and 6.4(5) of IEEE Std. 338-1987 are replaced by the following:

   Procedures for periodic tests shall not require makeshift test connections except as follows:

   (1) Temporary jumper wires may be used with safety systems that are provided with facilities specifically designed for the connection of portable test equipment. These facilities shall be considered part of the safety system and shall meet all the requirements of IEEE Std. 338-1987.

   (2) Removal of fuses or opening a breaker is permitted only if such action causes trip of the associated channel or actuation of the logic of the associated load group.

   (3) Test procedures or administrative controls shall provide for verifying the open circuit or verifying that temporary connections are restored after testing.

3. The description for a logic system functional test, as noted in Section 6.3.5 of IEEE Std. 338-1987, implies that the sensor is included. A logic system functional test is to be a test of all logic components (i.e., all relays and contacts, trip units, solid state logic elements, etc.) of a logic circuit, from as close to the sensor as practicable up to but not including the actuated device, to verify operability.

D. Implementation

The purpose of this section is to provide information to applicants and licensees regarding the NRC staff's plans for using this guide.

Except in those cases in which an applicant proposes an acceptable alternative method for complying with specified portions of the Commission's regulations, the method described in this guide will be used in the evaluation of submittals in connection with applications for construction permits and operating licenses. It will also be used to evaluate submittals from operating reactor licensees that propose system modifications voluntarily initiated by the licensee if there is a clear nexus between the proposed modifications and this guidance.

Value/Impact Statement

A draft Value/Impact Statement was published with the draft of this guide when it was published for public comment (Task DG-1028, September 1994). No changes were necessary, so a separate value/impact statement for the final guide has not been prepared. A copy of the draft value/impact statement is available for inspection or copying for a fee in the Commission's Public Document Room at 2120 L Street NW., Washington, DC, under Task DG-1028.

Footnotes

1. Copies may be purchased from the Institute of Electrical and Electronics Engineers, 345 East 47th Street, New York, NY 10017.

APPENDIX D

NRC Information Notice 92-33

Effect of Snubber on Response Time of Nuclear Plant Pressure Transmitters

Appendix D: NRC Information Notice 92-33 Effect of Snubber on Response Time of Nuclear Plant Pressure Transmitters

This appendix contains the NRC information notice on the use of snubbers in the sensing line of nuclear power plants.

Note: Due to the publication requirements, this document was retyped verbatim from the NRC version and formatted as closely as possible to the authentic document.

United States

Nuclear Regulatory Commission

Washington, D.C. 20555

April 30, 1992

NRC Information Notice 92-33: Increased Instrument Response Time when Pressure Dampening Devices are Installed

Addressees

All holders of operating licenses or construction permits for nuclear power reactors.

Purpose

The U.S. Nuclear Regulatory Commission (NRC) is issuing this information notice to alert addressees to increased response times for pressure sensing instruments that occur when pressure dampening devices are installed in the instrument sensing lines. It is expected that recipients will review the information for applicability to their facilities and consider actions, as appropriate, to avoid similar problems. However, suggestions contained in this information notice are not NRC requirements; therefore, no specific action or written response is required.

Description of Circumstances

On September 25, 1991, the GPU Nuclear Corporation, licensee for the Oyster Creek Nuclear Generating Station, initiated a plant shutdown after determining that seven of the eight isolation condenser line break pressure sensors did not meet the plant technical specification requirements for instrument response times. The licensee's trouble shooting determined that pressure dampening devices in the sensing lines for these differential pressure sensors had caused an increase in the response times. Furthermore, the licensee found that the time delay caused by these devices is significant at both low and high pressures.

Discussion

The pressure dampening devices (snubbers) which utilize sintered stainless steel elements are generally installed in instrument sensing lines to dampen pressure oscillations or to protect the instruments from particulate contamination.

When the sensors failed to meet the required response times during surveillance testes, the licensee reviewed the pressure sensing system and found that the snubbers were causing unacceptable time delays. The licensee later removed the snubbers because they are not needed when using the upgraded Barton pressure sensors that were installed.

Snubber time delay will not be detected in response time testing conducted directly at the sensing instrument (see Figure 1.) The effect of a snubber will differ from sensor to sensor because of differences in the volumetric displacement of fluid within the pressure sensing mechanisms. System response time can also be degraded by the accumulation of foreign material in sensing line snubbers.

The NRC's Office of Nuclear Regulatory Research is conducting a generic study on the performance of pressure instrumentation at nuclear power plants. The staff plans to publish the results in NUREG/CR 5851, "Long Term Performance and Aging Characteristics of Nuclear Plant Pressure Transmitters." The staff has completed a part of this effort, and the results of tests conducted on pressure sensor response time testing were published in an ISA (Instrument Society of America) transaction 91-720, "Response Time Testing of Pressure Transmitters in Nuclear Power Plants." This publication addresses various causes for delays in sensor response and documents that significant time delays can occur.

This information notice requires no specific action or written response. If you have any questions about the information in this notice, please contact one of the technical contacts listed below or the appropriate Office of Nuclear Reactor Regulation (NRR) project manager.

Charles E. Rossi, Director
Division of Operational Events Assessment
Office of Nuclear Reactor Regulation

Technical contacts: Thomas Koshy, NRR
(301) 504-1176

Iqbal Ahmed, NRR
(301) 504-3252

Attachments:

1. Figure 1. Test Configuration that Excludes the Effect of Snubbers
2. List of Recently Issued NRC Information Notices

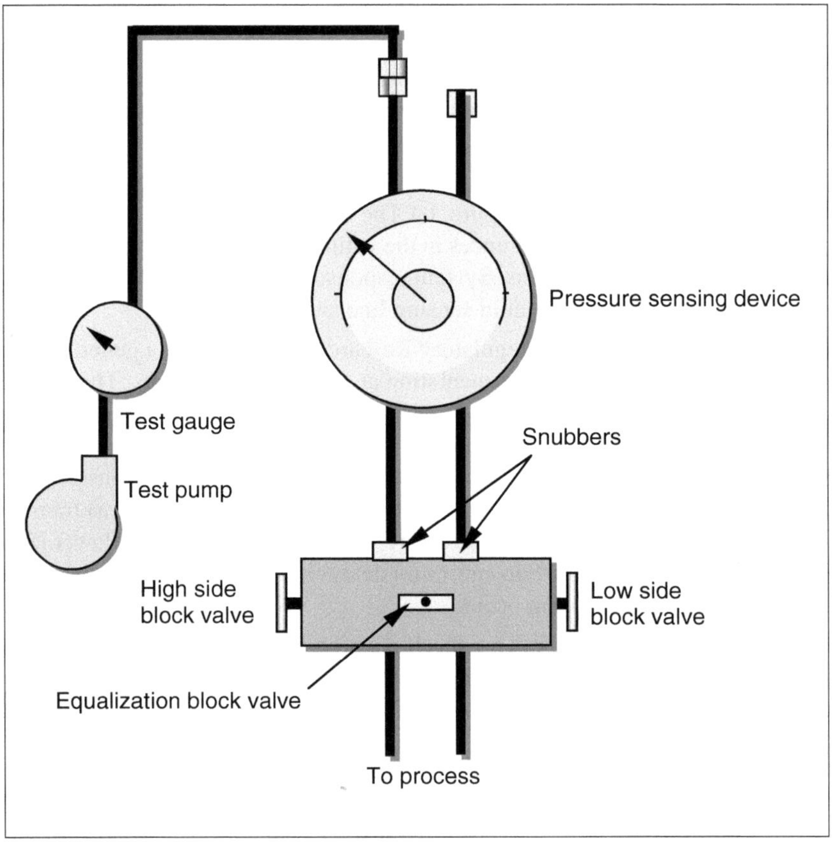

Fig. D.1. Test configuration that excludes the effect of snubbers
Note: Redrawn from original in the NRC Information Notice 92-33

Appendix E

NRC Documents on Oil Loss

Problem in Nuclear Power Plant Pressure Transmitters

Appendix E: NRC Documents on Oil Loss Problem in Nuclear Power Plant Pressure Transmitters

This appendix contains two NRC documents on the subject of oil loss in nuclear power plant pressure transmitters. These documents are:

(1990) – NRC Bulletin No. 90-01: Loss of Fill-Oil in Transmitters Manufactured By Rosemount

(2004) – NRC ISSUE 176 Document: Loss Of Fill-Oil in Rosemount Transmitters

The first document above (Bulletin No. 90-01) describes the nature of the oil loss problem and its potential consequences, and the second document (Issue 176) is provided by the NRC to close the oil loss issue for the nuclear power industry.

OMB No.: 3150-0011

NRCB 90-01

United States

Nuclear Regulatory Commission

Office of Nuclear Reactor Regulation

Washington, D.C. 20555

March 9, 1990

NRC Bulletin No. 90-01: Loss of Fill-Oil in Transmitters Manufactured by Rosemount

Addressees:

All holders of operating licenses or construction permits for nuclear power reactors.

Purpose:

This bulletin requests that addressees promptly identify and take appropriate corrective actions for Model 1153 Series B, Model 1153 Series D, and Model 1154 transmitters manufactured by Rosemount that may be leaking fill-oil.

Description of Circumstances:

NRC Information Notice No. 89-42, "Failure of Rosemount Models 1153 and 1154 Transmitters," dated April 21, 1989, was issued to alert industry to a series of reported failures of Rosemount Models 1153 and 1154 pressure and differential pressure transmitters. The reported failures occurred at Northeast Utilities' Millstone Unit 3 between March and October 1987. Subsequent investigation into the cause of the failures by Rosemount confirmed that the failure mode was a gradual loss of fill-oil from the transmitter's sealed sensing module.

Discussion of Safety Significance:

The performance of a transmitter that is leaking fill-oil gradually deteriorates and may eventually lead to failure. Although some failed transmitters have shown symptoms of loss of fill-oil prior to failure, it has been reported that in some cases the failure of a transmitter that is leaking fill-oil may be difficult to detect during operation. An undetected transmitter failure has a greater adverse effect on safety system reliability than a failure that would be readily detectable during normal operation. For example, electronic circuit malfunctions are routinely detected either by observing instrument channel readout or during periodic surveillance tests. Transmitter failures that are not readily detectable increase the potential for common mode failure and may result in the affected safety system not performing its intended safety function. This common mode failure potential is of increased concern when transmitter designs are particularly susceptible to loss of fill-oil.

Discussion:

Model 1151, 1152, 1153, and 1154 Rosemount transmitters are utilized extensively in nuclear power plants. Model 1153 and 1154 transmitters are supplied by Rosemount as both seismically and environmentally qualified equipment. Model 1152 transmitters are supplied by Rosemount only as seismically qualified equipment. Model 1151 transmitters are supplied by Rosemount as commercial-grade equipment.

Rosemount has indicated, to date, that failure of approximately 91 Model 1153 Series B, Model 1153 Series D, and Model 1154 transmitters due to loss of fill-oil from a glass to metal seal failure have been confirmed. Since the sensing module is sealed, loss of fill-oil cannot be visually confirmed without destructive analysis of the sensing module. NRC staff review of this issue has identified additional Model 1153 and 1154 transmitters with symptoms indicative of loss of fill-oil that may not have been brought to Rosemount's attention. Thus, the number of Model 1153 and 1154 transmitters that have experienced a loss of fill-oil may be even greater than that confirmed by Rosemount.

Model 1153 Series B, Model 1153 Series D, and Model 1154 transmitters, because their construction incorporates the use of a metal o-ring, appear to be particularly susceptible to loss of fill-oil due to a glass to metal seal failure. Accordingly, the NRC staff believes that the degree of susceptibility of these transmitters to loss of fill-oil warrants their being subjected to an enhanced surveillance program. In addition, certain manufacturing lots of Model 1153 Series B, Model 1153 Series D, and Model 1154 transmitters have been identified by Rosemount as having had a high failure fraction due to loss of fill-oil. Specific information needed to identify transmitters that are from these suspect lots has been provided to industry by Rosemount concurrent with Reference 4. Accordingly, the NRC staff believes that this additional degree of susceptibility warrants not utilizing these suspect lot transmitters in the reactor protection or engineered safety features actuation systems.

Rosemount has indicated that failures of Model 1151 and 1152 transmitters due to loss of fill-oil have also been confirmed. The construction of Model 1151, 1152, and 1153 Series A transmitters is similar to that of Model 1153 Series B, Model 1153 Series D, and Model 1154 transmitters (i.e., the utilization of a glass to metal seal) except the construction of Model 1151, 1152, and 1153 Series A transmitters incorporates an elastomeric o-ring instead of a metal o-ring. The NRC staff does not, at present, have sufficient information to effectively address the susceptibility of Model 1151, 1152, and 1153 Series A transmitters to loss of fill-oil. Therefore, in order to obtain relevant operational experience data, addressees are encouraged to report Model 1151, 1152, and 1153 Series A, as well as Model 1153 Series B, Model 1153 Series D and Model 1154 transmitters that may have exhibited symptoms indicative of loss of fill-oil or have been confirmed to have experienced a loss of fill-oil to the Nuclear Plant Reliability Data System (NPRDS). In addition, while enhanced surveillance of Model 1151, 1152, and 1153 Series A transmitters is not specifically requested by this bulletin, addressees are encouraged to undertake such efforts on Model 1151, 1152, and 1153 Series A transmitters utilized in either safety-related systems or systems installed in accordance with 10 CFR 50.62 (the ATWS rule).

Rosemount has indicated that they have instituted additional quality control and quality assurance steps in the manufacturing process and modified specifications on bolt torque to reduce stress levels. These changes should minimize the potential for Model 1153 Series B, Model 1153 Series D, and Model 1154 transmitter failures due to loss of fill-oil. As a result, Rosemount has indicated that transmitters of these types manufactured after July 11, 1989 are not subject to their May, 1989 10 CFR Part 21 notification. The NRC staff has not, to date, received operational experience data that indicates that Model 1153 Series B, Model 1153 Series D and Model 1154 transmitters manufactured after July 11, 1989 are as susceptible to loss of fill-oil as those manufactured prior to July 11, 1989. Accordingly, while enhanced surveillance of transmitters of these types manufactured after July 11, 1989 is not specifically requested by this bulletin, addressees are encouraged to undertake such efforts on these transmitters if they are utilized in either safety-related systems or systems installed in accordance with 10 CFR 50.62 (the ATWS rule). In addition, Model 1153 Series B, Model 1153 Series D and Model 1154 transmitters manufactured after July 11, 1989 that exhibit symptoms indicative of loss of fill-oil or are confirmed to have experienced a loss of fill-oil should be reported in accordance with the report requirements of this bulletin.

The NRC staff encourages utilities to work collectively under the guidance of a technical industry organization to develop and analyze an operational experience database concerning all models of Rosemount transmitters. The NRC staff will continue to obtain and analyze operational experience data pertaining to Model 1151, 1152, 1153, and 1154 transmitters. Further regulatory action, such as requesting expansion of enhanced surveillance activities to include Model 1151, Model 1152, and Model 1153 Series A transmitters and Model 1153 Series B, Model 1153 Series D and Model 1154 transmitters manufactured by Rosemount after July 11, 1989 or requesting replacement of additional suspect lot transmitters, may be taken if warranted.

Addressees may have obtained transmitters that were manufactured by Rosemount or that contain Rosemount manufactured sensing modules from a number of different sources. The following information is provided to facilitate addressee's identification of transmitters that were manufactured by Rosemount or that contain Rosemount manufactured sensing modules:

- Rosemount has indicated that unauthorized remanufacturers and refurbishers exist for Model 1151 transmitters. Unauthorized remanufacturers and refurbishers may also exist for Model 1152, 1153, and 1154 transmitters.
- All Model 1153 and 1154 transmitters, whether obtained directly from Rosemount, obtained through intermediary suppliers, or provided as an integral part of another component (such as an emergency diesel generator), should a) indicate manufacture by Rosemount, b) have a distinctive Rosemount model and serial number, c) have the physical profile characteristics of a Rosemount transmitter, and d) have a blue or stainless steel housing. Rosemount has indicated that Model 1153 and 1154 transmitters are not provided to other manufacturers for resale under a different brandname. In addition, a simplified diagram that describes the typical physical characteristics of a Rosemount transmitter is provided by Attachment 1.
- Model 1152 transmitters, except as noted below, should a) indicate manufacture by Rosemount, b) have a distinctive Rosemount model and serial number, c) have the physical profile characteristics of a Rosemount transmitter, and d) have a blue or stainless steel housing. Rosemount has indicated that they have supplied Model 1152 transmitter sensing modules to Bailey Controls (formerly Bailey Meter). Bailey manufactured transmitters that contain Rosemount manufactured Model 1152 sensing modules have gray housings that appear slightly different than Rosemount housings.
- Model 1151 transmitters, except as noted below, should a) indicate manufacture by Rosemount, b) have a distinctive Rosemount model and serial number, c) have the physical profile characteristics of a Rosemount transmitter, and d) have a blue housing. Model 1151 transmitters manufactured by Rosemount may have been

supplied for use in nuclear power plants by other original equipment manufacturers. These transmitters should have the physical profile characteristics of a Rosemount transmitter and have a blue housing. Fisher Controls may also offer for resale under their own brand name Model 1151 transmitters purchased from Rosemount. These transmitters should have the physical profile characteristics of a Rosemount transmitter, but have a green housing.

The earliest symptom a Model 1153 Series B, Model 1153 Series D, or Model 1154 transmitter may exhibit during normal operation prior to failure if it is leaking fill-oil is:

– a sustained drift

The symptoms a Model 1153 Series B, Model 1153 Series D, or Model 1154 transmitter may exhibit during normal operation subsequent or immediately prior to failure if it is leaking fill-oil include:

– a sustained drift
– an abrupt decreasing drift (for high range gauge or absolute transmitters)
– a change in process noise including amplitude variations, "one-sided-noise," or asymmetric noise distributions
– slow response to or inability to follow planned or unplanned plant Transients

The symptoms a Model 1153 Series B, Model 1153 Series D, or Model 1154 transmitter may exhibit during calibration activities if it is leaking fill-oil include:

– inability to respond over the entire design range
– slow response to either an increasing or decreasing test pressure
– a sustained zero or span shift

The NRC staff believes these symptoms can also be utilized to detect Model 1151, 1152, and 1153 Series A transmitters that may be experiencing a loss of fill-oil.

The NRC staff has reviewed the information which has been provided by Rosemount, including References 1, 2, 3, and 4, to assist industry in detecting transmitters that may be leaking fill-oil. The NRC staff has concluded that Rosemount has provided sufficient bases to support their proposed diagnostic procedures (trending calibration data, trending operational data, sluggish transient response, and process noise analysis) for detecting whether a transmitter may be leaking fill-oil. Accordingly, the actions requested in this bulletin are intended to reflect these diagnostic procedures. However, the NRC staff has concluded that Rosemount has not provided sufficient bases to support their proposed methodology (pressure versus time-in-service) for identifying which transmitters should be subject to an enhanced surveillance program. Specifically, the NRC staff believes that the methodology utilized by Rosemount to support their proposed pressure versus time-in-service criteria for identifying which transmitters should be subject to an enhanced surveillance program does not provide the necessary high degree of confidence that this failure mode will not occur.

Rosemount had initially indicated that Model 1153 Series B, Model 1153 Series D, and Model 1154 transmitters that were experiencing a loss of fill-oil should fail within approximately 36 months of in-service time. Recent information indicates that the rate of loss of fill-oil is application and pressure dependent. Although transmitters subject to continuous high-pressure (e.g. reactor operating pressures) may fail within this timeframe, transmitters utilized in low-pressure systems or not subject to continuous high-pressure may take longer to fail.

General Design Criterion (GDC) 21 “Protection System Reliability and Testability" of 10 CFR 50, Appendix A requires the protection system to be designed for high functional reliability and with sufficient capability to allow periodic testing of its functioning when the reactor is in operation in order to readily detect failures of subcomponents and subsystems within the protection system as well as loss of the required protection system redundancy as they occur. 10 CFR 50.55a(h) requires that protection systems meet the Institute of Electrical and Electronics Engineers Standard: “Criteria for Protection Systems for Nuclear Power Generating Stations" (IEEE-279). IEEE-279 states that means shall be provided for checking, with a high degree of confidence, the operational availability of each system input sensor during reactor operation. Thus, the NRC staff concludes that facilities that utilize transmitters that may be particularly susceptible to loss of fill-oil may not be in full compliance with these regulations because undetected transmitter failure could occur. Accordingly, the NRC staff requests that addressees take the actions requested below.

Requested Actions:

Operating Reactors

All holders of operating licenses for nuclear power reactors are, within 120 days after receipt of this bulletin, requested to:

1. Identify Model 1153 Series B, 1153 Series D, and Model 1154 pressure or differential pressure transmitters, excluding Model 1153 Series B, 1153 Series D, and Model 1154 transmitters manufactured by Rosemount subsequent to July 11, 1989, that are currently utilized in either safety-related systems or systems installed in accordance with 10 CFR 50.62 (the ATWS rule).
2. Determine whether any transmitters identified in Item 1 are from the manufacturing lots that have been identified by Rosemount as having a high failure fraction due to loss of fill-oil. Addressees are requested not to utilize transmitters from these suspect lots in the reactor protection or engineered safety features actuation systems; therefore, addressees are requested to develop and implement a program to replace, at the earliest appropriate opportunity, transmitters from these suspect lots in use in the reactor protection or engineered safety features actuation systems.
3. Review plant records (for example, the three most recent calibration records) associated with the transmitters identified in Item 1 above to determine whether any

of these transmitters may have already exhibited symptoms indicative of loss of fill-oil. Appropriate operability acceptance criteria should be developed and applied to transmitters identified as having exhibited symptoms indicative of loss of fill-oil from this plant record review. Transmitters identified as having exhibited symptoms indicative of loss of fill-oil that do not conform to the operability acceptance criteria should be addressed in accordance with the applicable technical specification. Transmitters identified as having exhibited symptoms indicative of loss of fill-oil that do not conform to the operability acceptance criteria and are not addressed in the technical specifications should be replaced at the earliest appropriate opportunity.

4. Develop and implement an enhanced surveillance program to monitor transmitters identified in Item 1 for symptoms of loss of fill-oil. This enhanced surveillance program should consider the following or equally effective actions:
   a) Ensuring appropriate licensee personnel are aware of the symptoms that a transmitter, both during operation and during calibration activities, may exhibit if it is experiencing a loss of fill-oil and the need for prompt identification of transmitters that may exhibit these symptoms;
   b) Enhanced transmitter monitoring to identify sustained transmitter drift;
   c) Review of transmitter performance following planned or unplanned plant transients or tests to identify sluggish transmitter response;
   d) Enhanced awareness of sluggish transmitter response to either increasing or decreasing test pressures during calibration activities;
   e) Development and implementation of a program to detect changes in process noise; and
   f) Development and application to transmitters identified as having exhibited symptoms indicative of loss of fill-oil of an appropriate operability acceptance criteria. Transmitters identified as having exhibited symptoms indicative of loss of fill-oil that do not conform to the operability acceptance criteria should be addressed in accordance with the applicable technical specification. Transmitters identified as having exhibited symptoms indicative of loss of fill-oil that do not conform to the operability acceptance criteria and are not addressed in the technical specifications should be replaced at the earliest appropriate opportunity.
5. Document and maintain in accordance with existing plant procedures a basis for continued plant operation covering the time period from the present until such time that the Model 1153 Series B, 1153 Series D, and Model 1154 transmitters from the manufacturing lots that have been identified by Rosemount as having a high failure fraction due to loss of fill-oil in use in the reactor protection or engineered safety features actuation systems can be replaced. In addition, while performing the actions requested above, addressees may identify transmitters exhibiting symptoms indicative of loss of fill-oil that do not conform to the established operability acceptance criteria and are not addressed in the technical specifications. As these transmitters are identified, this basis for continued plant

operation should be updated to address these transmitters covering the time period from the time these transmitters are identified until such time that these transmitters can be replaced. When developing and updating this basis for continued plant operation, addressees may wish to consider transmitter diversity and redundancy, diverse trip functions (a separate trip function that may also provide a corresponding trip signal), special system and/or component tests, or (if necessary) immediate replacement of certain suspect transmitters.

Construction Permit Holders

1. All construction permit holders that anticipate receiving an operating license within 120 days after receipt of this bulletin are requested to perform Items 1, 2, 4, and 5 of Requested Actions for Operating Reactors within 120 days after receipt of this bulletin.
2. All construction permit holders that do not anticipate receiving an operating license within 120 days after receipt of this bulletin are requested to, prior to the date scheduled for fuel loading, complete Items 1 and 4 of Requested Actions for Operating Reactors and to address the intent of Items 2 and 5 of Requested Actions for Operating Reactors by:
   a) Identifying and replacing Model 1153 Series B, 1153 Series D, and Model 1154 transmitters from the manufacturing lots that have been identified by Rosemount as having a high failure fraction due to loss of fill-oil that are installed in the reactor protection or engineered safety features actuation systems; and
   b) Documenting and maintaining in accordance with existing plant procedures a basis for continued plant operation that addresses transmitters that, subsequent to fuel loading, are identified as exhibiting symptoms indicative of loss of fill-oil that do not conform to the established operability acceptance criteria and are not addressed in the technical specifications covering the time period from the time these transmitters are identified until such time that these transmitters can be replaced.

Reporting Requirements:

Operating Reactors

1. Provide, within 120 days after receipt of this bulletin, a response that:
   a) Confirms that Items 1, 2, 3, 4, and 5 of Requested Actions for Operating Reactors have been completed.
   b) Identifies the indicated manufacturer; the model number; the system the transmitter was utilized in; the approximate amount of time at pressure; the corrective actions taken; and the disposition (e.g., returned to vendor for analysis)

of Rosemount Model 1153 Series B, Model 1153 Series D, and Model 1154 transmitters that are believed to have exhibited symptoms indicative of loss of fill-oil or have been confirmed to have experienced a loss of fill-oil. This should include Model 1153 manufactured after July 11, 1989.

c) Identifies the system in which the Model 1153 Series B, 1153 Series D, and Model 1154 transmitters from the manufacturing lots that have been identified by Rosemount as having a high failure fraction due to loss of fill-oil are utilized and provides a schedule for replacement of these transmitters which are in use in the reactor protection or engineered safety features actuation systems.

2. Model 1153 Series B, Model 1153 Series D and Model 1154 transmitters that, subsequent to providing the response required by Item 1 above, exhibit symptoms of loss of fill-oil or are confirmed to have experienced a loss of fill-oil should be reviewed for reportability under existing NRC regulations. If determined not to be reportable, addressees are requested to document and maintain, in accordance with existing plant procedures, information consistent with that requested in Item 1 b) above for each transmitter identified.

Although not required by this bulletin, addressees are encouraged to report information consistent with that requested in Item 1 b) above through the Nuclear Plant Reliability Data System (NPRDS) for all Rosemount Model 1151, 1152, 1153 and 1154 transmitters that exhibit symptoms indicative of a loss of fill-oil or are confirmed to have experienced a loss of fill-oil.

Construction Permit Holders

1. All holders of construction permits that anticipate receiving an operating license within 120 days after receipt of this bulletin are required to, within 120 days after receipt of this bulletin, provide a response that:

   a) Confirms that Items 1, 2, 4, and 5 of Requested Actions for Operating Reactors have been completed; and

   b) Identifies the system in which the Model 1153 Series B, 1153 Series D, and Model 1154 transmitters from the manufacturing lots that have been identified by Rosemount as having a high failure fraction due to loss of fill-oil are utilized and provides a schedule for replacement of these transmitters which are in use in the reactor protection or engineered safety features actuation systems.

2. All holders of construction permits that do not anticipate receiving an operating license within 120 days after receipt of this bulletin are required to, prior to the date scheduled for fuel loading, provide a response that confirms that Item 2 of Requested Actions for Construction Permit Holders has been completed.

3. Model 1153 Series B, Model 1153 Series D and Model 1154 transmitters that, subsequent to providing the response required by Item 1 or 2 above, exhibit symptoms of loss of fill-oil or are confirmed to have experienced a loss of fill-oil should

be reviewed for reportability under existing NRC regulations. If determined not to be reportable, addressees are requested to document and maintain, in accordance with existing plant procedures, information consistent with that requested in Item 1 b) of the Reporting Requirements for Operating Reactors above for each transmitter identified.

Although not required by this bulletin, addressees are encouraged to report information consistent with that requested in Item 1 b) of the Reporting Requirement for Operating Reactors through the NPRDS for all Rosemount Model 1151, 1152, 1153 and 1154 transmitters that exhibit symptoms indicative of loss of fill-oil or are confirmed to have experienced a loss of fill-oil.

As has been previously indicated, the NRC staff believes that the methodology utilized by Rosemount to support their proposed pressure versus time-in-service criteria for identifying which transmitters should be subject to an enhanced surveillance program does not provide the necessary high degree of confidence that this failure mode will not occur. Additional operational experience data, such as that to be generated in response to this bulletin, could be utilized by industry either to provide additional insight as to the appropriateness of Rosemount's pressure versus time-in-service criteria or to develop bases for staff consideration of an amendment to or termination of the actions requested by this bulletin. Accordingly, the NRC staff encourages utilities to work collectively under the guidance of a technical industry organization to develop an operational experience database concerning all models of Rosemount transmitters.

The written reports required above shall be addressed to the U.S. Nuclear Regulatory Commission, ATTN: Document Control Desk, Washington, D.C. 20555, and shall be submitted under oath or affirmation pursuant to the provisions of Section 182a, Atomic Energy Act of 1954, as amended and 10 CFR 50.54(f). In addition, a copy shall be submitted to the appropriate Regional Administrator.

Backfit Discussion

The objective of the actions requested in this bulletin is to ensure that transmitter failures due to loss of fill-oil are promptly detected. Loss of fill-oil may result in a transmitter not performing its intended safety function.

The actions requested in this bulletin represent new staff positions and thus, this request is considered a backfit in accordance with NRC procedures. Because established regulatory requirements exist but were not satisfied, this backfit is to bring facilities into compliance with existing requirements. Therefore, a full backfit analysis was not performed. An evaluation of the type discussed in 10 CFR 50.109(a)(6) was performed, including a statement of the objectives of and reasons for the actions requested and the basis for invoking the compliance exception. It will be made available in the Public Document Room with the minutes of the 179th meeting of the Committee to Review Generic Requirements.

This request is covered by Office of Management and Budget Clearance Number 3150-0011 which expires January 31, 1991. The estimated average burden hours are

6 person-hours per transmitter per licensee. This includes assessing the requested actions, gathering and reviewing plant records, analyzing the data obtained from the plant records, and preparing the required response. This does not include developing and implementing the requested enhanced surveillance program or replacing transmitters from the manufacturing lots that have been identified by Rosemount as having a high failure fraction due to loss of fill-oil that are utilized in the reactor protection or engineered safety features actuation systems. Send comments regarding this burden estimate or any other aspect of this collection of information, including suggestions for reducing this burden, to the Information and Records Management Branch, Division of Information Support Services, Office of Information Resources Management, U.S. Nuclear Regulatory Commission, Washington, D.C. 20555; and to the Paperwork Reduction Project (3150-0011), Office of Management and Budget, Washington, D.C. 20503.

If you have any questions about this matter, please contact one of the technical contacts listed below or the appropriate NRR project manager.

Charles E. Rossi, Director

Division of Operational Events Assessment

Office of Nuclear Reactor Regulation

Technical Contacts: Jack Ramsey, NRR, (301) 492-1167
Vince Thomas, NRR, (301) 492-0786

References:

1. Rosemount Technical Bulletin No. 1 dated May 10, 1989
2. Rosemount Technical Bulletin No. 2 dated July 12, 1989
3. Rosemount Technical Bulletin No. 3 dated October 23, 1989
4. Rosemount Technical Bulletin No. 4 dated December 22, 1989

Attachments:

1. Typical Physical Characteristics of a Rosemount Transmitter
2. List of Recently Issued NRC Bulletins

August 2004

Issue 176: Loss of Fill-Oil in Rosemount Transmitters

Description

Historical Background

The Rosemount Transmitter Review Group (RTRG) was established[1659] to perform an assessment of the actions taken to address Rosemount transmitter oil-loss concerns. This assessment included an evaluation of the adequacy of the information and actions specified in NRC Bulletin 90-01,[1658] Supplement 1, which informed licensees of activities undertaken by the NRC and the industry in evaluating and addressing loss of fill-oil in Rosemount transmitters manufactured prior to July 11, 1989, and requested licensees to take actions to resolve the concerns.

An action plan was developed by the staff and integrated the following RTRG recommendations to address Rosemount transmitter loss of fill-oil concerns: (1) conduct temporary instruction (TI) inspections to verify commitments made by licensees to address the requested actions of NRC Bulletin 90-01,[1658] Supplement 1, and to gather plant-specific data on Rosemount transmitter failures; (2) establish a dialogue with Rosemount, Inc., on Rosemount transmitter failure information; (3) review NPRDS data on Rosemount transmitter performance; and (4) review EPRI Report TR-102908, "Review of Technical Issues Related to the Failure of Rosemount Pressure Transmitters Due to Fill-Oil Loss," dated August 1994. This issue was identified in an NRR memorandum[1601] to RES in February 1996.

Safety Significance

Loss of fill-oil in Rosemount transmitters was determined to be a potentially undetected means of common mode failure. Such failures could result in loss of automatic reactor protection and engineered safety feature actuations.

Possible Solution

The staff determined that actions were needed by licensees to ensure that safety-related functions were maintained. These actions were first identified in Bulletin 90-01[1658] and subsequently modified in Bulletin 90-01,[1658] Supplement 1. The time frame for this action plan was based on the fact that licensees had implemented the requested actions of Bulletin 90-01,[1658] Supplement 1, and the plan was intended only as confirmation of the adequacy of the actions called for in the Bulletin.[1658]

The activities specified in the action plan were completed as a follow-up and verification of the implementation of the requested actions in Bulletin 90-01,[1658] Supplement

1. Licensees addressed the common mode failure concerns by either replacing affected transmitters with newly designed transmitters which corrected the oil leakage problem, or subjecting affected transmitters to enhanced surveillance monitoring to ensure their proper performance. A two-year period was established for completing the necessary verification activities recommended by the RTRG including TI inspections and reviews of recent Rosemount transmitter performance.

Conclusion

Temporary Instruction (TI) 2515/122, "Evaluation of Rosemount Pressure Transmitter Performance and Licensee Enhanced Surveillance Programs," was issued on March 17, 1994 and inspections were initiated in May 1994. Based on the results of the TI effort, the staff determined that licensees were effectively addressing the Rosemount transmitter loss of fill-oil issue by, in general, following the requested actions of Bulletin 90-01,[1658] Supplement 1, and the manufacturer's drift trending guidance.

The staff met periodically (between January 1994 and September 1995) with Rosemount, Inc. to exchange information on Rosemount transmitter performance. In addition, the staff completed NPRDS reviews for Rosemount transmitter failure information during the same period. Based on the information presented by Rosemount, Inc. and the results of the NPRDS reviews, the staff concluded that there was a significant decrease in the number of fill-oil failures since the issuance of Bulletin 90-01,[1658] Supplement 1.

On February 15, 1995, the staff completed its review of EPRI Report TR-102908 and confirmed that it was substantially in agreement with the previous conclusions, guidance, and requested actions contained in Bulletin 90-01,[1658] Supplement 1.

Based on the results of the above activities completed, the staff confirmed that all pertinent information regarding loss of fill-oil in Rosemount transmitters was contained in Bulletin 90-01,[1658] Supplement 1, and Rosemount technical guidance. Therefore, the staff concluded that the safety concern of the issue had been effectively resolved by the actions taken and no changes or additional actions were warranted. Thus, this issue was RESOLVED and no new requirements were issued.

Index

MIX
Papier aus verantwortungsvollen Quellen
Paper from responsible sources
FSC® C105338

If you have any concerns about our products,
you can contact us on
ProductSafety@springernature.com

In case Publisher is established outside the EU,
the EU authorized representative is:
Springer Nature Customer Service Center GmbH
Europaplatz 3, 69115 Heidelberg, Germany

Printed by Libri Plureos GmbH
in Hamburg, Germany